Felix Kaufmann's Theory and Method in the Social Sciences

BOSTON STUDIES IN THE PHILOSOPHY AND HISTORY OF SCIENCE

VOLUME 303

For further volumes:
http://www.springer.com/series/5710

Robert S. Cohen • Ingeborg K. Helling
Editors

Felix Kaufmann's Theory and Method in the Social Sciences

With an Introductory Essay by
Ingeborg K. Helling

Springer

Editors
Robert S. Cohen
Center for the Philosophy and History of Science
Boston University
Boston, MA, USA

Ingeborg K. Helling
Chemnitz, Germany

Methodenlehre Der Sozialwissenchaften by Felix Kaufmann
Originally published by Verlag Julius Springer
1936, Wien

ISSN 0068-0346 ISSN 2214-7942 (electronic)
ISBN 978-3-319-02844-6 ISBN 978-3-319-02845-3 (eBook)
DOI 10.1007/978-3-319-02845-3
Springer Cham Heidelberg New York Dordrecht London

Library of Congress Control Number: 2013956240

Printed on acid-free paper

Springer is part of Springer Science+Business Media (www.springer.com)

Felix Kaufmann

Source: Institute Vienna Circle. Taken from *Felix Kaufmann: The Infinite in Mathematics*, ed. B.F. McGuiness. Dordrecht: Reidel, 1978

Contents

Editorial Note

Felix Kaufmann (1895–1949) was a philosopher of science, especially a philosopher of economics and of the wider social sciences. He developed his clarifying, indeed critical and penetrating, thought in Vienna. He was an admirer of the work of Husserl, particularly the early *Logische Untersuchungen*, and he was closely associated with Hans Kelsen, the leading philosophical jurist of his time; later, in America after 1939, he was profoundly but critically a close collaborator of the sociologist Alfred Schütz. But Kaufmann was also devoted, in his philosophical discipline, to the standard of logical rigor, conceptual clarity, and even cognitive skepticism of the logical empiricism of the Vienna Circle, organized around Moritz Schlick, Rudolf Carnap, Otto Neurath and others. Kaufmann, self-described as the loyal opposition to any atomistic or reductive empiricism, remained in his Vienna years and after 1938 in his New York time a bridge, perhaps even a link, between phenomenology in Husserl's sense and the logical positivism of the Vienna Circle movement. Ernest Nagel summarized Kaufmann: "The central objective of his wide ranging studies was to make manifest the principles men employ when they succeed in making their experience intelligible."

The major work of Kaufmann's life was the treatise before us in this volume, *Die Methodenlehre der Sozialwissenschaften* of 1936 (new edition, 1999). His works were remarkably rich, and he had already published a phenomenologically developed philosophy of mathematics *Das Unendliche in der Mathematik und seine Ausschaltung* (1930). The latter appeared in English (*Vienna Circle Collection* vol. 9) as *The Infinite in Mathematics* (1978, edited by Brian McGuinness with an introduction by Ernest Nagel). All the many shorter pieces, whether mathematical or specifically concerned with economic theory and practice, are the result of Kaufmann's mature and detailed investigations. See for example his 'Soziale Kollektiva' (*Z. für Nationalökonomie 1*, 1929–1930) and his 'Juristischer und soziologischer Rechtsbegriff' in *Gesellschaft, Staat und Recht* (ed. A. Verdross), a Kelsen Festschrift of 1931. A classic exposition of his philosophical role was given by Kaufmann in his 'Phenomenology and logical empiricism' (1940) in a memorial volume for Edmund Husserl (ed. M. Farber). His later book with a nearly exact title, *Methodology of the Social Sciences* (1944, reprint 1958), was not a translation but a new book, for which he was considerably influenced by his coming to terms with the ideas of John Dewey. The book of 1936 was published in a Spanish translation in Mexico City in 1946. There is a fine bibliography of Kaufmann's published writings, prepared by Harry Reeder, appearing as a supplement to the book on the *Infinite in Mathematics*; 8 books, 47 papers, 32 critical reviews.

A fascinating account of the 1936 *Methodenlehre*, and of the subsequent personal and wider human influence of Kaufmann's life and teaching, are to be found in Ingeborg K. Helling's Introductory Essay. The impact of Kaufmann's creative achievement may be glimpsed in the following excerpt from his former student, the philosopher Reuben Abel: "Kaufmann believed that there was a direct intuition or apprehension of meanings, and of logical relations between them; these had to be presupposed in inquiry, and could not, like factual information, emerge within inquiry" (in his review of the Dewey-Bentley *Phil. Correspondence* in *Man and World* vol. 2 1968). The ensemble of the rational and the empirical, of the subjective and the objective, of Husserl and Schlick, was intended to be the deep construction of Kaufmann's life work.

This volume is the work of several contributors, mainly of course the book of Felix Kaufmann. Our translation of the 1936 book is based on a draft by John Viertel, critically reviewed by Carolyn Fawcett, and finally prepared and edited by R. S. Cohen and Ingeborg K. Helling. The Introductory Essay examines the work and reports the result of Helling's interviews with Kaufmann's colleagues, friends and students as well as with Kaufmann's son George Kaufmann.

Robert S. Cohen

Felix Kaufmann in Perspective: An Introductory Essay

Ingeborg K. Helling

As I recall the formative period of the Vienna Circle (1924–1930), Dr. Kaufmann was an especially valuable and stimulating member of that unique group precisely because he was a brilliant critical opponent in many fundamental aspects of the earlier forms of Logical Positivism. (Feigl 1981, p. 12)

Many followers of logical empiricism held the view that causal explanation of human actions has to be achieved by the use of terms which, like those of physics, refer exclusively to "objectively" observable traits of material objects. The fact that the few adherents of this philosophical school, for instance, Felix Kaufmann, who worked on methodological problems of the social sciences, raised basic doubts about the meaningfulness of this program, had no influence on the rigor of the philosophers of science, who by majority, were interested in the problems of the natural sciences. (Acham 1983, p. 201, my translation)

It is true that the balance between material research and methodological analysis is constantly changing. Fifty years ago the house of the social sciences was full of the echo of methodological discussions. Two thirds of Felix Kaufmann's famous book on the methods of the social sciences, which was published in 1936 and covered the previous thirty years, deal with the "*Methodenstreit*", the question of the difference between the natural and the social sciences, between sociology and psychology, and the status of values in the social sciences. There were few empirical studies in those days and, therefore, the discussion was concentrated on questions which, for the most part, we consider obsolete today (Lazarsfeld 1959, p. 225).

Have we really progressed much beyond Felix Kaufmann in the methodology of the social sciences? (Karl Acham, personal communication to the author).

I.K. Helling (✉)
Chemnitz, Germany
e-mail: ingeborg.helling@yahoo.de

R.S. Cohen and I.K. Helling (eds.), *Felix Kaufmann's Theory and Method in the Social Sciences*, Boston Studies in the Philosophy and History of Science 303,
DOI 10.1007/978-3-319-02845-3_1,

Introduction

The purpose of this introduction to the English translation of Felix Kaufmann's *Methodenlehre der Sozialwissenschaften* of 1936 is to fill in for the reader aspects of the theoretical and historical background presupposed by Kaufmann's work, with an eye to facilitating the reader's assessment of Kaufmann's contribution.

I proceed in the following way: After a biographical sketch, I describe Felix Kaufmann as a 'bridge-figure' with connections to a number of different approaches in philosophy and the social sciences. This will be done for both the Vienna period and the American period of his intellectual life, by tracing this position in the network of academic milieu and by pointing out substantial agreements and disagreements. In an introduction to a volume of the *Vienna Circle Collection*, it is Kaufmann's work in Vienna and his views concerning the Vienna Circle that are of course treated in the greatest detail: still, historical and systematic relations to phenomenology, the Austrian school of marginal utility theory, legal positivism, and pragmatism are considered too. A summary of Kaufmann's arguments will be presented and amplified by other published sources and unpublished correspondence and documents. The sections on logical positivism, Alfred Schutz, and John Dewey are based on Helling (1984, 1985, 1988a, b). For recent literature on Kaufmann, see Zilian (1990), Stadler (1997), Kaufmann (1999).

Biographical Dates

Felix Kaufmann was born in Vienna in 1895. He studied '*Staatswissenschaften*' (law, economics and sociology) and philosophy and received his doctorate in law in 1919, and a doctorate in philosophy in 1926. His academic position after his '*Habilitation*' (1922) in Vienna was '*Privatdozent*' (1922–1938) in the philosophy of law (cf. *Österreichisches Biographisches Lexikon* (1815–1950). pp. 271). He made his living through a managerial position as the Vienna representative of the Anglo-Iranian Oil Company.

Of liberal Jewish family, Kaufmann emigrated to the United States in 1938, after the *Anschluß* of Austria to Nazi-Germany. He became assistant professor of philosophy at the Graduate Faculty of the New School for Social Research in New York, and, in 1944, full professor at this institution. In 1949, he died suddenly at the age of 54.

Kaufmann's Milieus of Social Science in Vienna Between the Wars

In an essay on Austrian sociology from the turn of the century to 1938, R. Knoll and others (1981, pp. 59–101) distinguish the following schools of social science in Vienna between the Wars:

- political economy and sociology of the marginal utility school (F. v. Wieser, J. Schumpeter, L. v. Mises, F. Hayek, F. Machlup, G. Haberler, O. Morgenstern);
- neo-romanticism (O. Spann);
- left-wing Catholic social philosophy (A. M. Knoll, E. K. Winter);
- sociology of law and legal positivism (H. Kelsen);
- Austro-Marxism (M. Adler, F. Adler, V. Adler, O. Bauer, G. Eckstein, R. Hilferding, K. Renner, C. Grünberg);
- Vienna Circle (M. Schlick, O. Neurath, H. Feigl, F. Waismann, Ph. Frank, H. Hahn, V. Kraft, F. Kaufmann, E. Zilsel, K. Gödel, K. Menger, R. Carnap);
- psychoanalysis (S. Freud, A. Freud, A. Adler, W. Reich, A. Hartmann, S. Bernfeld, A. Aichhorn, E. Simmel, O. Fenichel, J.L. Moreno);
- empirical social research (K. Leichter, Charlotte and K. Bühler, E. Frenkel, E. Brunswik, M. Jahoda, P.F. Lazarsfeld, H. Zeisel).

Alfred Schutz is strangely omitted here.

In this account, Kaufmann figures as a member of the Vienna Circle. The reports about his membership are, however, ambiguous. Recalling the formation of the 'Thursday evening colloquium – the beginning of the Vienna Circle' in 1924, Herbert Feigl (1981, p. 60) lists Felix Kaufmann 'among its initial members' (together with Hans Hahn, Olga and Otto Neurath, and Viktor Kraft) and calls him 'an active and cherished member of the Vienna Circle' (ibid., p. 73). His name does not appear, neither among the members of, nor among 'those sympathetic to the Vienna Circle', in the pamphlet (1929) *Wissenschaftliche Weltauffassung: Der Wiener Kreis* ('The Scientific Conception of the World: The Vienna Circle', in Neurath, 1973, pp. 299–318) marking the official beginning of the Circle in other accounts of its history. However, Neurath mentions him among the 'researchers close to the Vienna Circle (Kaila, Kaufmann, Kraft, Menger, Reidemeister, Zilsel et al.)' in his 'Historische Anmerkungen' (1930/31 in Neurath 1981, p. 390) and comments on his work at other places.

> [Brentano's] pupil Husserl who was successful in Germany, was […] followed by metaphysical idealists strongly influenced by theology – Max Scheler, Heidegger, Conrad-Martius – whereas in Vienna his pupil Felix Kaufmann worked diligently on studies in logical analysis, which made him become close to the Vienna Circle (Neurath 1936, ibid., p. 673, my translation).

It seems to me that Neurath's numerous and varied statements on the members and on the persons with and affinity to the Vienna Circle are offered less as a matter of sheer description than as practical (and in Neurath's case, political) means of gaining acceptance for the Circle. Therefore, they change with the circumstances (place, time, and audience) in which they are formulated.

Some light can be shed on this matter by consulting the correspondence (Felix Kaufmann papers, *Sozialwissenschaftliches Archiv Konstanz*, hereafter KP), between Felix Kaufmann and Rudolf Carnap (cf. Helling 1985). It shows that Carnap, in preparing the aforementioned pamphlet of 1929, invited Kaufmann to submit a list of his publications and to subscribe to the publication. Kaufmann declined the invitation, asking Carnap to understand his reasons for doing so.

Carnap's reply to Kaufmann is congenial: "Your scientific and personal relations to our circle will not be affected in the least". Kaufmann's refusal seems to have been based on his substantial disagreement with some of the views current in the Circle (see 'Felix Kaufmann and the Vienna Circle', below).

In the summer of 1934, the American professor of sociology, Earle Edward Eubank, interviewed leading figures of European sociology (Käsler 1985). Felix Kaufmann appears in the protocol of an interview with Eric Voegelin:

> There are several people at Vienna University, who, while not working in sociology proper, have given us a lot of help; among others, I think of Felix Kaufmann and Fritz Schreier, both legal philosophers. Former students of Husserl, both have contributed to a theory of social action, and both have the right sociological perspective (cf. Käsler 1985, p. 144, my translation).

To whom does 'us' refer in the second line of the quote, and what exactly is the 'right sociological perspective'? Voegelin himself, like Kaufmann, was a participant in L. v. Mises' private seminar. Other members were F. Machlup, F. Hayek, O. Morgenstern, G. Haberler, K. Schlesinger, V. Bloch, A. Schütz (Schutz), S. Braun (Browne), I. Mintz, H. Lieser, M. Herzfeld, R. Strigl, W. Fröhlich, H. Fürth, E. Winternitz (cf. M. v. Mises 1981). In addition, both men met in a private discussion group, the *Geist Kreis*. The members of Mises' private seminar were not exclusively economists in today's technical sense and the subjects chosen for discussion were not always 'economics' in this strict sense either:

> [...] economic theory, methodology of the social sciences, economic policy [...] sociology, especially Max Weber's *'Verstehende Soziologie'* and the problems related to it, were favourite topics (G. Haberler in M. v. Mises 1981, p. 264, my translation).
>
> The year of methodological topics was particularly interesting, partly through Schutz's and Kaufmann's connection with Husserl's ideas and through Kaufmann's connection to the Schlick Circle. Sometimes Kaufmann brought other members from the Schlick Circle as guests, I particularly remember H. Feigl's visit (F. Machlup, *ibid.*, p. 261, my translation; see also Engel-Janosi 1974; Wagner 1983).

Kaufmann was also a member of the so-called Vienna school of legal theory, a group loosely organized around Hans Kelsen. It was due to Kelsen's influence that Kaufmann was appointed to the post of '*Privatdozent*' at the University of Vienna.

Kaufmann's Positions in Methodology

Before he was forced to emigrate in 1938, Felix Kaufmann had published books, articles and many reviews (e.g. on works by Carnap, Husserl, and Schutz) in philosophy of law, economics, sociology, and mathematics. In the foreword to his *Methodenlehre der Sozialwissenschaften* (1936a) he describes as the task of the book to show the interconnection of the problems of social theory with those of the general theory of knowledge, acknowledges strong influences of Husserl, v. Mises, and Kelsen on his thinking, and thanks A. Schutz and K. Bode. The structure of the book reflects these influences. Turning now to the main features of the *Methodenlehre*,

I shall introduce them by comparing Kaufmann's views with those of the Vienna Circle, Alfred Schutz and Phenomenology, and by summarizing his treatment of Kelsen's *Pure Theory of Law* (1936a, pp. 291–310) and the Austrian school of marginal utility theory (*ibid*., pp. 255–290).

Kaufmann's intellectual encounter with American pragmatism will be sketched on the basis of the unpublished correspondence between John Dewey, Arthur F. Bentley, and Felix Kaufmann. (For more detail, cf. Helling 1988b).

Felix Kaufmann and the Vienna Circle

In his *Methodenlehre* (1936a), Kaufmann agrees with the general position of the Vienna Circle on a number of issues:

1. Prediction is the aim of all empirical sciences, i.e. natural and social sciences do not differ in their task.
2. Laws and initial conditions are required for prediction in social and natural science.
3. Intersubjectively valid experience must be provided for the acceptance of theoretical statements.
4. Logical truth and factual truth must be distinguished.
5. Metaphysical statements can be discarded as setting pseudo-problems.

He disagrees mainly on the following points:

- Statements about psychical facts are *not* reducible to statements about physical facts (a matter of discussion within the Circle).
- The distinction between overt behavior and internal behavior does *not* coincide with the possibility of objective statements about observations and predictability.
- Theoretical statements can*not* be fully translated into elementary (atomic) sentences about observations of external events.
- The meaning of a statement is *not* identical with its method of verification.

Given that Kaufmann was in favor of empiricism in some respects but opposed to it in others, how are we to understand his position? I regard Kaufmann's phenomenological concept of experience as the basic element of his position and I shall develop an understanding of the above points of agreement and disagreement with empiricism by focusing on it. In his paper 'Phenomenology and Logical Positivism' (1940) he states that controlled examination of experience is the only way to knowledge and opposes, with the logical positivists, metaphysical speculation and intuition as sources of knowledge; indeed, metaphysical speculation and intuition have brought about muddles and much fuzzy thinking in philosophy, leading to confusion of logical truth with factual truth, and to the imputation of necessity to matters of fact. Or so Kaufmann, like the logical positivists, believes.

At the same time, he accuses the logical positivists of not going far enough: They stop short of an analysis of the structure of experience and settle instead for a sensationalist view. Conceiving of experience as a simple given, they come to think of verification as a simple structureless event. And, instead of analyzing the relationship between the *meaning* of a judgment and its *manner of verification*, as would be required, they equate the two.

Experience, in Kaufmann's phenomenologically inspired view, has a complex structure. The fact that psychologically it is a quick and simple given to experiencing subjects must not be misconstrued as structural simplicity. Experience is neither fully passive ('receptive') nor fully active ('spontaneous') but consists of elements of both kinds, which cannot be easily separated. The synthesis of elements of knowledge, which forms the content of judgments, occurs on different levels ('strata of experience'). Even a simple identification of the existence of an object of a particular kind involves recollection or activation of both former and future experiences (anticipations), which are directed at 'intertemporal, intersensual and interpersonal (intersubjective) validation' (1936a, pp. 11).

As a consequence Kaufmann's view with respect to *verification* is that a judgment of fact contains more than a registration of an isolated situation of experience and it therefore cannot be definitely and ultimately verified by a statement about such a situation. It can, however, be subjected to empirical control and can be changed. Note that Kaufmann uses the empiricist question of truth-criteria for statements in order to establish phenomenologically that there can be no isolated act of verification. Though verification is conceived of differently from the early Vienna Circle, it is still of extreme importance: Due to the open-endedness of every experience and its incorporation within contexts of experience (Husserl), the meaning of a statement always transcends a given method of verification, but verification is a necessary element in establishing its meaning:

> Included in the conception of Being as a horizon of open possibilities, are not only well-defined access points, but also *vacant positions*, which are open to further, as yet undetermined experience and which will, so it is anticipated, be congruent [*einstimmig*] with the context of the modes of comprehension already characterized. In this way Being 'transcends' any fixed number not only of individual lived experiences, but also of types of lived experience. And in the same way the meaning of judgments transcends any fixed number of truth-conditions [...] this 'transcendence' must not be misconstrued as a supersession of the correlation, i.e., not as an absolute transcendence [...] for each determination of possibilities that as yet remain open is in turn the result of the specification of a mode of comprehension, of a path of empirical access (1936a, p. 16).

Though Kaufmann argues rigorously against the identity of meaning and verification on the theoretical level, he does at times use them synonymously on the procedural level. His failure to introduce two different terms makes it difficult to appreciate his position, which forecasts the distinction between theoretical and observational concepts.

What, for Kaufmann, is the function of perception in knowledge? The amount of sense-material contained in a judgment marks the difference between direct experience and remembering, reality and fantasy. Perceptual judgments have a special status in the process of gaining knowledge, they form 'nodes' (*Knoten*) in the network of experiential knowledge. Yet they cannot be isolated from the pre-established habitualized (pre-predicative) knowledge into which they become incorporated in the process of judgment-formation: thus *no presuppositionless knowledge* is possible and knowledge of the world cannot be constructed from or represented in isolated *atomic sentences*. Note that Kaufmann attributes aspects to simple perception, i.e. every perception must be confirmed, this in contrast to other traditions, where confirmation is attributed only to scientific observation and thus, for perception, follows a principle of continuity between everyday life and science. Knowledge of the world thus has its basis in experience, but it is a changing basis. This basis, to use Popper's metaphor of the 'mud, on which we build knowledge like engineers build bridges', is in Kaufmann's analysis not treated as a non-examined entity but analyzed for its structure, its similarity and interaction with scientific knowledge.

But what is the effect of the open-endedness of experience on the level of scientific theory and research? Kaufmann formulates a principle of finite formulation, which is to prevent infinite regress:

> The horizon-character of experience corresponds to an *indefinite regress* in the verification of judgments it is indefinite, because there is no point within the regression at which its termination would impose itself logically, but it is not infinite because de *facto* it is always broken off at some point, and has to be. For anyone who has overcome the erroneous conception of knowledge without presuppositions, this state of affairs will no longer seem paradoxical (1936a, p. 18).

This statement asks for specification of the criteria of breaking off the series of verification. A descriptive or normative account of such 'closure' is not given in the 1936 book, where Kaufmann is concerned mainly with propositions and their empirical control. The central role given to empirical control of scientific propositions by the Vienna Circle is not challenged; rather, a different *meaning* of empirical control is provided: verification does not consist of a singular act but of series of controls which result in 'unanimous' experience (*'einstimmige' Erfahrung*). The truth of a perceptual judgment is dependent on the truth of other judgments; thus it is always possible to refuse an observational test when it does not fit the theory. Facts are theory-dependent. However, the implications of these views are not clearly worked out. (Cf. the later development such as Hempel's view (1952) of scientific theories as 'networks' which have connections with empirical reality at some points only, and Quine's (1960, 1961) 'contact-zones' between theory and reality.)

If one claims, as Kaufmann does, that every experience transcends its here and now, that every fact is theory-dependent, that experience is an everchanging basis and at the same time there is no object of knowledge that transcends experience, and that verification is necessary to *establish* the meaning of propositions, then more is required to explain scientific knowledge than a refutation of the possibility of its construction from atomic sentences and a differentiation of systematic and extra-systematic concepts.

For this Kaufmann concentrates on research procedure. Under the title 'universal methodological schema' (1936a, pp. 121–128) he names as one of the most important tasks for methodology the characterization of more or less general schemata of problem structures in research procedure. He looks at the functions served by elements of knowledge (knowledge of various levels of clarity and distinctness – of facts, essence, and laws) in research procedure in general and in the specific research procedures of various sciences. He suggests an ordering of the elements of knowledge within the temporal structure of the research process, the thematic relevance of research problems and heuristic principles of the selection of the tools of research, and an ordering or the relative status of factual knowledge in the research procedure. The latter is most important here:

> That a known fact is derogated means here that it is regarded as not decisive for the designated complex of knowledge; this does not necessarily mean, however, that this particular knowledge is negated, or even subjected to doubt, but merely that it may be reinterpreted; thus for example by drawing on either real or merely supposed facts (disturbances, changes of data) until then not considered, it can be brought into conformity with the known fact that derogated it [...] Known facts whose capacity for derogation within the framework of a procedure are excluded altogether we can designate as absolutely valid facts of knowledge *for this procedure* [...] We no longer see in certain known facts simply the solution of a problem but only a step along the path to the acquisition of knowledge (1936a, pp. 124–125).

In his 1944 book, Kaufmann develops this analysis. The acceptance or non-acceptance of observational statements *and* general statements is seen to be decided within a framework of 'basic' and 'preference rules of procedure', and propositions are distinguished with respect to their status in scientific procedure.

> Observation is but one way of verifying a proposition among the related controls [...] Observational results possess a certain primacy over theories but the very meaning of this primacy can be determined only within the framework of a system of rules of empirical procedure (Kaufmann 1944, p. 43ff.) [...] The untenable sensationalist view of a hierarchy of controls headed by the indubitable results of perception would lead us to believe that control works only from protocol propositions to universal propositions. As a matter of fact, however, it works the other way too: we may withdraw recognition from a protocol proposition because it cannot be brought into accord with an accepted synthetic universal proposition (empirical law) (1944, p. 58).

How does Kaufmann's position differ from the various reformulations of the verifiability principle discussed within the Vienna Circle between 1929 and 1935?

> At first such protocol statements (statements of perception) were considered as being of absolute validity: 'sentences which need no confirmation and form the basis for all other statements of science' (Carnap, *Erkenntnis*, 2). Neurath denied this absolute validity (*Erkenntnis*, 3). Even protocol statements can be declared invalid if necessary; for they are not free from elaboration, they are not more original than other empirical statements, they are as hypothetical and open to correction. Moreover, statements cannot be compared with data, with experiences, with anything extralingual, but only with other statements. This view of Neurath's was also accepted by Carnap. Protocol statements are not ranked above other statements. Carnap said (*Erkenntnis,* 3) 'there are no absolute original statements for the construction of science.' It is a matter of decision, of convention, where to stop [...] if protocol statements are no longer absolutely certain but open to correction, how can one decide when a protocol statement is to be dismissed, and when not? For Neurath the criterion was the harmony between empirical statements. If a protocol statement is in contradiction with the system of so far accepted statements, either the protocol statement has to be dropped or the system has to be changed so that with the addition of the new statements no contradiction remains (Kraft 1950, quoted from Neurath 1973, p. 50).

Thus the place Kaufmann attributes to protocol sentences and observation in scientific procedure is not very different from Neurath's or Carnap's versions of 1932/33. Kaufmann claims that the reason for the failure of the strong verifiability principle consists in a neglect of the structure of experience. The theory of science cannot take protocol sentences as irreducible units. The reinterpretations of the function and status of protocol sentences which occurred in the history of the Vienna Circle and later can and must, in his view, be integrated within a phenomenological theory of experience. Such a theory must be able to account for both pre-predicative experience and scientific theories. Incidentally, Kaufmann's opposition to the verifiability principle is nicely illustrated by the difference in two statements by F. Waismann. In 1930, Waismann formulates the verifiability principle in this way: "A statement which cannot be verified conclusively is not verifiable at all; it is simply devoid of any meaning" (1930/31, p. 229, my translation). In 1949, without mentioning either Kaufmann or phenomenology, he introduces the concept of the 'open texture' of empirical concepts, "a factor not discussed thus far", as a reason why statements cannot be conclusively verified.

How does Kaufmann's concept of experience relate to his position on behaviorism? The following statements of his analysis are necessary for an appraisal: subjective and intersubjective elements of natural science, internal and external experience, content of assumptions in natural and social science, and the structure of social facts.

Kaufmann accuses both behaviorism ('naturalism') and introspectionism ('anti-naturalism') of drawing false conclusion from the fact that inner experiences are given to only one person: behaviorism concludes that such experiences by virtue of not being externally observable, are not controllable intersubjectively and, therefore, cannot be dealt with scientifically. Introspectionism concludes that knowledge of inner experiences is of specific evidence and requires an altogether different methodology. Both share false conceptions about the elimination of subjective elements in the epistemic process of the natural sciences and about direct observation.

Replying to these positions, Kaufmann argues as follows: Every experience of external events contains inner experience (the spontaneous elements of experience described above). By using instruments of measurement, natural sciences do not reach objectivity independent of the 'subjective' feelings of warmth, length etc., but still use such feelings. The very concept of intersubjective control presupposes the existence of an alter ego which is not reducible to external observation. The meaning of signs and language in science presupposes a relationship between psycho-physical subjects which is not a property of signs as such. The difference between the method of the natural sciences and the method of the sciences of the mind is a matter of whether general propositions about correlations between physical (external) and psychological acts are used in the process of incorporating observations into general contexts of experience. Physical objects correspond to external experience, psychical objects to internal experience. However, external and internal experience cannot be totally isolated, since they are intimately tied to each other. The contents of inner experiences such as remembering and fantasy are external facts (sensations) or facts of one's own body (feelings). External experience, on the other hand, by presupposing identification and differentiation of objects, presupposes inner experiences.

Thus, the problem of relationships between physical and psychical objects of knowledge requires an analysis of 'The Strata of Experience' and their joint implications, (cf. Kaufmann's paper, 1940).

As to the nature of psychological and social facts, while we cannot strictly contrast experience of physical and psychical objects, we can contrast experience of physical and psycho-physical objects. The isolation of the psychical from the psycho-physical requires an abstraction similar to the abstraction of color from its foundation network of bodily things. The intentionality of all acts of consciousness is of importance in this connection. An analysis of an act of thought, for instance, can focus either on the elements of this act itself, or on its intentional object or content. The isolation of psychical phenomena must strive for the latter, and presupposes the isolatability of act and object of act. However, for the objects of the social sciences the interconnection or unity of psycho-physical phenomena is the adequate level. For instance, action is a psycho-physical phenomenon and must never be seen as a physical occurrence only.

How is this general account (1936a, pp. 87–91) related to psychology and sociology? Kaufmann defines as the core of scientific knowledge of social facts the meaning-interpretation of actions as 'symptoms'. (This concerns sociology and psychology, insofar it is concerned with human action.) Meaning-interpretation does not require special sources of knowledge. Rather, it is a synthesis of external and internal experience; accordingly, concepts of social facts can be constructed from concepts of physical and psycho-physical facts. *Social facts* are *not intramental*. The peculiar nature of psycho-physical facts, in contrast to physical facts, is that in addition to the syntheses of consciousness which constitute them, another level of syntheses is involved: they are interpreted as symptoms of acts of consciousness of other persons. Here we have the core of Kaufmann's position with respect to the differences of social and natural science: They are not totally different because their objects are constituted in consciousness through spontaneous and receptive elements, though to be sure, these objects are not contained by consciousness, and they are different because the syntheses involved are of a different kind. Observable, physical and psycho-physical (social) facts are always perceived within some context of experience. To paraphrase an example dear to analytical philosophy of action (Anscombe 1953), in perceiving 'John's waving his arm' as a 'greeting', we employ an interpretation that adds to the identification of a moving object at some point of space as a moving human arm. The physical object then is interpreted as a *symptom* for an act of consciousness. There is nothing private about such regular coordinations of physical objects and acts of consciousness (facial moment, gestures, signs of language) because they are incorporated within shared schemes of interpretation. As for controllability, we do make predictions on their basis in daily life (cf. Ryle 1949). For the inferences involved, Kaufmann introduces the terms 'symptom–relation' and 'inference on epistemic grounds'.

> A state of affairs S_1 is called 'symptom for state of affairs S_2' if from the presence of S_1 the – past, present, future – existence of S_2 can be inferred. But that such inferences can be made indicates nothing else than that between S_1 and S_2 a real relation – an empirical connection – exists [...] this real relation must not necessarily be of the kind that the symptom ('epistemic ground') of a state of affairs coincides with one of its causes ('real ground').

Instead it could also be the effect of this state of affairs, or as a consequence of shared causes, a regularly accompanying phenomenon [...] the *significance of the symptom* S_1 is nothing else than *the meaning of the judgement about* S_2 *which is based on knowledge of* S_1. We call such a judgement *an interpretation* of S_1; the concepts 'symptom meaning' and 'interpretation' are thus correlative [...] symptom relations are neither one-to-many relations nor many-to-one relations [...] on the one hand a state of affairs S_1 can function as a symptom for various states of affairs S_2, S_3, ... S_n, and on the other hand each of this states of affairs S_2, S_3, ... S_n can, if need be, also be inferred from other states than S_1 [...] every inference from one fact to another fact presupposes *general assumptions* about the connections between facts. The real relation, from which the symptom-relation arises, is not a relation between two specified individual facts as such, but between any two *arbitrarily selected* facts of a given kind. The general assumptions on which the interpretation is based are called *schemata of interpretation* [...] the assertion, S_1 is a symptom for S_2, is incomplete as long as a schema of interpretation establishing the symptomatic connection has not been furnished. However it is to be noted that S_1 can function as a symptom for S_2 on the basis of *various* schemata of interpretation.

If, given a schema of interpretation, S_2 cannot be inferred from S_1 alone but only from the combination (coexistence or succession) of various facts (S_1, T_1, U_1), then we want to say that the facts S_1, T_1, U_1 stand in a *meaning-context*.

[...] by the 'meaning of a fact' within the framework of an epistemic process nothing else is to be understood than the position (function) of that fact within this process. [...] The fact asserted in the judgment can either be a physical or psychical (psycho-physical) fact (1936a, pp. 153–156).

In the interpretation of the meaning of action and products of action we are concerned with inferring psychical (intentions, projects) and psycho-physical facts (actions) from the physical facts, bodily movement. The distinction between real grounds (causes) and epistemic grounds (symptoms) is important for the methodology of the social sciences, because interpretation is concerned with finding the latter. For the methods of an empirical science the difference is not crucial. Phenomena of meaning do not constitute an autonomous domain of knowledge independent of facts. That the thinking another person is engaged in is analogous to one's own thinking can be assumed as a hypothesis and be confirmed indirectly in the same way as a hypothesis about not directly observable objects, e.g. atoms, in physics. In 1944, Kaufmann expresses this view in language, responsive to the earlier concern in the Vienna Circle over the problem of protocol sentences:

[...] it is erroneous to conceive of human actions as observable physical facts, and this applies to artifacts and institutions. In all these instances there is indeed reference to the observation of physical facts, but we do not observe actions *qua* actions, artifacts *qua* artifacts, institutions *qua* institutions and therefore we cannot say that they are given in observation. Accordingly, we may state that every interpretation of social facts presupposes a fundamental interpretation, namely that of the underlying physical fact as a social fact (1944, p. 166).

The chief difference between rules of procedure concerning propositions about the psycho-physical world, and those concerning propositions about the physical world is that the protocol propositions are of a different kind. In the psycho-physical domain they imply interpretations by which psycho-physical facts are correlated with physical facts. But the two kinds of protocol propositions have an essentially similar status in scientific procedure. Both can be sufficient conditions for the acceptance or elimination of singular propositions. This is one reason why they are seldom properly distinguished. Another reason is that particular sets of observational data are often 'automatically' interpreted as psycho-physical phenomena (*ibid*., p. 126).

But if there is a regular correlation between psycho-physical and physical facts, why not use physical facts alone in science? This question leads us to Kaufmann's position with respect to physicalism as argued for by Neurath, Carnap and Hempel in the early 30s. Kaufmann examines two formulations of physicalism:

Hempel's version of 1935, which argues for the identity of meaning of a psychological sentence and its physical translation on the grounds of identical truth-criteria expressed by observational sentences about external events.

Carnap's version of 1935, which abolished the belief in atomic sentences as the basis of all knowledge and the meaning-identity of psychological and physical sentences, i.e. the belief that psychological sentences have cognitive content only insofar as they can be expressed in space-time terms. Instead, direct *and* indirect verification of psychological sentences is argued for: a sentence about a psychological state can be verified directly by the person experiencing it without using space-time external terms, and can be verified indirectly by an observer by means of space-time terms. Since the fact referred to in both kinds of verification is identical, the direct verification coincides with indirect verification. For the purposes of science, indirect verification can be substituted for direct verification and physical sentences can be substituted for psychological sentences.

Kaufmann insists that by translating psychological sentences into physical sentences, the former are not eliminated from the system of scientific propositions. As is often true of Kaufmann's criticism of ideas proposed by members of the Vienna Circle it is less a matter of objections on Kaufmann's part to possible procedures they propose than it is objections to the underlying logical and empirical grounds. In the first case, he claims that not all the control sentences of psychological sentences are physical sentences, e.g. in order to verify the statement 'B understands A's question' one could in most cases use as a control sentence 'B will give correct answer'. The 'accepted scientific propositions' which are required for a deduction of the control-sentence are knowledge about the relationship between thought and language, or, if one isolates the external facts (body movements) contained in the act of speaking, between psychological and physical facts. Similarly, in reply to Carnap's 'equipollence' of concepts, Kaufmann by no means denies a regular correlation between psychological and physical facts; rather, he denies that indirect verification can be established as equipollent with direct verification without making use of a general empirical correlation between the two kinds of facts, from which correlation psychological terms cannot be removed. This time, Kaufmann's procedure of eliminating 'exaggerated claims' by means of clarification of what is claimed and on what grounds misses the point for neither Hempel nor Carnap had made the claim he criticizes them for, namely that psychological sentences could be eliminated altogether. Kaufmann's position by no means precludes the use of natural-scientific methods in the social sciences; rather their fruitfulness becomes an empirical question.

As a result of his critique, Kaufmann draws the following conclusion about the relation of direct and indirect verifiability, and about operating with physical and psycho-physical terms in social science:

> insofar as we are in possession of an empirical procedure which permits us to assume an unequivocal coordination between psychic events and space-time events – those accessible to external observation – we can first carry out these observations and put them in order – and

thus for an extensive phase of the procedure operating only with space-time facts – and only at the end interpret the result psychologically (1936a, p. 142).

This position is clearly as far away from radical interpretive approaches, which would not operate with an assumption of a general correlation between observable external and non-observable internal events. To place Kaufmann's views into a historical context, let us briefly look at Hempel's later view (1972, p. 14, my translation) on the reduction (translation) of psychological terms into physical terms. Human action, according to this view, is explained by motives understood as dispositions which are not reducible to physical terms:

> The psychological attributes in question [...] are not be understood as strictly physicalistic or behavioristic dispositions [...] The explanation of human action by means of psychological factors presupposes [...] a complicated network of law-like connections between psychological attributes and physical attributes.

The discussion of Kaufmann's view of physicalism – he insists on the non-reducibility of sentences about psychological and social facts to sentences about physical facts *and* stresses the necessity of both kinds of facts for the verification of psycho-physical sentences – should have provided some of the elements required for placing Kaufmann's concept of '*Verstehen*' within the framework of his methodological analysis. In a formulation close to the spirit of the Vienna Circle but rather foreign to contemporary followers of Schutz in what is called 'phenomenological sociology', Kaufmann introduces Schutz's analyses of the structure of the social world as the answer to the question: what are the truth-criteria of propositions about the meaning of actions of other persons? "The manner of verifying these judgments and therefore also their meaning, depends on the relative spatio-temporal situation of those making the judgment and of those acting" (1936a, p.157). (Here we have one of the occasions (cf. also p. 171, p. 173) on which Kaufmann equates meaning and verification despite his explicit rejection (pp. 10–12) of their identity.) Truth-criteria of statements about the meanings of action vary with respect to the directness/indirectness of the symptoms used to establish the validity of judgments. The concept of 'objective meaning' provides no procedural significance without the specification of a scheme of interpretation. For Kaufmann, this variation of the meaning of judgments relative to interpretative schemes is in principle, though not in practice, also true for the natural sciences (where an established hierarchy of ways of incorporating data into contexts of experience makes the variation less visible and less important.)

> Thus a major part of the judgments in the human sciences, as well as a major part of the statements in the natural sciences, are valid only *relative to a reference system that must be specified*, and the problem of the discovery of universal invariants, i.e., propositions the truth of which is not affected by the transition from one reference system to another, has become just as acute in the one domain as in the other (1936a, pp. 160–161).

'*Verstehen*' a method of the social sciences has the following features:

Understanding of other minds and all forms of meaning-interpretation presuppose as a basic assumption the existence of fellow men.

The specific evidence of understanding does *not* provide a *truth-criterion* for scientific sentences about social facts.

Again, Kaufmann argues for a modified unity of science approach, claiming a similarity of understanding and explanation. Just as there is not *one* and only one *explanation* of a fact, there is not one understanding of social facts. In both cases, the following questions (1936a, p. 164) have to be asked: What data (physical data/psycho-physical facts) are to form the basis of explanation/understanding? Under which conditions will the explanation/understanding of an object of experience be regarded as successful? The definition of both explanation and understanding is the 'incorporation of facts into general contexts of experience' and in both cases 'laws and singular facts (initial conditions)' are required.

The link between claiming, at the same time, the non-reducibility of social facts and the unity of scientific procedure is provided, in Kaufmann's pre-American period, by a 'conventionalist' conception of laws:

> Every empirical law can [...] thus be put into the form: if facts of the kind *E, F, G*, ... will appear in one given domain, then facts of the kind *M, N, P* will appear in a determinate environment of the first named facts (1936a, p. 58).

Temporal succession of cause and effect is not a necessary element of experiential laws. The mathematical precision of laws is not a property of reality, but a matter of choosing a convenient language for expressing laws. Absolute validity of experiential laws can only be established by convention.

A conventionalist conception of laws is also held by Otto Neurath: for Kaufmann, laws are rules of inference, for Neurath, they are 'not to be seen as proper statements but as directives for finding predictions of individual courses of events' (1931; quoted from 1982, p. 52). It might be noted, too, that Kaufmann explicitly rejects Neurath's physicalism but agrees with his conception of protocol sentences. While Neurath rejects '*Verstehen*' as a method, Kaufmann sets out to reformulate it.

Since 'laws are nothing but rational constructions', there is no reason why such constructions cannot be applied to social facts. The prototypes of such laws are, for Kaufmann, Weber's ideal types:

> As quite generally, laws are nothing else than general assumptions – thus 'rational constructions' – which have been set up on the basis of prior experience and have to continue to be confirmed by the facts, so ideal typical interpretive schemata are 'rational constructions of meaningfully understandable' action [...] for the human sciences [...] the lawfulness which is crucial for the sociological rules [is] a lawfulness of understanding (1936a, p. 228).

Kaufmann defines the concept of society as 'a field of application of certain schemata of interpretation for social relationships' (1936a, p. 208). The concept of a 'social relationship' between persons was introduced by Max Weber as consisting fully and conclusively in the *chance* that the social action of a specifiable kind will occur. In reformulating this notion, Kaufmann again combines the phenomenological concern with interpretation (physical movements become action through interpretation, i.e. incorporation within a context of experience) and an empiricist question of truth-criteria and prognosis:

> That such a chance exists, however, means nothing else than that under the assumption of the existence of such a mutual orientation, the actions in question can appropriately be interpreted, where the most important criterion of an appropriate interpretation lies in the confirmation of the prediction of the course of future actions based on this interpretation (1936a, p. 207).

Weber requires of ideal types of social action and products of social action adequacy on the level of meaning *and* causal adequacy. Kaufmann fully accepts the different structural forms of understanding fellow-men proposed by Schutz (1932) as a clarification of judgments of the adequacy of meaning in social science. This may be the place to illustrate Kaufmann's numerous assertions to the effect that several strata of meaning have to be distinguished in methodological analysis. Because Weber's meaning adequacy refers to the goal-directedness of action, meaning adequacy is an empirical relation, subject to historical changes. Therefore, ideal typical constructions involve assumptions and verification of factual relations – both on the level of 'meaningful comprehension' of actions as well as on the level of their 'causal' validation, the latter by establishing the actual occurrence of actions through prognosis or retroduction. Those who posit rationality of goals as the paradigm-case for understandability should be able to answer the question whether they see it as an anthropological constant of human action and its understanding, or as a specific historical type of action that has become dominant in modern societies; they should be aware, too, of the limits they thereby set for social analysis. Though Kaufmann accepts the dominance of goal-directed action in social science, he is prepared to see it as an empirical and historical question.

As for Weber's exposition of the nature and the function of ideal types in the social and natural sciences, Kaufmann argues that Weber made too much of the differences. Using the case of the ideal laws of free fall in empty space as an example, he points out that in operating with social laws (ideal types, interpretive schemes) as in operating with idealized laws in the natural sciences, it is necessary, to decide whether we want to apply them to reality in isolation, i.e., use them to make predictions or whether we want them to be conceived as partial laws requiring supplementation (1936a, p. 229).

For Kaufmann, the similarity holds. It holds, notwithstanding his view that within the framework of Weber's theory of ideal types, the postulate of causal adequacy – and thus also statistical observation – does not have the function of a point of departure for induction, as is the case in the statistics of processes which have no meaning, but merely an (accessory) control function. It has to be pointed out, however that in the actual process of pursuing knowledge these two functions cannot be sharply separated from each other (1936a, p. 228).

In stressing the similarities of ideal types in the social and natural sciences, Kaufmann's arguments are similar to those offered by Hempel in his analysis of ideal types in the social sciences (1963). (It is very likely that Kaufmann knew the manuscript of Hempel and Oppenheim's work on types, published in 1936. In a letter of July 20, 1934, Hempel writes: "I would be very pleased if on that occasion [the Philosophy Congress in Prague] I could discuss various questions with you on the work I am doing here together with Mr. Oppenheim." KP, my translation).

Kaufmann's examples for interpretive schemes are the rules of card-playing (taken from Weber), and rules of organizational and legal behavior. The rules of a game of cards can be used as an interpretive scheme by means of which the behavior of the players can be understood and predicted. They are invariant with respect to individual players and time, place, and other properties of the game just as the meaning of a judgment in the objective sense is invariant with respect to the occasional aspects and elements of judging. The rules are *not* ideal objects, that realize themselves in an instance of card-playing, rather, they are products of abstraction from intended meaning, in which the occasional data of intensions (who, when, where) are left open. Husserl's analysis of the meaning of judgments, a central and constantly recurring tool for Kaufmann's attempts at the clarification of meanings, is also applied in the analysis of concepts of social collectivities:

Careful analysis of the concepts of social science shows, that while the assumptions that there are social realities or ideal social essences, which are independent in the sense that to them correspond specific sources of knowledge transcending physical or psycho-physical experience, is entirely erroneous, yet such an independence must be attributed to them as is due a product of abstraction, in contrast to the exemplary experience which forms the basis for the abstraction (1936a, p. 205).

While the view may be adequate for an understanding of the content of the concept of a 'legal person' and some aspects of the holism-individualism debate. I do not see – nor does Kaufmann intend – that it can be used for distinguishing between macro-concepts and micro-concepts in sociology. A particular family or group, for example, could also be described as a field of application of interpretive schemes. Further, Kaufmann's use of examples is unfortunate because he prefers examples of activities for which the rules are constitutive *and* known. But the field of regularities of meaningful behavior, i.e. action, is much broader.

Felix Kaufmann and the Austrian Theory of Marginal Utility

Kaufmann regards his chapters on 'the *Methodenstreit* over the theory of marginal utility' in economics and on Kelsen's 'pure theory of law' in jurisprudence as applications of the results of his general philosophical analyses. These applications should, therefore, show the merits and limits of his technique of 'clarification'. It is important to note that Kaufmann always performs the extraction of 'intended meaning' from various ambiguous formulations of the analyzed concepts and theories with an eye to their procedural function for gaining specific knowledge. His style of reasoning is a remarkable, and rare, combination of tolerance towards different scientific views, and of belief in the power of reason and logical analysis. In this concern with scientific procedure and rational reconstruction of existing theories I see an essential continuity between the *Methodenlehre* (1936a) and his later *Methodology* (1944), where the rules of procedure become central.

Kaufmann's method of analysis is, in both cases, to separate the content of claims from the grounds explicitly offered for them or implicitly presupposed in them. In particular, he considers whether the connection between a claim and its ground is of a logical or empirical kind or whether both are confused. If he finds that a claim is made on false grounds, he does not reject the claim altogether but looks to see whether it might be justified on different grounds. For instance, it may be valid for a narrower field than claimed or as a heuristic postulate rather than as a statement of fact. In this procedure of clarification, he combines the Vienna Circle's emphasis on the difference between factual and logical analysis with the wider Husserlian approach in separating out what is presupposed and what is posited in a judgment. (For an excellent analysis of 'epistemic claims' in this tradition see R. Zaner (1970).) It appears, then, that in his methodological analyses Kaufmann uses the distinction between synthetic and analytic propositions not as a general context-free criterion.

In summary, the main results of Kaufmann's analysis are:

1. There is confusion about the status of the principle of marginal utility. If taken to be unmistakably true, it is not an empirical hypothesis but the explication of a definition: it then cannot be falsified by experience because it makes no claims about reality. If understood as a proposition about actual behavior, it loses its incontestable validity and becomes an empirical hypothesis which can be falsified by research.
2. It is a methodological mistake to infer the intensity of needs on grounds of a choice between goods *if* there is no independent evidence of a hierarchy of needs. The elements of a theory of an economy without exchange do *not eo ipso* serve as the elements of a theory of an exchange-economy (the latter are not logically contained in the former).
3. Consumer demand is *not* the dominant factor in the determination of prices.
4. The theory of prices can*not* be *logically* deduced from value theory.
5. The quarrel about the possibility and fruitfulness of mathematical methods in economics reflects mistaken assumptions by both parties: indirect measurement cannot be rejected on grounds of intrinsic qualities of the objects to be measured, but the use of mathematical operations as such does *not* produce new empirical knowledge.

Let us briefly look at the recognition Kaufmann's analysis of marginal utility theory received. In his book *Positivism* (1939, 1952, p. 251) Richard v. Mises starts his chapter on marginal utility and mathematical theory with a reference to Kaufmann: "On the whole we agree with the careful analysis that Felix Kaufmann gave in his *Methodenlehre der Sozialwissenschaft* [...]." In a report on 'The Austrian School of Economics' (KP 000 626–648), the American economist Alan Sweezy describes the internal structure of the group in the 20s and 30s and refers to Kaufmann's critical position.

> I should like to distinguish between three groups within what is generally known as the Austrian School:
>
> (a) Mises, together with Hayek, Machlup (swinging over to the Kaufmann group) and the rest of his closest personal disciples Strigl and Robbins – between these first two there is no essential difference of opinion on fundamentals.
> (b) Mayer, Schönfeld, Rosenstein, and as far as his published works are concerned, Morgenstern. The chief interest of this group has been to develop utility theory further along traditional lines. In their methodology they are emphatic enough in rejecting the Mises standpoint but then seem to me to stop half-way and to remain in a highly ambiguous position.

(c) Kaufmann and Haberler, in which combination Haberler quite frankly takes the part of pupil [... Kaufmann's] chief interest in economics [is ...] one field of application for his more general work in sociological methodology [...] a methodological confusion lying at the heart of Strigl's *Kategorienlehre* [...] has been expounded with admirable clarity and thoroughness by Kaufmann in his two articles 'Die ökonomischen Grundbegriffe' (1923) and 'Logik und Wirtschaftswissenschaft' (1925).

Felix Kaufmann and the 'Pure Theory of Law'

Kaufmann believes that Kelsen's results in his early work are valid independently of their Neokantian philosophical presuppositions, namely, the assumption of a dualism of *'Sein und Sollen'* [fact and norm] and the doctrine of a basic norm. According to Kaufmann, there is no specific normative method, because normative sentences can be handled, just like other propositions, by appeal to their truth values. In his analysis of the 'value-problem in the social sciences' (1936a, pp. 169–193), in the general part of the book, he states:

> In normative considerations, two kinds of ascertainments must be distinguished; first, ascertainment of goals, second, ascertainment of the conditions of practical correctness with respect to these goals. Once the goals have been set, then the question of the efficacy of a certain action with respect to the goals (correspondence to norms) is a question of fact (1936a, p. 175).

This – Weberian – view is applied to legal theory, because "dogmatic legal science as the theory of the interpretation of law, never asks its questions concerning the law as such, but always about the 'law of a certain legal order'" (1936a, p. 296). It is simply not necessary to assume a hypothetical 'basic norm', which for Kelsen, mediates between the spheres of the factual and the normative.

> As our analysis of the concept of norm has shown, this assumption is not tenable; it owes its origin solely to the ambiguity of the term 'norm', where the aspect of issuing a command is confounded with the aspect of 'correctness' (itself in turn conceived in a confused manner) – which points to unspoken underlying goals. After the elimination of the ambiguity a specific kind of positing as criteria for the validity of law results – i.e. a positing where the two statements, 'The sentence L – which in its content is shown to be a legal sentence – is valid' ('is a component of a certain legal order', 'is' 'positive law') and the sentence L was posited 'in that specific manner' are *per definitionem* equivalent. Therefore there is no state of affairs at all that would require a justification to be carried out by a hypothetical approach (1936a, p. 299).

Similarly, on the basis of logical analysis, Kaufmann denies that the *'context of validity of legal sentences'*, i.e. their validity on the basis of the validity of other sentences, is specific for normative relations. In separating Kelsen's results from their philosophical basis, Kaufmann again uses the empiricist criterion of meaning (despite his theoretical reservations) and Husserl's ideas about the nature of judgments.

He engages in 'clarification of meanings' by separating the different meanings intended in the use of a concept and produces logical schemata to identify the logical structure of expressions. Independently of whether he succeeds in doing so or not, the limits of this kind of approach are clearly visible: by separating justice and command as two meanings of 'norm', the pressing questions of what norms should be set in a community are pushed outside of philosophy.

Felix Kaufmann and Alfred Schutz

According to Schutz's memorial speech (1949) for Kaufmann (*Sozialwissenschaftliches Archiv Konstanz,* Schutz-Papers, hereafter SP, 6379–6416), and his collection of 'Husserl's Influence on Me' (SP 6827–6836), Kaufmann was an older friend and a kind of tutor in their Viennese days; he brought him to read Husserl:

> In those days I was very close to the late Felix Kaufmann who at the same time worked on his first book, *Logik der Rechtswissenschaft* in which he successfully attempted 'to recast Kelsen's pure theory of law by substituting for its neokantian assumptions [...] a phenomenological epistemology'. He encouraged me to read [Husserl's] *Logical Investigations* and the first volume of [Husserl's] *Ideas*.

Schutz and Kaufmann spent long and regular hours reading Husserl together. On Kaufmann's request, Schutz included a methodological note in *The Phenomenology of the Social World* (1932), in which he clarified by which of Husserl's methods (transcendental or eidetic reduction) he had reached his results (1967, pp. 43–44). They had somewhat different interests: according to Schutz (SP 6379–6414), Kaufmann's interest in Husserl was not like Schutz's in 'the problems of *noema* and *noesis* and transcendental logic but rather in formal logic as an analytical *a priori*, the idea of a *mathesis universalis*, signification and meaning'.

What emerges from the published texts of Schutz and Kaufmann is a lifelong 'division of labor' between the two scholars: Schutz, writing about methodology, frequently quotes Kaufmann; Kaufmann, in turn, relies on Schutz's analyses of the structure of social action and the different perceptions of the social world. For instance, in Schutz's (1932/1967) *Phenomenology of the Social World*, publications by Kaufmann hold the fifth rank of references (after Weber, Husserl, Sander and Scheler), and in Kaufmann's (1936a) *Methodenlehre der Sozialwissenschaften*, references to Schutz (1932) hold the fourth rank (after Kant, Weber and Husserl). In Schutz's *Collected Papers* (1971, 1973) and in Kaufmann's *Methodology of the Social Sciences* (1944), both writers still refer to each other's writings when dealing with the issues named above. On the basis of a comparison between Schutz's and Kaufmann's publications, Schutz's views on social science can be read to be much less a foundation of 'alternative sociology' than is commonly assumed. In addition, the comparison shows that both men choose the same social theories, namely marginal utility theory and the pure theory of law for discussion which, considering the rich field of social inquiry in Vienna at the time (see above), is a very selective choice indeed. Neither the Lazarsfeld-type of empirical social research nor Max Adler's philosophical Marxism are so much as even mentioned (the former is perhaps understandably neglected because it had only just been started.) Finally, it shows the difference between the context of social science within which Schutz and Kaufmann worked in Vienna and the fully professionalized theoretical and empirical social science disciplines in the post-war United States, in which Schutz was to become famous (posthumously) in the late 60s.

A phenomenological concept of experience is central to both men's work. Kaufmann uses it to fight the logical positivists' early versions of logical atomism and the strong verifiability principle. In turn, in *The Phenomenology of the Social World*, Schutz often quotes Kaufmann's paper 'Soziale Kollektiva' (1929/30) and his writings in legal philosophy, mathematics and logic; especially important is the reference to Kaufmann's application of the concept of invariance to key concepts of legal philosophy, sociology and economics. Kaufmann's interpretation of ideal types as 'theoretical social laws' has already been referred to. I regard the concept of invariance of meaning, motives, etc., as essential to Schutz's methodology, that is to that element of methodology that I have called the element of 'scientific rigor', as opposed to the 'adequacy' element (cf. Helling 1979). Schutz writes, following Kaufmann:

> We have shown how the two most advanced 'theoretical' social sciences – pure economics and jurisprudence make use of ideal typical constructs (in our sense) in order to delimit their subject area and establish an objective context of meaning. What is true for the theoretical social sciences is generally true for all social sciences. Subjective meaning contexts are comprehended (scientifically) by means of a process in which that which is scientifically relevant in them is separated from that which is irrelevant. This process is made possible by an antecedently given highest interpretive scheme which defines once and for all the nature of the constructs which may be used (1932, p. 283; 1967, p. 248).

The problem of adequacy of methods to their objects of inquiry in Schutz's methodology can be approached from this angle. In this view, Schutz's call for the adequacy of methods in the social sciences is counteracted by his view that the objects of the life-world become objects of social science in a process governed by scientific perspectives, and described by Kaufmann as 'the process of the transformation of an object of experience into an object of inquiry'. There is, then, for Kaufmann and Schutz, no simple dominance of the objects of experience over methods, but an interaction between them.

Empirical social sciences, in Schutz's and Kaufmann's view, work on the level of objects of inquiry. For the rules governing this level, Schutz (in his American period) uses Kaufmann's (1944) concepts: The process of selection of observations for an ideal type is governed by the 'scientific problem', which in turn is determined by the 'scientific situation', i.e., the stage of development of a discipline. Work in the social sciences has to follow the 'rules of procedure' of a discipline, which are established methods, concepts and lines of argument of that discipline. In this interpretation of Schutz's methodology, the description of *The Structures of the Life-World* (Schutz and Luckmann 1974) therefore belongs not to sociology in the narrow sense, but to a 'protosociology' (Luckmann 1973).

The propositions of the social sciences can be verified empirically, but to this end one must use the phenomenological concept of experience, in which neither physical nor social objects are given in perception 'immediately'. Kaufmann's description of 'spontaneous' and 'receptive' elements of experience, and the incorporation of individual perceptions into a context for both kinds of objects and of protocol sentences in the social sciences are relevant here.

Schutz and Kaufmann do not propose a total difference between the natural and social sciences. Rather, on the level of the logic of explanation, both take a moderate unity of science view. On the level of the establishment and control of propositions, they do see considerable differences.

Whereas no methodological disagreements between Schutz and Kaufmann can be inferred from the published writings, cues to differences of opinion and interest can be found in their correspondence (Helling 1984). Such differences exist for the concept of '*Verstehen*', the scope and limits of methodology, and Schutz's use of Husserl's term *epoché* in his conception of the '*epoché* of the natural attitude'.

On understanding: In a letter of August 27, 1930, Schutz refers to

> some principal reservations concerning your concept 'of understanding' (causal reduction to psychic objects) [*kausale Reduktion auf Psychisches*] which are, I think, closely intertwined with the problems of Husserl's position that have become questionable to me.

On the natural attitude: In a letter to Kaufmann (17 September 1945) Schutz defends his conceptions of the natural attitude and its role in the systems of knowledge.

> I fully agree with what you say about the relation between *epoché* and skepsis, the difference between systematically uncovering the various strata of the meaning of existence and the presupposed doubt in existential positing. But this is a problem in the theoretical sphere, a problem of science, of logic, [...] of phenomenology. Indeed, the paragraph of *Formal and Transcendental Logic* you quote refers, as is evident from its context, to the scientist's concept of reality and truth. It refers to scientific apophantic judgment [...]. On this level of the problem everything you say is correct. Granting you this, I cannot see why it should be incompatible with my conception of the natural attitude and of *epoché*. The natural attitude refers to the life-world, which, being one and unified, is also the substratum of apophantic, and, possibly, critical scientific judgment. But in the natural attitude the concept of reality is not gained through judgment. From the start the life-world is taken for granted in the way in which it appears, unless motives appear which run counter to this general supposition. But just this general positing of the life-world as given, taken for granted, something beyond doubt I have called the '*epoché* of the natural attitude' which involves refraining from doubt, not from belief. The point is the naive attitude of man *in* the world, who poses its existence simply as real. It may be that calling such an attitude '*epoché*' is incompatible with Husserl's terminology.

On the proper scope of methodology: Congratulating Kaufmann on his book *The Methodology of the Social Sciences*, Schutz writes (21 October 1944):

> I believe that arguments can be raised only against the principal premises of the book. These accepted that the argument is stringent and for the most part, cannot be challenged [... My arguments relate] to problems not treated in the book, problems which I regard as essential elements of a methodology of the social sciences. I can't see how a specific methodology of social science and its themes can be developed without intensively treating

the problems of action, communication, intersubjectivity, of subjective and objective meaning, and the structure of the formation of types in the social sciences; and in particular, without going into the relationship of the interpretation of the social world by those living and acting in it to the interpretation of this same world by the social scientist.

Kaufmann (29 October 1944) answers this sharp critique:

> You have surely understood that I was troubled by the question of how deep to dig the foundations and how broad to build the structure, and that compromises between different perspectives are inevitable [...] I hope that I was able to make my main point sufficiently clear, namely the difference between deductive logic and methodology, so that it can serve as an orientation for the future research [...] in order to clarify the relation between your and my analyses it is important to see that I deal exclusively with the structure of scientific research, and therefore the problems of social interaction [*Umwelt*] with their constitutive strata do not become thematic on this level. The types of social science, on the other hand, are theoretical laws in my framework. The question of the formation of types in my book appears as the rational reconstruction of rules of higher order: none of the problems you raise for the 'objective stratum' is lost this way. However, this stratum [of meaning] cannot be transcended in methodological analysis. The methodologist too has his 'brackets' and must obey this limitation.

On the natural and scientific attitude: On the differences between the natural attitude and the scientific attitude which Kaufmann seems to have questioned, a year later Schutz (25 September 1945) sums up his position as follows:

> [The question of] the degree of coherence [in both systems of knowledge] is variable in both cases. I should think that its limits depend on the scope of the projects involved [problems posed] and the systems of relevance emerging from them. This essentially comes to the same thing as your formulation: that they depend on the practical aims *and* on the stage of development of science. (I would prefer to simply speak of the stage of 'knowledge', respectively – in the theoretical sphere – of the stage of development of science.) So far we are in agreement, except that even in the case of identical or typically identical selections the different interpretations of meaning involved would still justify distinguishing a particular theoretical province [of meaning]. But the selections *cannot* be identical, since in the world of working the selecting factor is the pragmatic motive with all its systems of relevance emerging from the fundamental anxiety. The theoretical thinker, on the other hand, is free of the fundamental anxiety but, for this very reason, also set apart from the 'world within reach', the time dimension of the pure [face-to-face] relationship, etc. His system of relevance is exclusively determined by the aporetic general situation as pregiven in a preconstituted science by the procedural rules governing any possible solution of his problems. You yourself have described and analyzed this situation with incomparable clarity. This is the starting point of all the problems you deal with in your book, and the manner in which you unfold these problems and carry through their analysis is in my opinion not invalidated by any of my approaches.

The differences that emerge from the correspondence and from Schutz's autobiographical statement quoted at the beginning of this section require some discussion since for sociology, they are not of solely historical interest.

First, they show that the labels 'Husserl' and 'phenomenology' can be used to pursue different projects. In their early writings, both Schutz and Kaufmann state the relationship of their writings (1932, 1936a) to transcendental phenomenology. Kaufmann remarks that though he has been deeply influenced by Husserl, his book is not to be read as a *phenomenological* theory of the social sciences, the methodological analyses contained in it do not pursue the problems of transcendental phenomenology, their aim is a 'formal', not a 'transcendental' critique (cf. the 'foreword' of 1936a). Schutz, in the 'appended note' he wrote on Kaufmann's suggestion to clarify the methods used in his (1932) study (1967, pp. 43–44), is more explicit. On the one hand he believes that the phenomenon of 'meaning' and 'understanding' can be comprehended by a transcendental analysis of the structure of internal time-consciousness, i.e., 'the bracketing (disconnection) of the natural world and therewith the carrying into effect of a complete change of attitude (the *epoché*) toward the thesis of the 'world-given-to-me-as-being-there (*als daseiende gibt*)''.

He then moves from that type of analysis to 'phenomenological psychology' which does not require the transcendental-phenomenological reduction, because, 'in ordinary social life we are no longer concerned with the constituting phenomenon as these are studied within the sphere of the phenomenological reduction'.

Still, Schutz is convinced that the results of transcendental analysis can be applied to 'the phenomenon of meaning in ordinary (*mundanen*) social life'. The aim of such phenomenological psychology is to create

> not [...] a science of the facts of this inner sphere of appearance, but a science of essence (*Wesenswissenschaft*) [...] thus the invariant, unique, *a priori* structure of the mind, in particular of a society composed of living minds.

Schutz and Kaufmann concentrate on different elements of Husserl's work: Though Kaufmann also stresses the 'foundational nexus of experience', his recurrent tool of methodological analysis is the analysis of judgments. Whereas Schutz seeks the basis for the comprehension of phenomena of meaning in a transcendental analysis of the structure of consciousness, Kaufmann holds that the core of the concept 'meaning' is to be found as it is used in the expression 'meaning of an act of thought', i.e., the objective meaning of judgment, which involves an abstraction from the occasional moments of the act of judging (cf. 1936a, pp. 153, pp. 38). That Kaufmann chooses judgments in the objective sense as a model for the meaning of understanding the actions of fellow-men seems by no means a compelling choice of a basis. His choice shows Husserl's strong rationalistic inclinations. On the other hand, his choice seems more justified for the level of methodology, because the social scientists' understanding of action in propositional form does claim to be valid independently of 'who, where and when' makes the judgment. The possibility of such judgments in social science is real because in everyday-thinking that abstraction is, according to Schutz's analysis, also performed in its anonymous ideal types. I spoke of a 'division of labor' between Kaufmann and Schutz because Kaufmann integrated Schutz's results into his methodology, which, because of Kaufmann's conception of methodology is restricted to the analysis of given scientific concepts, theories and procedures, their foundation in the world of everyday life being none of its business.

Since Kaufmann works on the level of scientific research, the finished goods, as it were, there is not as much need for him to analyze the differences between scientific and everyday-thought as there is for Schutz, whose project it is to ground the scientific concepts of social action and their products on the corresponding concepts in everyday-thought.

The general features of perception and experience which, taking them from Husserl, Kaufmann uses to argue against logical positivism, encompass the experience of natural *and* social objects and therefore, he stresses the similarities of explanation and understanding. Surely, the 'life-world' which according to Husserl is the forgotten ground of the sciences, refers to the way in which natural *and* social objects are given in our experience. In this sense, Schutz's project is a more restricted one than Kaufmann's.

Kaufmann's concentration on judgments may explain why he was attracted by *logical* positivism. It also could be a reason why, though in both books (1936a and 1944) it is one of his recurrent techniques of 'clarification' to specify the function of propositions, for the goals of knowledge of the theoretical schools he investigates; he is not very much interested in the practical methods used in the social sciences (the methods of observation, interviewing, conversational analysis, etc.). This restriction limits Kaufmann, as he recognized himself (cf. the foreword of 1936a and the Dewey-correspondence below), to methodology in the sense of 'a theory of scientific criticism' rather than of 'a generalized description of the behavior of the scientist (qua scientist) *in toto*' (the project he ascribes to Dewey). He believes that considerations beyond this limitation 'however interesting they may be from the point of view of the history of theories, the psychology and sociology of knowledge' (1936a, foreword) cannot be dealt within this framework.

In order to prevent the reader from drawing any conclusion to the effect that Kaufmann was 'a cold objective type', preoccupied with judgment, with Schutz for his part believing in 'humanistic sociology', I would briefly like to point to 'rationalistic' elements of Schutz's philosophy. By choosing 'the constituting process in internal time-consciousness' as the level at which the analysis of understanding and meaning has to commence, Schutz comes to describe action as gaining its meaning only through a reflexive grasp; while going on, it is meaningless. Though he insists that methodology has to account for communication and the life-worldly basis of theorizing, he himself splits the scientist into a 'solitary' thinker who by definition cannot communicate and the scientist who engages in 'scientific work' and in so doing communicates with fellow scientists (and the subjects of social research). Thus the legacy of traditional European philosophy – pure thought, clarity and distinctness – as transmitted by Husserl, is equally vivid in Schutz and Kaufmann.

Historically, Schutz's later fame is due to the fact that his writings were used – beginning with the early ethnomethodological writings, of Garfinkel (1961) and Cicourel (1964) – as arguments against the dominance of survey-research claiming the identity of measurement and explanation in the social and natural sciences.

For this purpose Schutz's 'phenomenology of the natural standpoint' was 'put on its feet' by treating the science of essence Schutz attempted to achieve by the methods of transcendental reduction and eidetic variation as factual statements, as a 'sociology of everyday-life' – considered as an alternative to sociology within the unity of science approach. The recognition Kaufmann received in sociology is restricted to his (1944) *Methodology of the Social Sciences* and, for the most part, to those of its parts that were taken up by Schutz, especially the 'rules of procedure' and 'the corpus of a science' at a given time (cf. Cicourel 1964; Garfinkel 1961). While Cicourel refers to these concepts as reconstructions of social science, Garfinkel also applies them to everyday knowledge of social structures. He explicitly refers to Kaufmann in introducing two ideas central to his work, the 'corpus of common sense knowledge' and the 'documentary method of interpretation': "The former has been developed from Kaufmann's concept of science" (1961 fn 7), and in arguing for the dominance of the latter in common sense *and* scientific interpretation, he uses Kaufmann's insistence on the dependence of subjective *and* objective meaning, on interpretative schemes.

After the *Anschluß*, in 1938, Kaufmann left Austria, emigrating to the United States. He obtained a post as assistant professor of philosophy at the New School for Social Research in New York, and later, in 1944, a professorship. In a letter to his friend A. Schutz, he expressed his happiness to be able, at long last, to conduct scientific research professionally, and he expressed gratitude to the Americans for having saved him and his family from the Holocaust (cf. Helling 1984a). In Vienna Kaufmann had been named on the antisemitic '*Proskriptionslisten*' (cf. Stadler 1979) shortly after his *Habilitation* (1922) and his '*venia legendi*' (formal certification as a university lecturer) was withdrawn immediately after the '*Anschluß*'.

Kaufmann's intellectual reference-groups shared his fate: Most of the Vienna Circle members (except for Neurath and Popper), the economists of the Mises group, and Schutz all emigrated to the United States. But, in contrast to their earlier concentration in Vienna, they now lived in different places and therefore required different modes of communication.

In the 10 years before his death in 1949, Kaufmann published in the following areas: contributions to the 'Phenomenological Movement' in America (Kaufmann was on the board of the newly founded journal *Philosophy and Phenomenological Research*), debates with Ernest Nagel and Rudolf Carnap on the notion of 'truth', general philosophy of science, philosophy of the social sciences, John Dewey's *Logic: The Theory of Inquiry* (1938). Economics and jurisprudence recede into the background, possibly due to the differences between Roman and Anglo-Saxon law and to the greater specialization of the social sciences in the United States. Whereas there is a strong continuity with the Vienna endeavours in his other interests, Kaufmann's preoccupation with John Dewey was new, and central for his self-perception. This development was described by Feigl in the following way (see also the interviews with Abel, Altmann, Dworkin, and Nagel below):

> A distinctly pragmatist note emerged fairly late in Dr. Kaufmann's work, possibly coming from the influence of John Dewey's thought. His reluctance to accept the tools of pure syntax, semantics and pragmatics; his insistence that the semantical concept of truth involves a realistic metaphysics and his identification of truth with warranted assertability, all this manifests the curious (but not necessarily paradoxical) fact that an outspoken phenomenologist and Kantian can in this respect be more positivistic than the positivists themselves (1950, p. 12).

Felix Kaufmann and John Dewey

In his introduction to *Methodology of the Social Sciences* (1944) Kaufmann stressed its difference from the *Methodenlehre* (1936a) (published in this volume) and his obligation to John Dewey. He dedicated it to Alvin Johnson:

> Shortly after the publication of my *Methodenlehre der Sozialwissenschaften* (Vienna 1936), it was suggested that I write a similar book in English, and I started to work on it. But gradually it became a very different book. This is largely due to my study of Dewey's *Logic – The Theory of Inquiry*. While I was strongly impressed by Dewey's analysis of scientific procedure, I could not accept his theory of meaning. This led me to a reconsideration of the problem how the logical analysis of scientific procedure (methodology) is related to deductive logic. I came to the conclusion that methodology must be clearly distinguished from deductive logic and recognized as an autonomous rational discipline. This distinction dominates the argument throughout the book (1944, vii).
>
> The (1944) book is dedicated to Alvin Johnson, the organizer of the 'University of Emigrants' who led scores of European scholars to the shores of freedom and guided their first steps in the new world with subtle wisdom."If it were not for Dr. Johnson this book and many others would never have been completed" (1944, viii).

I do not consider the *Methodology* 'a very different' book from the *Methodenlehre*. First, many elements are the same (especially the emphasis on logical clarification of concepts and the application of the general analyses to the social sciences). Second, the emphasis on grounds for decisions within scientific research procedure is already developed in the *Methodenlehre* and third, the rephrasal of 'truth of an empirical statement' in terms of 'warranted assertability' may be seen as already contained in Kaufmann's earlier treatment of the function of observation in science. However, his ideas are sharpened in the direction of methodology as an autonomous discipline, and he develops a more refined vocabulary for a descriptive and normative account of scientific research.

Next, I shall discuss the American reception of the 1944 book by briefly pointing to some reviews, and, in more detail, describing John Dewey's reaction. The basis of my description (cf. Helling 1988b) is the unpublished correspondence between John Dewey, Arthur F. Bentley, Jules Altmann, and Felix Kaufmann and the published correspondence of Dewey and Bentley (1964).

The book was reviewed most favourably by C. Wright Mills, most critically by George Lundberg. Mills wrote (1945, pp. 470–472):

> The secret of Felix Kaufmann's success as a methodologist is the live connection he tries to maintain with research procedure. On the one hand, he translates logical and epistemological issues into questions about the rules of procedure and tries to solve them in those terms. On the other hand, he explicates the meaning of procedural difficulties by restating them in terms of larger philosophical traditions. His book is a two-way shuttle and in both directions it is immensely clarifying [...]. The book as a whole is one of the best statements I have seen of the relations between physical science and social science and of just what a social research man can accept to learn from the older disciplines [...]. Since Dr. Kaufmann's chief technique of analysis involves the translation of general problems into questions concerning rules of procedure, the logical status of these rules is of central importance. His position on this point is quite similar to that of Dewey's, and [...] I believe that on several

> points Dr. Kaufmann's closeness to social science enables him to clarify and make more explicit what is involved. It is clear that he has been influenced by logical positivism, but he has by no means accepted the more radical statement of this position. In trying to make a unity of certain German and Americans traditions, Dr. Kaufmann's work represents, at several points, an advance over both of them [...]

Lundberg's (1945) review expressed the view that the identity of methods employed by the natural and social sciences need not be based on elaborate philosophical analyses, thereby implicitly referring to the growth of statistical methods in American sociology. Other reviews state that the book contained too much by way of philosophy and too little by way of concrete methods for the practicing social scientist.

After Ernest Nagel, Morton G. White, and A. F. Bentley told Kaufmann that they could not see as close a connection between Dewey's *Logic* and Kaufmann's *Methodology* as Kaufmann himself had, Kaufmann sought a direct response from Dewey. The contact was arranged by Bentley, who had met Kaufmann in Vienna and had written about Kaufmann to Dewey in 1940 (Dewey and Bentley 1964, p. 74):

> I am inclined to say of Kaufmann that I know of no one who is a better prospect to carry on your torch. He had gone quite a way along your line under his own power before he came to this country.

Kaufmann explained the relationship between his *Methodology* and Dewey's *Logic* in an appendix (of 20 pages) to a letter to John Dewey (2 February 1945). He started by summarizing Dewey's book and offering an interpretation of it, as follows:

> Some Tentative Suggestions for an Interpretation of Dewey's *Logic*
>
> I. Problems treated and interrelated in Dewey's *Logic*
>
> (A) Interpretation of inquiry as an organic activity in a cultural environment and scientific methods as habits of inquiry
> (B) Analysis of the general criteria of warranted assertability, including interpretation of deductive logic in terms of its function in inquiry
> (C) Critical reinterpretation and reconstruction of traditional epistemologies in methodological terms. Considering B as the chief objective of the book we are led to interpret A as an introduction to (B) whereas (C) has to be interpreted as an application (which is to my mind most important and promising) of some of the chief results of (B) to the history of philosophy. The full significance of (B) cannot be grasped unless it is isolated from (A) and (C), particularly from (A). (It might be convenient to have the term 'methodology' cover (B) exclusively, whereas 'Inquiry into Inquiry' would encompass (A), (B), and (C).
>
> II. A suggestion concerning formulations in B
>
> All statements in (B) should be formulated in non-genetic terms. The postulational (normative) character of methodology will then become more apparent and it will be easier to apply to it some results of the analysis of postulate systems in mathematical logic.

III. Fundamental points in (B) formulated in non-genetic terms

1. Knowledge of fact has to be defined in terms of rules of empirical procedure (criteria of warranted assertability). A definition of 'knowledge' in terms of a notion of immutable truth is not in accordance with the use of the term 'knowledge' in inquiry.
2. These rules determinate what is understood by 'sufficient ground for a change in the body of accepted propositions'. Such a change may be either the acceptance of a proposition or the elimination of a previously accepted proposition.
3. Whether a change in the body of knowledge is warranted or unwarranted (in terms of the rules adopted at the time under consideration) is completely determined by the (cognitive) situation at that time, i.e. by the body of accepted propositions.
4. It follows from (3) that the rules do not establish relations between propositions as such, but rather relations between a subset of the body of accepted propositions, called 'sufficient ground', and changes in this body.
5. To solve a problem in empirical science means to give a warranted answer to a given pertinent question. (Hence 'problem' and 'solution' have to be defined in terms of criteria of warranted assertability.)
6. Given a problem – which implies according to (5) the 'givenness' of criteria of warranted assertability – we may judge a procedure, i.e. a series of changes in the body of knowledge, as presumably relevant (irrelevant) or as presumably more or less relevant for the solution of the given problem. (These judgments presuppose criteria of presumable relevance respectively, e.g. specific rules of procedures).
7. It is not in conformity with the adopted rules of scientific procedure to isolate within the framework of this procedure immediate experience (knowledge by direct acquaintance), and to contrast it with mediate inferential knowledge. In the first place observation is not a passive perception of data but involves a selective process. In the second place we cannot grasp the role of the observational test within the context of inquiry unless we determine its interrelation with other rules of procedure.
8. The system of the rules of procedures provides for the possible elimination of any accepted proposition. This is tantamount to saying that no accepted proposition is exempt from possible invalidation in scientific procedure.
9. Rules of procedure too are subject to changes. These changes are not determined 'from the outside' but within inquiry, which means that (a) the criteria of warranted assertability and (b) the patterns for the solutions of given problems are defined in terms of other rules of procedure.
10. In order to understand the forms and principles of deductive logic we have to determine their place within the system of procedural rules.

Relations between the interpretation of Dewey's *Logic* as outlined in III. and my *Methodology*.

Kaufmann continues:

I agree with points 1–9. The sentences in III within parenthesis indicate points which can be 'read into' Dewey's *Logic*, though they have not been actually made by him. The introduction of the terms 'scientific decision' for any change in the corpus of a science and 'scientific situation' for the body of propositions accepted (at the time of a decision) might also be traced to the *'Logic'* as soon as B has been isolated from A and formulated in non-genetic terms.

The distinction made between

(a) basic rules (criteria of warranted assertability) and preference rules (pattern of the solution of problems)
(b) rules of the first order (criteria of the correctness of scientific decisions and rules of higher order (criteria of the correctness of changes in the rules)
(c) empirical laws (synthetic propositions that may be falsified by one negative instance) and theoretical laws (specific rules of procedure which are criteria of warranted predictions) can hardly be 'read into' Dewey, but they seem to be in harmony with his view. The same is true for
(d) the introduction of the notion of 'step' (indivisible correct scientific decision)
(e) and the distinction between single rules of procedure and common properties of all systems of rules which are implicitly referred to when one speaks of scientific method as such.

(While any single rule may be altered, the common properties of the system of rules are invariable for scientific procedure, which is tantamount to saying that 'scientific procedure' is defined in terms of them).

I cannot accept point 10 and consider this disagreement, which concerns the nature of discourse (analysis of meanings) as the only fundamental difference between Dewey's conception of methodology and my own [...]

From the beginning of the correspondence, Kaufmann's separation of genetic analysis and analysis of the meaning of scientific procedure is rejected by Dewey (26 February 1945):

> I should have a great difficulty in improving on your A), B), C) summary [...] It shows a clear grasp of what I was trying to do. But then I come to the statement that the significance of B) cannot be grasped unless it is isolated from A) and C). My immediate reaction is to the contrary, that its significance can't be understood if it is isolated, since such isolation cuts the vital nerve of my whole treatment.

Kaufmann, in contrast, regarded his logic of procedure as a development of (level B of) Dewey's theory. He felt justified to do so on the basis of Dewey's remark (in the letter quoted above): 'Given the full recognition of the connection of B (relation of criteria of warranted assertability and deductive logic) with A), the material of B) can *then* be developed on its own account [...]', and, of course, from his understanding of such passages from Dewey's *Logic* as the following:

> To engage in an inquiry is like entering into a contract. It commits the inquirer to observance of certain conditions. A stipulation is a statement of conditions that are agreed to in the conduct of some affair. These stipulations are at the first implicit in the undertaking of inquiry. As they are formally acknowledged (formulated), they become logical forms of various degrees of generality [... A postulate] is empirically and temporally *a priori* in the same sense in which the law of contrasts is a rule regulating in advance the making of certain kinds of business engagements. While it is derived from what is involved in inquiries that have been successful in the past, it imposes a condition to be satisfied in further inquiries, until the results of such inquiries show reason for modifying it [...] Only after inquiry has proceeded for a considerable time and has hit upon methods that work successfully, it is possible to extract the postulates that are involved (1938, pp. 16–18).

As the correspondence develops, all the concepts introduced by Kaufmann and regarded by him as elaborations in Dewey's spirit came under attack from Dewey and Bentley:

1. His use of the expression 'autonomous logic of science' which he employed in order to distinguish the rules of procedure from deductive logic was rejected. It was understood as an introduction of a system of rules separated from the process of inquiry and its connection with practical life, and it raised the suspicion that he attached to logical analysis (analysis of meaning) a higher dignity than to empirical analysis.
2. His use of the expression 'intuition of meanings' had to be, understandably so, defended against the suspicion that it involved the assumption of some mental power, and had to be reformulated.
3. His concepts 'scientific situation', 'scientific problem' and 'scientific decision' came to be regarded as breaking up the continuum of inquiry and involving a judge of some higher competence.
4. His concept 'presupposition' was understood to imply a researcher who consciously presupposes rules, whereas for Kaufmann, the presuppositions are reconstructed by the methodologist and not necessarily conscious to the researcher.

These differences are reflected in letters such as the following (Bentley to Kaufmann, 22 February 1945):

> You go along with Dewey – or rather you develop in parallel with Dewey, much of it under your own power before you found him – until near the end. You then say suddenly: Behold the Son of God born among men. Behold the rules which validate all things and without which there would be no breath of life. Good little Jesus [?] is all right with us, but we can get everything out of him as a natural born human that the others can get as an immaculate conception. From my point of view you simply do not need at all your fiat at the finish. If you want to use the word 'validation' the way you see it, we agree. J. D. does not attain it. But we also add no such thing is found in the heavens or on earth or in the waters under the earth. We can [...] get everything human nature can use in 'situations' and we can always be broadening the situations. It is the breath of life to Peirce, to Dewey and to me (in the most characteristic phases of work) that we are not inclined to tell the universe just where it gets off in any final way; we are willing to let it run on.

In the last instance, Kaufmann understood by 'methodology' something different from Dewey and Bentley. His judgment in the following letter, is, I think, correct (Kaufmann to Dewey 2 August 1946):

> [...] you understand by 'logic of science' a generalized description of the behavior of the scientist (qua scientist) *in toto* whereas I mean by it a theory of scientific criticism. In formulating a question of scientific criticism – such as whether it is correct for a person to accept a certain statement on the basis of given evidence – we presuppose (postulate) that all the relevant words and sentences have unambiguous meanings, even though actual scientific work is progressing on different levels of clarity. But to state this is not to deny the possibility of a shift in meanings (of a change in definitions) as inquiry

goes on. In looking upon a particular question of scientific criticism as a logical unit I do not mean to sever the scientific decision at issue from the potentially endless series of scientific decisions [...]

Clarifications of mutual misunderstandings do occasionally occur in the correspondence, but they never resolve the basic differences. In order to document these and the style of the discussion, I include two complete letters by Dewey and Kaufmann.

504 South St., Key West, Fla. April 2, '47
Dear Dr. Kaufmann:

It has been a long time since I wrote you, and I owe you an apology for seeming neglect. But a good many things having interfered with continuity in my efforts – not the least, presumably, being my marriage in Dec. just before we came down here. We shall return somewhere about the middle of this month and I hope Mrs. Dewey and I may have the pleasure of seeing you and Mrs. Kaufmann at our apartment, 1158 Fifth Ave.

But what I started to write about was a question you asked in a letter to Bentley some time ago. He had evidently said something about "observation showing us something". You asked what does observation show? The point as I got it since you mentioned the need of prior specifications of limitations or conditions. What can observation show us? At all events, my great difficulty in our prior correspondence has been to locate the difference between us – it has been an elusive thing. Now if I were to say that your question seems to assume that observation is some kind of a faculty and that its workings have to be examined before any statement about what it 'shows' can be accepted; I doubt if you would accept that statement of your position – though it seems to me to be implied – if you did accept it, it would be a pretty good specification of the kind of things which separates us. So I'll put it this way. The answer to the question what does observation shows us, it's that [it] shows us what precisely it *does* show us. This answer is not an evasion nor yet a mere tautology. For the body of accepted scientific conclusions at a given time is that which observation shows us. Observation to us in other words is a name for inquiries it comes to a head. However, I want to go further than this. Our position, at least as I see it, to us, a fact, rests upon the assumption that we have no way of telling what observation is save in terms of scientific conclusions connected with the procedures that render them scientific. Now, if your assumption is that we have any other way of assigning significance to observation that would mark the kind of difference there is between us. Of course Kant's *Critique* represents the basic systematic assumption that there is another way and that it must be used before any given conclusion can be accepted as properly meeting the criteria for 'science'. However, in spite of your occasional favorable reference to Kant I don't wish to assign this view to you gratuitously. I am writing to ask if the difference between us is something of the kind herewith indicated in the case of observation and what it shows.

With regards,
Sincerely yours,
JOHN DEWEY

April 8th, 1947
Dear Professor Dewey:

Many thanks for your letter of April 2nd. Permit me to repeat my warmest wishes on the occasion of your wedding and to express my hope that you are enjoying your stay at Key West.

In my current attempts to develop the points made in the first part of my *Methodology* I am constantly and gratefully aware of how much your and Bentley's searching questions help me in reformulating my view. This applies to the question raised in your present letter. In replying to it I should like to state, first of all, that nothing could be more dissonant with my view than an interpretation of scientific inquiry in terms of a faculty psychology, and a conception of knowledge *a priori* as prior in time (innate) and superior in validity to empirical knowledge. While I admire Kant's *Critique of Pure Reason* immensely I cannot help regretting that he leaned so heavily (at least in his terminology) on the faculty psychology in the form it had been given by Tetens. Misunderstandings of his doctrine by contemporaries which he branded as 'almost intentional distortions' are largely due to this fact. As to his conception of knowledge *a priori* he was opposed to the doctrine of innate ideas; but I admit that some of his phrasings are apt to convey an impression to the contrary.

In asking what observation can show us, I am suggesting a systematic clarification of the function of observational tests in inquiry; i.e. in the process of validating, or invalidating, assertions. Two main questions emerge in this context, viz.

(a) What principles of structure (from organization) are implied in the statement of an observational finding, irrespective of the function which such a finding may have in testing a given assertion?
(b) What does it mean to say that an observational finding provides a test for a given assertion?

I am primarily concerned with question (b) which has to be discussed within the frame of an analysis of warranted assertability. When we state that a given observational finding warrants a given assertion (more precisely, the acceptance of this assertion) we must be prepared to defend this statement against anyone who challenges it; i.e. we must be able to demonstrate that the acceptance of the assertion on the strength of the observational finding, is in conformity with implicitly adopted standards of warranted assertability. An integral part of such a demonstration is the explicit formulation (clarification) of these standards.

The problem may be elucidated by comparing it with the issue of determining the standards of rigorous mathematical proofs. We cannot even accurately formulate, let alone answer, the question whether a mathematician's claim that he has presented a rigorous proof of a mathematical proposition, is tenable, unless we lay down transformation rules in terms of which 'rigorous proof' is to be defined.

Similar considerations apply to the far more complicated and diverse criteria of warranted assertability. The very establishment of the notion of warranted assertability which I regard as a milestone on the road toward a rigorous theory of science suggests the explicit formulation of standards (canons) of validation and invalidation. Let me develop this point at some length.

The notion of warranted assertability warns us against defining 'empirical inquiry' in terms of attainment of or approximation to truth where truth is conceived as an unalterable property of propositions unrelated to the knowledge situation at a given time. It suggests a re-definition of such fundamental terms of the logic of empirical science as 'problem', 'solution', and 'ground' without reference to a notion of transcendent truth. This holds irrespective of whether truth is interpreted in terms of correspondence, or of ultimate coherence. While it does make a difference whether we conceive of factual truth as an ideal of inquiry or as pre-established irrespective of inquiry, this difference does not affect the point at issue. For the very conception of scientific inquiry as a self correcting process implies that the ideal of ultimate coherence of our scientific findings is a transcendent ideal. We can never claim that we have attained, or approximated to ultimate coherence, since we can never know whether the coherence established at a certain stage of inquiry will not be upset by future inquiry. We say that progress in science has been made, if there is more coherence at a given stage of inquiry, than there was at a preceding stage; but we cannot, strictly speaking, interpret this gain in coherence as an approximation toward the ideal of ultimate universal coherence. Peirce's 'faith of the logician' may be strengthened by progress in inquiry, but it is not vindicated by this progress. It is no more adequate to interpret progress in inquiry as progress toward an infinite goal of ultimate coherence, than to interpret progress in counting as a progress toward the infinite; and to state, for instance, that 100 is closer to the infinite than 10.

Your 'finitistic' interpretation of scientific procedure appears to me as an outstanding achievement. It is, indeed, impossible to 'bring logical theory into accord with scientific practice', unless we dispose of transcendent concepts in our interpretation of inquiry. My own approach follows this clue.

The program of bringing logical theory into accord with scientific practice demands (as I understand it) that we should define methodological concepts in accordance with their use in scientific criticism. I shall illustrate the point for the pivotal concept of warranted assertability. Its definition should imply reference to the criteria, and to nothing but the criteria, for the distinction between warranted and unwarranted assertions. This distinction is made in terms of 'rules' which establish 'conditions' for the acceptance of assertions and for the reversal of the acceptance.

No matter how widely these canons may differ, they have some traits in common. First of all they relate the acceptance (or the elimination) of the assertion under scrutiny to the 'scientific situation' that is to the body of assertions accepted at the critical time. Referring here, for the sake of simplicity, only to the procedure of acceptance of an assertion, we can say that acceptance (in conformity with the rules) depends upon whether or not a set of specified assertions ('grounds') belongs to that body of accepted assertions. A definition of 'warranted assertability' (in general) must therefore refer to rules of scientific procedure which establish conditions for the acceptance of assertions in terms of previously accepted (and retained) assertions. And the meaning of 'warranted assertability' in a specific case must then be determined by listing specific rules of this kind (which I called 'basic rules of scientific procedure'). Only in defining 'warranted assertability' in this way shall we comply with the postulate that the definition of a scientific term should be in conformity with the scientific use, for the only use which the scientist makes of the notion of warranted assertability is in distinguishing between warranted and unwarranted assertions. Now it seems to me that this demand cannot be satisfied unless we detach the logic of science from a causal explanation of the process of inquiry. Otherwise 'extra-systematic' definitions of procedural terms will be allowed to enter into the logic of science. This is the point from which our discussion started, and which I should like to re-emphasize. But I feel that your *Logic*, which abounds in valuable suggestions for the future analysis suggests also the just outlined approach. At any rate I do not think that the finding of this approach was unrelated to my previous study of your great work.

We secured for Dr. and Mrs. Bentley a quiet room in the Mayflower Hotel.

Best regards,
Yours sincerely,
(Felix Kaufmann)

Though Dewey and Bentley accuse not only Kaufmann, but also other authors such as Russell, Nagel, Carnap, and others of establishing logical analysis as a new kind of metaphysics (cf. Dewey and Bentley 1964, p. 68, p. 451), their correspondence with Kaufmann contains what Kaufmann might call 'extra-systematic typifications'.

This correspondence is, apart from its substantive value, a document of the difficulties of communication between a European emigrant and American scholars, for Kaufmann's statements are interpreted by Bentley and Dewey against the background of Dewey's (1915, 1942) views of *German Philosophy and Politics* (cf. Helling 1988b).

(I wish to thank Jules Altmann for making the following letter to him from A. F. Bentley, 13 May – or March – 1945, available to me.)

> I mean it literally when I say [...] over and over, that Kaufmann in action is German-language-in-action and that German-language-as-in-Kaufmann is such that the wave motions of Dewey-Bentley-language simply disappear in the sea. He never catches them at all. He stands blank. I on my side am blockaded by this complete absence of recognition on my part of a spark of acceptance on his part of the points on my part that are advanced [...] Kaufmann (the man as a walking organism) is so affected by his German vocabulary – the meaning of words as he uses them – the shades of meaning – that we don't get together [...] K. sees *Geist, Vernunft, Verstand* – all these as mental operations in a nonmental world (I don't give a damn what he says as creed about it – I am reporting what he does in speech) [...]

Dewey and Bentley suspect that there lies at the basis of Kaufmann's theory of scientific criticism yet another 'quest for certainty'. For Kaufmann, clarification of meanings and procedures by rational reconstruction is an instrument for settling the various methodological battles which had recently been fought in the German scientific literature of many disciplines (sociology, economics, history, mathematics) and to facilitate scientific progress. For Dewey, there is no need for clarification and foundation of the sciences in this sense; his aim is to analyze 'science as the most developed form of free human intelligence' in order to apply its methods in everyday life.

Despite the fact that the schools of thought of post-World War I Vienna treated in this introduction – logical positivism, phenomenology, pure theory of law, and the Austrian school of marginal utility – produced radically different solutions to the problem of finding a basis for knowledge, their effort to solve this problem is indeed their common trait. All of them seek to analyze scientific knowledge and to establish rules for gaining and criticizing it. Among them, Kaufmann was the great mediator.

A Selection from Interviews (1982) and a Letter on Kaufmann

Family, Friends, and Colleagues

George Kaufmann

H.: I suppose you don't remember much about the old country. I'm really interested in how your father managed to have this job and at the same time moved in these [intellectual circles].

G.K.: My father had incredible energy, to the point that it killed him. He had a tremendous capacity to work. But on the other hand, his *Stanzerl* are very much an element of his personality. He was very successful as a manager of the Anglo-Iranian Oil Company. For example, this is the only thing I know about it, he competed with other oil companies for the contract with the Austrian Federal Railways and he got the contract for the Anglo-Iranian Oil Company, which aside from being Jewish and an intellectual were reasons why he was obviously a target. My father had anticipated the problems with the Nazis by 1931. And he used to tell a story about how he had just bought a radio and the first thing that they heard was something that the Nazis had done. And they never listened to the radio again. But despite that my father waited till almost too late because he felt that he had to write this one more book, the *Methodenlehre* to make his reputation. He was in all these circles because that was what he was really interested in. And he was in the business because he felt that his first responsibility was to earn enough money for his family. He had grown up in a family who originally were fairly affluent Jewish businessmen. But they lost everything in the inflation and so he felt he had to do this and he did it. But his real interest was intellectual work. He had his three books on philosophy of law; then he wrote the *Methodenlehre* and articles on this or that or the other. He was in the economics circle despite the fact he was not a believer in Mises' philosophy. Hayek was a close friend to him. And he was in the logical positivists circle despite the fact that he was not a logical positivist. There is a story about that: When I was in college I went to the library and somehow I would gather there was some reference to my father and made some crack about him not being a logical positivist but a logical negativist. But he wrote me back saying that he was glad that I hadn't said he was illogical […]

The assassination of Schlick was to my father one of his horror stories, kept coming back […] A great tragedy happened to him: He started to lose his hearing during the emigration. He was for ten days in London and all of a sudden, my father began to lose his hearing. It deteriorated and he was very nervous about it. He was sensitive about it. It was a long time before my mother did get a hearing aid those days, these things didn't work very well. And the result was that he was much more cut off. He and John Dewey, after all, were in the same city, right? But I don't think they met more than twice […] because it was too much of strain for my father […]

In my father's system mathematics was the architectonic. I remember him using that very phrase. And his basic view in general was that the philosophers didn't know enough mathematics. But he was as open-minded as one can be in intellectual matters – he certainly wasn't open-minded in terms of his judgements about the upbringing of children. And it was for that reason, he was the only one of the refugees, European scholars, who came over here, who took Dewey seriously […] Dewey himself said that and even wrote a letter to him […] He was prepared to take it seriously; he reconsidered his own system, responded to problems, and to Carnap and all these people it was the same and it was, I think, a part of the same aspect.

Of course, now, in feeling the sense of the essence of his approach to being a scholar he was open to all these other meanings. Equally he was critical; he wasn't accepting things; but he was totally open-minded.

One of the things that always concerned me was that he was a little bit – for somebody of his intelligence and with his background – he was a little bit naive in this respect. He was always offended by other people's careerist interests. For him the subject matter was the key. And the last thing that he wrote, in the sense of being public, John Dewey had his 90th birthday in 1949, the year my father died, and my father did three or four, I think it's four substantially revised drafts of the speech that he gave. The first one was a '*Kritik*' [Kaufmann 1959] of Dewey's system, and then he successively cut it down, so that at the end the celebratorial speech remained and he was geniously concerned about what was the appropriate way of doing this.

His standards for other people as for himself were too high. And so that in that way he would cut himself off despite the fact that on a personal basis he could be, he wasn't that kind of personality. He was not a personality who goes out, who invites people – quite the contrary. He was charming but I do think that the hearing problem had been a very serious psychological impact on the whole emigration; the tragedy for him.

In this particular era work was everything for him and it took priority over everything; took priority over his family, although in the old country, I mean, he was in the Mises and so and so circle. He had this job […] so basically his work was all his meaning. He cared about his family, loved his family and he was quite interested in most of my education, but his work had priority for him

He came over here and he dedicated his *Methodology* to Alvin Johnson. And I remember we talked about this: "Why don't you dedicate it to your family?" I said. But my father said: "No, this is the right thing." […] He was a Dewey disciple in the sense of Dewey's concern for education, that part of progressive education and Johnson had the idea of starting a new school; adult education, people coming in the evening and people who had been too poor to go to college now taking adult education courses. And then he had that great idea, [of the University in Exile] fascinating idea.

I didn't know Johnson that well; my father regarded him as a demigod. But Johnson, on the other hand, had the highest regard for my father. He kept his letters from Johnson which is his response to the dedication in which Johnson says he regards my father as the greatest methodologist […]

My father was very much a believer in this '*mens sana in corpore sano*' and so in the old country, skiing, mountain-climbing, and till the end my father was a walker, very much. And he did some of his best things when walking. So that, for example: the New School was on 12th street; the Columbia Library was on the 116th street; we lived way, way up, on 232nd street [...] The north–south streets go on twenty miles. My father, you couldn't do that now for security reasons, but in those days, when my father had a class – these classes were all in the evenings – he would write in the mornings, write during the day and then teach in the evenings. But many times, regular mornings, my father walked from 12th street [...] as far as 200th [...]

After my father died we were short of money. Bentley would send me checks. Bentley made it possible for me to finish university. Bentley used to visit us. Bentley would come to New York and they would walk. I mean, so my father wasn't completely unwilling to talk to him. So he would talk to Bentley [...] the main point was, he hardly talked to anyone because of the hearing problem [...] terrible. Look, he was very young, when he died, he was 54 when he died. But when it happened he was 43, and I think it was something that made the auditorium nervous, so that he would be uncomfortable at meetings; he wouldn't attend all these conferences. This is where the real personalities exchange, particularly in this country. One to one he could deal with and, for example, he met with Nagel [...]

In teaching he worked hard; he took it seriously; and some of his students obviously loved him [...] He had basically a double standard. People who were just taking a course here and there, in grading them etc., he was generous. On the other hand, people who, you know, particularly, the doctorate candidates, those he was tough with. And I remember discussing this with him, about the grades. And his viewpoint was: "Look they want to go out and they themselves want to teach other people." [...] And I used to say to my father: "How come you are never dean?" And my father said he didn't want to. There were two reasons apart from the hearing. He had trouble with the hearing. But his basic reason, his basic reason was the take away from his work. And he wasn't interested. And, I mean, it was 500 dollars extra; and 500 dollars was a lot of money. But he made a judgment that he didn't want to be dean. And I was pushing him because it was a form of ambition; but we had this conversation and he was always quite clear, he didn't want to. But he was, and this is very important, he was the chairman of the Scholarship Committee. And this he took very seriously [...] He did to some extent, even in basic courses prepare; he had basic undergraduate courses. The basic courses he was teaching, even to the non-doctorate people, he would work on and try to freshen them up and learn from what difficulties the students had and so on; in spite of the fact this wasn't what he was really interested in.

I know nothing of the substance. I can't help you on that. Indeed, to me it's a great tragedy because he died exactly in the vacation after my first philosophy course; so we were about to start talking about this. But one thing I can tell you, too. The first is: He said to me more than on one occasion that he was at the point where the substance of the problem wasn't what created the difficulty for him. What created difficulty for him was to try to convey this so that other people will understand it. And he even chose the title of this document 'The Pursuit of Clarity' which he wrote and he rewrote in order to make these difficult concepts accessible. That was going to be his next book and the very title of it is very much symbolic of what he had in mind […] But I do think that he thought a lot of disagreements were unnecessary, were due to failures of understanding, mutual understanding. Then he thought there were people who were still locked into positions, who were unwilling to understand the contrary position […] they weren't willing to consider etc.. I mean, in his mind, he never put it this way. I think he would have been the last to put it this way, but it was *Die Sünde gegen den Heiligen Geist* [the sin against the Holy Spirit]. This was the ultimate, to a scholar this was the ultimate sin; and so he took extraordinary pains to refine the ideas, to try to make them more […] understandable. But he even had the idea of writing a book, the title of which was going to be *Philosophy for George*. Of trying to take some of the ideas, obviously not all, and writing in such a way that children can understand. I don't think he meant children; I guess I was at high school at the time; he wanted to be able to talk about some of these things in such a way that people can understand. And indeed, we did. I mean, I remember a discussion, some play of Shakespeare and some very popular amusing play that he and I had seen together as a movie. So he thought that it ought to be possible to take some of these basic problems, value and so on, and put them in such a form that a layman could understand that anybody could understand. Obviously he never got anywhere; but forbidding as I assume a lot of his stuff is, when you read it, he took great pain and regarded this as an important part of his function not to be up there, but to try to be clear and he was against jargon […] and sloppy thinking was to him just anathema […]

I'll give you another insight, it is personal. My father enjoyed smoking. But my father didn't think it was a good thing […] He decided that he was not going to smoke until he published a book. When he published his first book which must have been in the mid '20s, he began to smoke three cigarettes a day. And whenever he published another book he would add one cigarette to his quota. So that at the time that he died – he had six published books – and he would smoke eight cigarettes a day […] He might sometimes smoke that last cigarette five minutes past midnight, but he never deviated from that; and I would ask him about that. He said: "Boy, you know,

I'm enjoying this" [...] But when I said: "Look at all the articles you've published"; he would say: "Well, he smoked Kingsize cigarettes [...]" It's symbolic, but he really had extraordinary self-discipline [...]

My father was very musical and he loved to sing which after he got hard of hearing he had to stop doing. But my mother would play the piano and I was brought up on '*Die schöne Müllerin*' [Schubert] and Löwe, '*Heinrich der Vogler*' and these songs. He never played an instrument, but he did sing; and he loved to go to concerts, etc. He was also very widely read, he had a tremendous memory and could recite from '*Faust*' and so on at any occasion. And there was a sort of family story: I don't know whether it's true or not but between them, my father and his brother could recite the whole '*Faust*'. He placed a great store on this and also on knowledge of languages. He used to criticize himself because he had never learnt Greek, because he wanted to be able to read Aristotle, Plato in original [...] I can't remember it exactly, but one of his slogans [in Latin] was that a person is worth as many languages as he knows. And, of course, he had been brought up on this classical education with a lot of Latin, etc. [...] he was a voracious reader.

H.: I remember invitations in the old country where he asked friends to come and somebody would sing. So that had to stop in this country, too? [...]

G.K.: We were very well off [in the old country] because of my father's job. Over here we were poor, not poor, well, but money was a problem. It was a problem for two reasons. First because we weren't able to bring over any money; second because his earnings as a professor, you know, and third because we contributed to other people, gave them money when they came over. We had to support my grandmother, my mother's mother; we had to help to support her and after the war sent money to all sorts of people. I remember, I mean, my father's secretary and, I guess, his assistant had been very decent to my father when we left. And my father considered this to be a moral obligation on his part; so he would sent these people care packages and things like that. I was able to go to school because of scholarships and so on. They were very careful about the budget, which was completely different from the way they had grown up. Now, obviously, there were a lot of people who were worse off, poor; well, on the other hand there were people that came over and they made a lot of money. But they never complained about it; they were grateful to be alive. But I remember, for instance, talking to my father about, you know, making money on the book, the *Methodology*; he didn't make money on the book. He had to pay for it because of all the corrections he had to make in the proof. He taught summer school to bring in extra money; not every year, but some years [...]

H.: Schutz said in his memorial speech that your father was very happy to be in this country not only because that enabled him to get away from the holocaust but also because it enabled him to be a full time scholar.

G.K.: I think that's absolutely right. That's really what he wanted to do. On the other hand if things had worked out differently, if there hadn't been the whole '*Anschluß*', etc. [...] I mean we never discussed this, but I'm quite sure it's true: there would have come a time when my father had earned enough money and, presumably, being independent that he would have given up his job. And then he would have devoted himself to scholarship, presumably. At that age it turned out he died. If he had been fit by the time he got to fifty he certainly would have had enough money. Perhaps earlier. Now the Anglo-Iranian Oil Company had offered him a job in London. I don't know what the terms of that were; but he opted out of this. Now I think there were two considerations. I think one of them was that he was going to take this opportunity to be a scholar and I think that he probably anticipated that there would be a war and England would be in it. And then it would be better and safer to be in the United States.

H.: I interviewed a man named Rosenstein-Rodan whom he knew in the '20s in Vienna. He told me that he was in London. And one day in the '40s he suddenly, got a telegram from New York, from Felix Kaufmann offering him a job in this country. And he said Felix Kaufmann saw that England was going to be taken by the Nazis and he was looking after his old friends and worked hard to get them, bring them to safety.

G.K.: Well, I'm glad you mentioned that because there was another thing. An enormous amount; I mean, I said before how devoted he was to scholarship, etc., but the thing that took absolute priority at this time, of course, was trying to help people to get out, friends and relatives, and get them jobs, to the point that this was so much a matter of the conversation that the key document was the *affidavit*, which was a testimony of somebody here saying that they would in fact finance you if you couldn't get a job. And I, I mean in the first summer, in 1938, he went to Ithaca. How my father came to that, I don't know. Except that he knew there was a library, there at Cornell University, and he had decided he had to publish an article right away. I don't know why; he had a job, but anyway. He had decided [...] We rented an apartment and we spent the summer there. And we would eat lunch. – My mother had never cooked, she could cook, but we had a maid in the old country. We had a maid and I had a governess. Also had an English nurse up to the last week. So we used to eat at home because we couldn't eat out all the time. But Sunday as a special treat we would go to this one place and we would have a 'blue plate special' [...] this was towards the end of the depression in this country, the 'blue plate special' was 35 cents, it was a three-course meal. We would order two of those and split them three ways. And then, since there were only two desserts I would be given a nickel to get an ice-cream. And we thought this was fabulous and we had lunch for 75 cents for the family. The point of the story was that while my parents were having their coffee and desserts, I got into a conversation with these two Cornell students and then they came up and I introduced them to my parents and so on. And then I said: maybe these people can help to get an affidavit. And, actually, it did turn out that they got to be best friends with my parents. And they introduced them to somebody else, who got an affidavit [...] this was, you know, the absolute concern getting everybody out and then helping them [...]

H.: What about his friends from the old country like, for instance, Carnap? Did he meet them in this country?

G.K.: Oh, sure. Well, I already told you they had this elaborate correspondence and that was over here [...] In 1948, I think, surely and largely at Carnap's instigation my father had a visiting professorship during the summer of 1948 at Chicago in the philosophy department. And we lived in the Carnaps' apartment [...]

H.: Well, you see, in today's academic scene it would be impossible for two persons of so different convictions, one being a famous logical positivist, Carnap, and the other describing himself as opposed to some central ideas of that school to have a friendship of that kind.

G.K.: This was my, I mean it sounds terrible, almost; this was my father's special thing. It didn't matter to him that somebody believed, had different views as long as they were willing to deal with the problems in an honest matter and that they worked on them, etc. That's all that mattered to my father [...]

Herbert von Fürth

H.F.: [...] Ja, also ich war ja im Hauptberuf Jurist und im Nebenberuf Nationalökonom und war daher hauptsächlich an seiner [Kaufmann's] Rechtsphilosophie interessiert.

H.: Sagen Sie, war die Emigration nicht besonders schwierig für einen Juristen wegen der Unterschiede zwischen römischem Recht und amerikanischem?

H.F.: Ja. Ich war im Jahre 1931, 1932 hier als Rockefeller Fellow in social science. Infolgedessen hat sich die Rockefeller Foundation in der anständigsten Weise darum bemüht, mir gleich eine Stellung zu kriegen.

H.: Übrigens können wir Englisch sprechen, ich muss sonst alles nachher übersetzen, weil das Buch auf Englisch erscheint.

H.F.: Bitte, bitte. When I came here I had great luck, first of all, that Professor Haberler was here at Harvard and . . .

H.: When did he come? When did he come to this country?

H.F.: Professor Haberler came here about 1930, I think. First, again, as a Rockefeller Fellow and then in 1934 as assistant or associate professor and then in 1937 or '38 he became full professor at Harvard. And, second, that as a former Rockefeller Fellow I had not only my contacts which I made during my Rockefeller year, but also the official protection of the Rockefeller Foundation, so that after one year as a special student at Harvard I immediately got a job as a professor of economies at a small Negro college in Pennsylvania; and four years later then I was called to the Federal Reserve as an expert on international finance and stayed with the Federal Reserve until my retirement 1966, 1967 and then went on to the Foreign Service Institute as a faculty member and so my only title is Faculty Associate Emeritus of the Foreign Service Institute; and I'm still giving lectures there. But therefore my problem was really very simple. By the way, when I arrived here for the second time in 1938 I stayed for the first few weeks with Dr. Kaufmann in New York before I went to Cambridge, Massachusetts. And I had known Dr. Kaufmann from my first student days on and I was a lawyer of his firm, of the Austrian subsidiary of the Anglo-Iranian Oil Company, and was, as you perhaps know, together with Fritz Hayek, the founder of this little Viennese circle, of whom Felix Kaufmann was one of the most prominent members. So we had a very close contact.

H.: That's *Geist Kreis*, is it?

H.F.: Yes, yes. But, please, this *Geist Kreis*, as you know, was a nickname to us by a lady economist who was incensed that we didn't accept ladies [In colloquial German at that time, *Geist*, spirit and mind, was something reserved for the male sex – H.].

H.: And you didn't have an official name then?

H.F.: No, no. There was no official name, there was no official organization. There was nothing at all. You are familiar, I assume, with [1974] Professor Engel-Janosi's book on that subject, his autobiography […]

H.: So you met Kaufmann in […]

H.F.: 1918, after the war, when we both came back. He was a few years older than I was; but I was in the war only for two years; and he was in the war, I think, for four years and that equalized our studies.

H.: I see. And then you studied law?

H.F.: I studied law and economics. That was combined in Vienna and he did exactly the same; only that he was primarily interested in philosophy of law.

H.: Did he study before the war, too?

H.F.: I really don't know exactly when he started. Perhaps he started already in 1913, I simply don't know.

H.: What sort of a life was that: studying law with Kelsen at that time in Vienna? Who else was there?

H.F.: Oh, it was extremely stimulating because the law faculty was excellent; but the economics faculty was also excellent and enlivened by the fact that the two major professors or three major professors were completely at loggerheads with each other. There was old Friedrich Wieser, of course, one of the founders of the Austrian school, who was, by far the most eminent of them; but then there was Othmar Spann, who had later got a bad reputation; but at that time although we all knew he was not a great economist he was probably the best teacher I have ever had or seen in all my life, fantastic teacher [...] and he was a great nationalist, although not a Nazi at all, on the contrary, as far as I know, the Nazis put him in a concentration camp when they came; but he was very nationalistic and therefore antagonistic to the more international orthodox Austrian School. And the third, a Mr. [Carl] Grünberg, was a Marxist. So that one really had a lot of stimulus.

In fact, our little circle, as I had written to Engel-Janosi and he repeated in his autobiography, was really an offshoot of the Spann Seminar because one day, at one of these seminar meetings, we were just graduated, but were still attending what one would call in this country 'post-graduate seminars', Spann made a statement, I don't know exactly what it was; and Hayek and I both objected violently. And after the seminar was over Spann called Hayek and me and said: "I'm deeply disappointed that my two favorite students opposed me on such a fundamental question"; whereupon Hayek and I told each other that we didn't like a seminar in which one was not supposed to speak one's mind. And that we would start a kind of a seminar of our own and, as Spann, not just on economics but on all social sciences because Spann was really a universal man and that was the origin of our circle. And we took all the promising youngsters whom we knew, about twenty of them, all together: art historians and historians and statisticians and all that; and started a little circle which met once every two weeks, completely without any organization; we always met, the landlord presided and that was that; and I was the official secretary of the whole group, you see. And the interesting thing is that of these twenty odd people half of them became internationally famous. It's quite fantastic. Hayek himself, of course, and the economists Haberler, Machlup, Morgenstern. Hayek and Morgenstern were the two outstanding economists of the group, and Voegelin in political science, whom I considered then and still consider today the only real genius of our group, and Kaufmann and Schutz and the art historians Benesch and Wilde, who are both still today considered. Just the other day, I read in *The Times Literary Supplement* a review of a book by a British Art historian and it said there: Well, this fellow studied under the great old master, Johannes Wilde. So, I mean, he is still today considered. And Engel-Janosi was a very good historian, and the statistician Karl Menger, the son of the founder of the Austrian school, of course; so all these people became internationally famous. And that's a fifty per cent ratio, which I think is really without precedent in the history of such circles.

H.: It's, of course, due to Hitler, too, isn't it? Due to the catastrophe: I suppose, if they all had stayed in Vienna.

H.F.: Well, no. The really good ones all left Austria before that. They were called away. Hayek was the first to be called to London in 1931; Haberler was called to Harvard, Menger was called to Notre Dame at that time; Machlup was called to Buffalo; Morgenstern to Princeton, well, [...] And there were four extraordinary outstanding economists alone, you know [...] And there were quite a few others who were of lesser ranks. I mean, obviously, the best ones left first [...]

H.: So Kaufmann was a member of your circle and?

H.F.: And of the [Ludwig v.] Mises Seminar. You are acquainted with his biography [Mises 1978], are you? That's important. There Kaufmann is also mentioned quite frequently.

H.: And also he attended the . . .

H.F.: Kelsen

H.: The Kelsen Seminar, I got that. And the Vienna Circle. So how did he make that – for me it seems too much.

H.F.: Well, don't forget, for instance, the Mises Seminar was really an offshoot of our circle. More than half of our members became members of that seminar and more than half of the members of this Mises Seminar had been members of our circle, you see.

H.: But was it at the same time? It was not one after the other?

H.F.: It was founded 2 years later, 2 years after ours. And the so-called philosophical Vienna Circle, [soon the accepted title, coined by Otto Neurath for the circle around the philosopher Moritz Schlick – H.], you know that Mrs. Haberler and Mrs. Fürth are sisters of the founding members of the Vienna Circle? There were three cousins of them. So it was practically a family affair.

H.: But that is very interesting.

H.F.: You see, that is one of the things. The intellectual circle in Vienna was extremely narrow. Once when I went through my family papers I found an invitation to a session of the Vienna Branch of the *Verein für Sozialpolitik*, with which you are familiar, too. The chairman of the meeting in the 1890s, the chairman of the meeting was Haberler's father, the two speakers were my father and the man who later became my father-in-law. I didn't meet Gottfried [Haberler] until 1918 and while I had met Mrs. Haberler in 1916 – she happened to be a High School class mate of an old friend of mine – I didn't meet my wife until 1924. So it wasn't that the families were so closely connected, but they were just members of the same [milieu].

My parents knew Ludwig Mises, of course, and it was all the same people and completely interracial [...] One third of our membership was [so-called] Aryan, one third Jewish and one third mixed, either of mixed descent or of Jewish descent married to non-Jews or non-Jews married to people, to women of Jewish descent. And that reflected really the composition of this intellectual society in Vienna, what I call the 'liberal establishment', consisting of the university and the professions and the bureaucracy – in contrast to the 'conservative establishment' consisting of the army, the church and the aristocracy.

H.: Yes, I see. Did you mention just now that your wife and your sister-in-law were related to members of the philosophical Vienna Circle?

H.F.: Yes.

H.: Who were they?

H.F.: Hans Hahn was the son of a Jewish journalist who had married an aunt of Frieda and Emmi.

H.: I see.

H.F.: And his sister was married to Otto Neurath. And these three, the two Neuraths and Hans Hahn, were [later in the '20s] really together with Schlick the pillars of the philosophical circle [with Neurath, the 'first' Vienna circle before 1914].

H.: I'm really glad I could come to talk to you because I did miss a lot.

H.F.: And the thing is between the Hahns and the rest of my wife's family there was a big family feud – so she met Hahn through me.

H.: Was Kaufmann related to other members of the Vienna Circle, too?

H.F.: No, no. I don't know anything about his family, I must say.

H.: What about the Gomperz Kreis?

H.F.: Well, you mean Heinrich Gomperz, the philosopher? Well, I knew him, but I know nothing about – I was never really interested in philosophy; somehow these things, never understood what this means. Still today when I read something of Alfred Schutz, very, very intelligent, of course, and especially, since Eric Voegelin thinks it's highly important, I'm quite willing to agree it's highly important but I don't see the point, that's my fault.

H.: Well, maybe you are more of a practical man?

H.F.: Well, yes, you are quite right, you are quite right. I consider also economic theory useful only in so far only as it is policy [–related] and well, if you are interested in my attitudes at all [...] The last thing I wrote in Europe was something where the first quotation is from Kaufmann. In this article as a kind of a joke – silent joke – I made it a point to quote only members of our circle.

H.: Oh, I see. Yes. Was that a general policy of your members?

H.F.: Oh no. I never told anybody. It was just a private joke of mine.

H.: Well, you see, Schutz and Kaufmann quote each other very much.

H.F.: Yes. We had, of course, our discussions. You see, here in that paper I quote Kaufmann, Voegelin, Fröhlich and then again Kaufmann and Schutz, and then again Voegelin, Morgenstern, Haberler, Hayek. I couldn't quote Menger, of course, because it had nothing to do with economics. But, you see, I've always been interested in these borderlines between law and the social sciences. And that was also my work here as a Rockefeller Fellow. And that's been practically my work ever since; it was my work at the Federal Reserve; and now the only scientific task I'm still doing is to teach a course on history of economics at the Foreign Service Institute and, as I tell my students every time, more thought and history than economics. But obviously, since we [at the *Geist Kreis*] were discussing these things all the time, well, if they wrote them down, they quoted each other.

H.: What sort of things did you discuss?

H.F.: Everything, everything under the world. Everybody was permitted to talk about whatever he wanted, and, of course, he usually talked about the things about which he was writing something and, I think, most of the published works of the members of our circle had first been discussed in our group, you see? Menger has a list of all the papers given in this group [...]

H.: So you had a very broad education, as the Americans would say, in the University and also ...

H.F.: Yes, yes. Also at the University and through our group. And the fact is that we had art historians and literary historians, you see.

H.: And Kaufmann was different with respect to the fact that he also was interested and knew about philosophy and mathematics?

H.F.: Well, Karl Menger, of course, was a mathematician. So he was not the only one. The fantastic thing about Kaufmann was that he was a philosopher and an eminently practical businessman. You know, he was the head of the Austrian branch of the Anglo-Iranian Oil Company.

H.: Yes, I heard that but I never understood how that was possible.

H.F.: He was an excellent businessman and he once told me:"You know, philosophy of law is so much more difficult than selling oil. I can do everything I have to do in my office in four hours a day and can spend my time on the things I'm really interested in. And in these four hours," he said it himself, "I can outcompete all these poor businessmen" and he did. He got for Anglo-Iranian a monopoly of selling oil to the Austrian Federal Railroads, which was a fantastic thing, you see. And he just did it with the help of one secretary and one office boy. He said: "What there is to it? There is nothing to it at all. One just has to be honest and know one's subject and not to be too greedy, and then one can outcompete all these people from Poland and Romania. So that is all." Because I once told him: "Is it not a waste of your time to spend time selling oil when you could spend it on much more important problems?" He said: "I'm not wasting my time."

H.: But wasn't it, I remember from Popper's autobiography [1976], that he said at that time and you said that before, too: there just weren't any academic jobs?

H.F.: That's right. So he was an unpaid instructor at the university.

H.: Yes. So I'm asking, did he really have an alternative?

H.F.: Well, he could have probably gotten a job somewhere else, you see, if he had published a little bit more. I think through the influence of Husserl and this thing: he probably could have gotten academic jobs abroad. But he didn't want one. And, you see, as soon as he emigrated, he immediately got this job in New York, 1938. Now, I came here – he came here in the summer of 1938 and I came in the fall of 1938. And by that time he had already his professorship at the New School.

H.: What was his relationship with Kelsen then?

H.F.: Kelsen? Oh, he was a tremendous admirer of Kelsen, tremendous admirer of Kelsen.

H.: I want to get a clear picture of life in those days, in particular of Felix Kaufmann's life. So he went to all of these four circles all the time. Is that right? Because that's a lot of activity, to be a member of four circles.

H.F.: Well, I assume he had one or two meetings a week, you see, that was all, if that much. I don't know how often the Mises Seminar met because I was not a regular member, I don't know whether they met every week or every two weeks. We met every two to four weeks and I'm sure the Vienna Circle didn't meet more often than that. So probably, he didn't have meetings more than once a week. So that's not too much.

H.: What about the political situation?

H.F.: The political situation was in so far very good as the only parties that counted were the Socialists and the Conservatives, Catholic Conservatives. And none of us was either a real socialist or a Catholic conservative. And therefore none of us was really much interested in practical politics. Hayek and I, when we came back from the war, we founded an Association of Democratic Students at the University of Vienna in 1918. But while it had a little bit of an influence while we were there, it completely deteriorated after we left. And we never bothered with politics after that. I was sorry about it because my family had been politically interested for generations. Our grandfather had been member of the Austrian parliament from 1860 to 1888; and my father was a member of the Vienna City Council; my mother was a leader of the Austrian Women's Movement, so I was reared with politics; but liberal politics had no future in Austria after 1918, so . . .

H.: Was that true for Kaufmann, too?

H.F.: I don't know, as I said I know nothing whatsoever about his family, about his background.

H.: No, I mean his politics.

H.F.: I think he was not at all interested in politics. And I think, that's true of all of us. Eric Voegelin, at one time, felt that perhaps the Schuschnigg government had some ideas that could be elaborated and put into some theory; and he got a little bit too much entangled with the people, but I don't know how he feels about that today. At that time, I wasn't happy about it and I'm not happy about it today because – while Schuschnigg was possibly the only alternative to Hitler, and therefore from my point of view, of course, preferable, certainly, from an absolute point of view, he wasn't what an old liberal like I would have wanted either. So . . .

H.: It now seems to me that the fact of these circles could have been one reason why Kaufmann was able to publish in such different fields.

H.F.: Oh, yes, sure. You're completely right. And that was the idea. We didn't want to become narrow-minded specialists and, especially, I didn't want it because for purely material, financial reasons I had to be a lawyer although my interest was in social science and so we purposely tried to become as, as you said, broadly educated as possible.

Now another thing is, of course, I once told Fritz Hayek: "Without knowing it we grew up among giants." Because, you know, that general intellectual atmosphere in Vienna centered around Freud and Adler in psychology; Wittgenstein and Popper in philosophy, Klimt and Schiele in art, Mach and Boltzmann in the philosophy of science;

Wagner and Loos in architecture; I mean, unbelievable, absolutely unbelievable, not to speak of the Austrian school, after all, I still knew Menger and Wieser personally; and my father had been a friend of Böhm [–Bawerk] and Philippowich. So in such an atmosphere one had to grow up as an intellectual. You are, I'm sure, familiar with Schorske's [1980] little book on [...] Vienna at the turn of the century. He maintains that the reason for this intellectual brilliance was that the more or less liberal bourgeoisie didn't have any political influence and therefore turned to intellectual pursuits. Now, I don't know whether that's true or not, but, in any case, that's his idea. And his idea is that breakdown of the whole liberal philosophy led to this exciting explosion. One can more simply say with Hegel that 'the owl of Minerva' starts to fly at dusk, I don't know.

H.: It's tragic, too, isn't it.

H.F.: It's tragic, yes. Oh yes. Tragic.

H.: What about Max Weber's influence? Do you recall any of that?

H.F.: It was very great. Max Weber, of course, had been in Vienna but before 1918, so I never met him; my mother was a friend of Marianne Weber's and I still remember an evening at the opera with Marianne Weber, who had come to visit us; and in 1920 I felt I wanted to study one semester in Germany and I wasn't sure whether I wanted to go to Munich, to Max Weber, or to Heidelberg, to Alfred Weber. And I decided to go to Alfred Weber because I said Munich is so close to Vienna I can always see Max Weber later. Of course, the next year he died.

H.: And what about Kaufmann and Weber?

H.F.: Kaufmann never met Weber personally but, of course, Max Weber was for all of us the one great sociologist and philosopher of social science; so his influence was enormous.

H.: I see, yes. Was that true for – but, of course, the division into sociology, economics, psychology wasn't as strong as today, was it?

H.F.: Well, sociology was considered, if you pardon me, a little bit an out-of-step science. Something which is very interesting but which doesn't have the rigor of economics, so that economics was supposed to be the social science par excellence; and I think, Alfred Schutz in Vienna would have called himself more a philosopher than a sociologist.

H.: I see, yes. Which he was, too, because he didn't do any empirical study.

H.F.: No empirical work at all. And if you ever get hold of a copy of the *Zeitschrift für Öffentliches Recht*, you'll see that I try desperately to find something to do for the sociology of law. In fact, I was quite amazed when I was a Rockefeller Fellow here to find that the sociology of law was taken far more seriously in the United States than in Austria or Germany. Although it was imported from Austria and Germany more or less. But still, here it was taken very seriously; in Austria and Germany it was not.

H.: Yes, I have often wondered about Felix Kaufmann's conception of methodology because in the German book of 1936 his concrete methods are Weber's 'ideal types'.

H.F.: Yes.

H.: And then he has this chapter on Kelsen and one on 'marginal utility'.

H.F.: On what? On?

H.: 'marginal utility in economics'

H.F.: Yes.

H.: But he did not in that book discuss any actual social theories.

H.F.: No, no. I think we were utterly unfamiliar, let's say, with Dürkheim and functionalism in general […] These things were completely alien to us.

H.: So it was really Weber?

H.F.: Was really Weber.

H.: Weber and in Kaufmann's case also Kelsen and economics and that was the social sciences?

H.F.: Yes, yes.

H.: What about the Bühlers? The psychologists in Vienna at that time?

H.F.: Well, I knew of him; I didn't know him personally. We had among our members one psychoanalyst, Robert Wälder, who then came to this country, and became quite a prominent theorist of psychoanalysis here; and we, of course, all knew of Freud and Adler. Well, as you know, Hayek was an absolutely fanatic opponent of psychoanalysis and I have the feeling that we really didn't think that that was a very interesting thing; and, obviously, it influenced us subconsciously a lot. But I don't think we would have really wanted to consider psychoanalysis a social science on par with the others.

H.: Did Kaufmann know Neurath at that time?

H.F.: I'm sure he did, of course. Of course, he must have known him […] Of course, Neurath alienated most of us by his radical socialist attitude. I personally never met him although he was a cousin of my wife's.

H.: I see, yes. Because of that reason?

H.F.: His wife knew my mother, but I never met him.

H.: Because of this reason?

H.F.: Well, the whole interest and, especially, his participation in the Bavarian Communist attempt was considered very ill-advised [Neurath did not consider himself a communist]. We were certainly all dedicated anti-communists.

H.: So Kaufmann did his law degree in 1920? His degree in law, the doctorate, he did this in 1920?

H.F.: Yes, yes.

H.: Then he did another one in philosophy in 1922 and then he got the "Privatdozentur"?

H.F.: Yes.

H.: Well, then maybe we could turn to Kaufmann's life in the United States.

H.F.: There I know very little because, after all, I never lived in New York. So we corresponded a little bit, but that was all.

H.: He did not really make an impact in social science in this country.

H.F.: I doubt it, I doubt it. I think he was really overshadowed by Schutz.

H.: Yes, but that was much later, too, because Schutz did not, I mean, Schutz became famous after he died.

H.F.: Oh really?

H.: Yes, so. And I wonder why that was.

H.F.: Well, perhaps his whole thinking was too much anchored in these old Viennese concepts. After all, don't forget that Austrian economics after its initial impact at the time of Böhm-Bawerk then started to make an impact again only after Schumpeter came to Harvard in 1932. That was a renaissance of Austrian economics; and as far as I'm concerned theoretical economics in the United States.

H.: Did Kaufmann participate in economic circles in this country. Do you know?

H.F.: I doubt it very much. I doubt it very much.

H.: Because when you say he was probably – in order to be accepted in this country he was probably too much rooted in the Austrian concepts. What do you mean by that in concrete terms?

H.F.: Well, in the United States at that time certainly the prevailing philosophy was pragmatism [...] and in order to impress his students, [Kaufmann] I think, had to start really from the roots to tell them what theory meant, what philosophy meant, etc. etc. And I'm sure that quite a few, at least those who wanted to have practical influence in this country probably felt that it was far too theoretical and too foreign to everything that people are really discussing, subjects that get you a good professorship somewhere. But that's my interpretation; I never talked to him about it and don't forget, those of us who emigrated after 1938 were so happy to have a decent job that we really didn't bother very much with the question of whether we are having a great impact on the public in the United States. That was the least of our worries […]

H.: Back to Vienna – was Kaufmann considered a phenomenologist there or a logical positivist – or?

H.F.: No, a phenomenologist, oh yes, certainly. You know, he and Schutz, of course, battled the orthodox positivists very much and, you know, both of them were absolutely adhered to Husserl […] I don't know whether they ever met Husserl personally […] But he mentioned Husserl all the time.

H.: Yes, the curious thing about that is that in the old book [*Methodenlehre*] he says: Even though I regard Edmund Husserl as a very great philosopher and even though I owe very much to him this book stays outside of transcendental phenomenology. So.

H.F.: Yes, because he really wanted to be a methodologist rather than a basic philosopher but as far as, since after all, all methodology must be founded on some philosophy, certainly the philosophy on which it was founded, was Husserl's.

H.: I see.

H.F.: Now, I suppose, Schutz was more outspoken a supporter of Husserl.

H.: What do you think was Kaufmann's function in your discipline, in law, if there was any? […]

H.F.: I think he was among, together with Voegelin, Hayek and Schutz, he was probably the most influential on the people around our circle. And I know he had a tremendous influence on the people around Mises and he had a great admiration for Mises, much greater than I had because for me Mises was a dreadful dogmatist. I never completely understood why he should be really ranked in the same rank, as people, let's say, Schumpeter. But he was a very decent person and a very good economist, of course.

H.: What about the normal sociologists like Paul Lazarsfeld at that time?

H.F.: Well, Lazarsfeld was one year below me in high school, in the same high school; but I suppose none of us would have ever gotten the idea of asking Lazarsfeld to join our circle. None of us. I mean, the success of Lazarsfeld in this country still amazes me. I still don't understand it.

H.: And when you discussed social science problems you did not talk about methods in the narrow sense of [practical research] methods?

H.F.: Well, Kaufmann talked about methods a lot. I never was particularly interested in it; but, of course, we respected him and so if he thought it was important then we were willing to discuss it.

H.: Yes. But didn't he talk about methods in a philosophical sense and not in the sense of how do you get your data?

H.F.: Oh no, oh heavens, nobody talked about that [laughter]. That was much too mundane.

H.: I see. But that's what, since I'm a normal sociologist, this is what I have to.

H.F.: Well, sure.

H.: Well, my own position is, I don't think one can separate methodology totally from the actual practicability of doing research.

H.F.: I know that's a modern point of view, I know that. Well, I'm getting more and more skeptical about the practical applicability of economics anyhow. Conventional economists are completely helpless in the face of problems today whatever they are: Marxist or Keynesians or classicists. None of them has any real answer.

H.: Yes. That's what the layman says, but it's interesting to hear that from you.

H.F.: Oh, Schumpeter said that seventy years ago. Eighty years ago. Now at that time economics was really in its infancy one could have thought that by now it could have, but it simply didn't. [...]

Gottfried von Haberler

H.: [...] When you first met him [Kaufmann], when was it?

G.H.: Oh, that must have been in the middle 20s at the University of Vienna. I cannot place the year, but the middle 20s. We had a so-called *Geist Kreis*, you may have heard about that, in which he was a member and I was a member and Dr. Fürth was a member and during that total time I knew him very well. And I saw a good deal of him also in the United States when he was at the New School. I was at Harvard at that time; and he once came to Harvard for a seminar. I knew him really quite well. I really helped him to [come to this country]. And a friend of ours gave him an affidavit to come to the United States.

Mrs.H.: And then he was here and he was afraid Hitler might kidnap him here, so, you see. Well, people were very, at that time, people were very nervous.

G.H.: But now you'd better ask concrete questions.

H.: Yes I would like to know more about Felix Kaufmann's relationship to these various circles in Vienna.

G.H.: Yes, he was a member of this *Geist Kreis*; he was a member, I believe, of the *Nationalökonomische Gesellschaft* and he was a member of this Schlick Circle, at least he was well connected with that group. He was a student of Kelsen, you know that probably?

H.: Yes.

G.H.: So he really was by profession, he was a lawyer, and legal philosophy was his first field and then it branched out to the social sciences; but he had his hands in economics, social sciences, philosophy. He wrote a book on the infinite in mathematics [Kaufmann 1936]. You know about that that has recently [1978, Engl. tr. of 1936] been published [...]

H.: How was it possible to be in so many circles?

G.H.: Yes, yes. He had his hands in many fields and his business at the same time. He was really a most remarkable man. He was a poet. As you can see from his *Stanzerl* [poems] here.

Mrs.H.: He also was a musician.

H.: Was his book well received, the 1936 book [on methodology]?

G.H.: Which one is that?

H.: Methodenlehre der Sozialwissenschaften

G.H.: Yes, it was quite well received and we all read it and, I must say, I learned a lot from it. Yes. As you will see from the other papers which his son should give you – he had his hands, let me say, in philosophy, pure philosophy; there are some of them on Neokantianism and on Husserl. Yes, that is very important. You probably know that. That he was an admirer of Edmund Husserl.

H.: I know that but I don't know how visible it was that he was. Did he talk about Husserl, for instance, in the Geist Kreis or did he teach about Husserl?

G.H.: Dr. Fürth knows most about the *Geist Kreis*. He, I think, kept a sort of record; he probably can tell you; but Kaufmann was very close to Husserl and a great admirer of Husserl's phenomenology, and he was between phenomenology, Husserl, on the one hand and the Vienna positivists on the other. So he was in between all these schools and the social sciences and economics [...] But some of [the poems] are really very excellent [...] on Edmund Husserl I remember one by heart:

> Was Edmund Husserl mich gelehrt,
> kein Wort vergeß' ich je,
> doch aller Weisheit Basis
> bi, ba, Basis, das ist die Epoché

You see, it's a very nice one [cf. Kaufmann 1992, p. 38] It goes on through many lines [...]

> What Edmund Husserl taught me
> of this, I will never forget a single word.
> Of all wisdom though the basis,
> be, ba, basis
> this is the *epoché*

[my translation].

H.: Did he teach legal philosophy in Vienna? Is that what he taught?

G.H.: Kaufmann didn't teach in the philosophy department, but in the legal department, *juristische Fakultät*, and this is rated not actually the faculty of philosophy.

H.: So, he must have taught legal philosophy strictly?

G.H.: Yes, I think that was the subject he was teaching. But he was not a professor; he was a *Privatdozent* [...] But he was a member of the Mises-Kreis.

H.: And did you have any controversies, academic controversies about methods and so on in the Mises-Kreis?

G.H.: Yes, to some extent.

H.: Because I'm trying to understand the atmosphere in Vienna at that time.

G.H.: Yes, you know, Mises was, well, yes, this is a technical thing. Mises was an apriorist of the logic of economics. Kaufmann did not agree with that. But they never had a main disagreement or controversies, but he has one of these poems which relate to that [...] He called Mises an economist in paradise. There he says: "The poor man here, I cannot function", because nothing is short, you see. To economize means to economize short things, but in paradise nothing is short. And then I remember this verse:

> Da sprach der Herr: Du sollst nicht klagen derfen, (bitte, Sie verstehn Deutsch).
> Du kannst Dir eine Theorie entwerfen, das macht den Menschengeist erst so erlaucht,
> daß er zum Denken nichts zu wissen braucht.
> Zwar kannst Du niemals einen Satz erproben,
> doch ewig drum sollst Du mich stündlich loben,
> So bleibt die Lehre aufrecht unentwegt,
> Wo nichts erprobt wird, wird nichts widerlegt.

> Then spoke the Lord: 'My son thou shouldst not grumble
> Wherefore should there upon thy theories crumble.
> The human mind with such mobility is blessed
> That without knowledge it can function at its best.
> True thought canst not subject laws to confirmation

> But that itself should be a cause for jubilation
> The theories thus may ever stand unmoved.
> What can't be tested cannot ever be disproved.
>
> [cf. Kaufmann 1992 p. 30, Engl. tr. p. 43].

H.: That's wonderful!

Mrs.H.: Witty, yes.

G.H.: And that, of course, angered Mises, because Mises – but this is a fine point, I don't know if you want to hear of that – because Kaufmann really disagreed with Mises [...] most angry relations. Mises was also in New York [...]

H.: And what about his relationship with Alfred Schutz in Vienna?

G.H.: Well, they were members of the *Geist Kreis*, they were personal friends. Schutz was also an admirer of Husserl [...]

H.: Did Kaufmann bring his phenomenology into the Mises Circle, too, or was that different? Did he talk about phenomenology?

G.H.: No, no. He mainly talked about economics, and, philosophy of economics, scientific philosophy but not phenomenology. He was, I would say, he was not an outright phenomenologist. He was somewhere in between phenomenology on the one hand and positivism on the other.

H.: And was considered as such?

G.H.: Yes

H.: I see – I had often wondered why he didn't make more of an impact in the social sciences in this country.

G.H.: Yes, I think that's true, he did not make an impact. He taught at the New School, but he was not quite healthy anymore. His hearing was very bad; and here on the whole people were not so much interested in philosophy of science. I mean, the economists were not interested as much as they were in Europe. Because this was a special thing in Vienna: this group of economists, philosophers and you had this common interest in philosophy of science [...] Yes, you can say that economics has become much more specialized. Most of them had no interest in philosophy of science. At that time in Vienna economics was not specialized; it was not quite developed. But you are quite right: his impact here was not very strong . Now who else, is there anybody alive who was his colleague at the New School? [...] I'm sorry. I'm not quite well prepared.

H.: It's very kind of you to see me.

G.H.: Well, he was a very close friend of ours. I much admired him [...] His hearing problem was very bad. I remember when he came to Harvard to have a seminar, and it was difficult [...]

H.: What do you consider Kaufmann's function within your discipline? See, you said I learnt a lot from him; I would like that a little more concrete.

G.H.: The distinction between the empirical sciences and logic and mathematics, you see, which I think this is something which I found very, very useful. Well, you can learn it from other people, too, but I learnt it from Felix Kaufmann [...]

H.: Kaufmann did not talk to economists in this country then? He became more of a strict philosopher?

G.H.: Well, at the New School, you know, there were the economists, he probably talked to them; but I don't think that they were very much interested in his philosophy. As I said we invited him once to Harvard: and we had a seminar, but it did not go off very well; and it was [...] his bad hearing. He certainly was a very remarkable person [...] I'm very, very glad that someone is interested in him; you see, it's very worthwhile [...]

Friedrich August von Hayek

F.v.H.: I knew Felix Kaufmann well in Vienna in the twenties. I did, however, leave Vienna already in 1931. Kaufmann was unusually versatile as a philosopher of law and as a mathematician, professionally he worked for the Anglo-Iranian Oil Company. He was a member of our *Geist Kreis*, of the Mises [private] Seminar and of the Schlick Circle.

H.: In the *Geist Kreis* and the Mises Seminar was Kaufmann regarded as a phenomenologist, or as a positivist, or as none of these two?

F.v.H.: Clearly as a positivist. He transmitted the discussions of the Schlick Circle to the Mises Seminar and the *Geist Kreis*. That was interesting for us. Well, Kaufmann was seen as a positivist, the phenomenologist for us was Schutz.

H.: If Kaufmann was seen as a positivist, how could he – at the same time – be a member of the Mises Seminar?

F.v.H.: Well, he was a positivist one could talk with, a reasonable positivist.

H.: Were there any other personal interlinkages with other discussion groups, e.g., the many Austro-Marxist groups?

F.v.H.: No. Kaufmann was the only personal parallel with the Schlick Circle. The *Geist Kreis* and the Mises Seminar were personally almost identical. An Austro-Marxist would by this very token not be a member of the Mises Circle and the *Geist Kreis*, though Kaufmann may perhaps have had social-democratic attitudes.

H.: What was the function of these circles and private seminars?

F.v.H.: The function of the Geist Kreis for us was to continue the discussions of the university. It was named thus ironically by the female members of the Mises Seminar: because the meetings took place at our parents' houses, we could only invite men.

H.: Many members of the Mises Seminar and of the discussion circle held academic positions and became famous in the countries to which they emigrated, while in Vienna they worked as professionals outside of the universities (e.g.; Schutz, Popper, and Kaufmann). Is this related to anti-semitism in Austrian universities?

F.v.H.: No. In the human sciences there were no assistant positions, one could not make a career this way. Therefore, we had to work in a practical field. This held for myself as an Aryan as much as it did for many Jewish friends.

Ernest Nagel

E.N.: I went to Vienna in the summer of 1934 and this is where I met Kaufmann. I knew about him through a book of his *Das Unendliche in der Mathematik und seine Ausschaltung* [Kaufmann 1930], so I had this kind of interest. I was not familiar with his interest in legal philosophy. But eventually I got to know that, too. Our first meeting was really in connection with the philosophy of mathematics. Now at that time he was also the Viennese representative for the [Anglo-Iranian, later to become] Shell Oil Company. I first met him at home and then he took me to his office on the *Ring* and showed me around. He really was a wonderfully lively person. His duties as a representative then weren't very heavy, his office was filled with books. And he really enjoyed life very much. Mrs. Kaufmann at that time just had the baby. I saw him quite a number of times during my stay in Vienna, since I had relatives in Vienna, too, at the time. So it was very convenient. Life was very different from what it is now, we met in coffeehouses and spent the whole evening with him just talking. That was the first occasion. I was particularly interested in Kaufmann's attitude, sympathetic attitude, towards the intuitionistic philosophy of mathematics, such names as Brouwer. So there would be mutual interest in these things. I don't think I corresponded with him when he was in Vienna. At least I don't remember that, but then I saw him quite often.

As a matter of fact, I think the very first week he came to New York I met him and his wife. We did frequently meet on a number of friendly occasions even though we disagreed on a number of philosophical issues. But you see, he was obviously very sympathetic to the *Wiener Kreis* at the same time he regarded himself as a loyal opposition to it. And he felt it a very serious obligation not to rest on his laurels but to continue writing and producing books in this country. And as you know the book *Methodology of the Social Sciences* [1944] that was intended as an adaptation of the *Methodenlehre* [1936]. He was very much worried that people at the New School would think that he had not done enough justice to his reputation as a scholar, but I told him that was a needless fear because he wrote quite frequently and published. There was one symposium that was published in the periodical *Philosophy and Phenomenological Research* and this involved a notion of truth, as I recall it. In several volumes [1943, 1944, 1945] he and I had a kind of controversy. You know, things looked so, I mean despite what was going on in Europe, and there was kind of a hope in the atmosphere that finally you were going to do philosophy in a way that would make it a really respected discipline. The combination of his interest in phenomenology, his interest in Kelsen, too and his interest in the philosophy of mathematics, this was all the beginning of a new day of philosophical thought. It was usually very instructing and inspiring to talk to him. He was full of energy and a great deal of hope what the future would bring about.

And then, I think to his own surprise and to the surprise of many of us who knew him he became very much under the influence of John Dewey. Now this was true for a number of other German refugees at the New School. It was true for Schutz, too. I mean for example that he thought that William James' psychology was really a contribution to phenomenology. I am not sure about that, but that is neither here nor there. There was a kind of receptive attitude toward American thought on the whole unusual for European scholars. American thought was usually regarded, at least before the great exodus of German scholars, as a kind of provincial thought. But this was not true for the people who came to this country. I myself think that Kaufmann exaggerated the similarities between Dewey's views on method and the phenomenological approach. He read Dewey because in a way Dewey was a person whose ideas were not entirely strange to European people. After all he started life as a Hegelian and that remained in his system.

We also had a kind of philosophy group meeting at least once a month in New York. And Kaufmann came quite often to that; and also he took part in something called the Conference on Methods in

Philosophy and Science. This met twice a year, usually at the New School and Kaufmann took part in that pretty regularly. He presented papers there. That group was started sometime in the early 40s as a reaffirmation of the naturalistic outlook. And of course, Dewey was very sympathetic to that. Dewey himself came to the meetings quite often. I recall there was a heated discussion at one of these meetings because of the way some of the phenomenologists began to interpret Dewey. Many of us thought this was not quite correct. But there was kind of receptivity, to people who came to the New School [emigrés] as well as to people who were in New York and got to know some of these. There really was an exchange of ideas. This, I think has not continued, for some reason, I don't know why, but there is [now] a kind of gulf between existential philosophy and American naturalism. At least during the long period including the years of the Second World War there was no reason at all of thinking that you could not talk to one another. Which has since been always a great difficulty for us and probably in sociology too. As far as I know Kaufmann was obviously interested in certain types of sociological thought: Max Weber, surely, and he had a great admiration for Schutz, but he did not show any familiarities, at least as I recall, with the work of people who were interested in empirical sociology, people like Lazarsfeld. There was no communication there to my knowledge […] I knew Lazarsfeld quite well and we gave a joint seminar on methodology of the social sciences and the emphasis was at that time on types of empirical analysis. But there was no sign of interest in Kaufmann as I knew him. Maybe he was, I really don't know why, there is an obviously very important transition between sociological thought which is based essentially upon a philosophical approach and the frame for the empirical approach, people like Lazarsfeld and Merton […] I want to be sure that we don't misunderstand one another. The monthly seminar, that was first led by Lazarsfeld and myself, I think.

There is a system at Columbia University: university seminars. It met once a month. Now, when the seminar first started people usually came from Columbia University. Later on it became much more inclusive; there were people from New York University, from the New School, eventually our university seminar became most exclusively interested in mathematical models of social research. So we had a number of mathematicians, for example I do remember that Abraham Wald, he was a European, he was Hungarian, he managed to escape and came to this country and he was a very distinguished mathematical statistician […]

H.: This seminar was different from the one you mentioned earlier, when you said you had a group of philosophers meeting once a month?

E.N.: Oh yes, it's very different maybe I ought to explain: This philosophical group originally started because we had a friend in common who was unable to leave the house. So the ones who knew him used to meet once a month at his house to bring the outside world to him; he became really too ill to do even that and so we used to meet at other places. The New School was one place, New York University was another place and this consisted very largely, not exclusively but primarily of philosophers [...]

H.: You say he did not mix or get involved in groups who actually were doing the methodology or the actual social sciences at this point?

E.N.: Yes, that I think is pretty certain. At least my memory is pretty strong on that, he did not. I don't know, have you seen some of the symposia [Symposium on Meaning and Truth, PPR 1943–1944, 1944–1945; Symposium on Probability, PPR 1945–1946] that were run?

H.: Yes. I wonder why this was the case. Was it his conception of methodology or was it a political conception or maybe something private?

E.N.: Well, my guess is that he was primarily a philosopher and to do empirical work was not according to his conception of what social science should be. I mean after all Schutz's book on the *Phenomenology of the Social World* does not depend in any significant way on empirical research. So Kaufmann's model of what a social science should be was very different from this crude empiricism.

H.: When you were in Vienna, did you see him act in some of the Vienna Circle discussions?

E.N.: Yes. I think I spent two months in Vienna and I went to a meeting of the *Wiener Kreis* and there it was quite clear what his position was. I think he felt greatly attracted to people like Schlick and Carnap, but he couldn't go all the way so he regarded himself in a sense as a critic of it, and also to the professor of mathematics [Hans] Hahn, no longer alive. When I was in Vienna, Kaufmann had a great deal of interest in set-theory, you know the Cantorian theory of types and so he was clearly influenced by people like Hahn and people like Menger, Karl Menger. It's there where I think his sympathy for the Vienna Circle was closest. And at that time the influence of Wittgenstein was getting great. The other way I can think of it, he was sympathetic to intuitionist mathematics, to which many of the people like Friedrich Waismann were attracted, who was a great protegé of Wittgenstein and when he finally published, he showed a great deal of understanding for Brouwer's approaches. And that thing is very close to Kaufmann's view.

One thing I never had the chance to discuss with him for some reason: see in the infinity book [1930] he took the view that arithmetic was complete. In the technical sense I mean, the axioms of arithmetic were sufficient to prove any arithmetic statement or disprove it. Then in 1931 when Gödel published his work which showed that arithmetic was not complete, that you could not develop a purely mechanical method for establishing it, I never understood whether this influenced Kaufmann's view or changed his view on this matter enough to rewrite *Das Unendliche* in English. A translation of that [appeared] and I wrote an introduction for it [*The Infinite in Mathematics*, 1978] and if he changed his mind about this point I don't know.

H.: I would love to get an idea how these discussions were actually happening in their natural situation in the Vienna Circle.

E.N.: Well, when I was there of course Schlick was the dominant figure. Carnap was no longer in Vienna, he was in Prague. It was Schlick and Waismann and a lot of others, Hahn was a member of that group, who else? Of course, what was so characteristic whenever a foreign logician came to Vienna, he would attend the meetings. I think I met Tarski at the *Wiener Kreis*. Now on those two occasions when I attended a meeting, Schlick held forth and there was a discussion of some of their views […]

But there is one other point, maybe this is relevant. You know he was a great admirer of Kelsen and he took essentially Kelsen's position in the philosophy of law, essentially it was a Kantian view that there were some ultimate first principles, a fundamental norm by which, according to Kelsen, he saw the legal system, as a system he regarded as positive law, as an establishment of norms and of values. But it would have been out of question for Kelsen to think of empirical research in terms of legal thought. It was for the same reason that he was not interested in empirical research in sociology.

H.: I suppose it's that 'objective spirit' hovering about.

E.N.: That's right, that's right. Another thing, this is kind of a long afterthought: In this country there was a movement in the philosophy of law called legal realists. It was in a way inspired by Justice Oliver Wendell Holmes, who was at the supreme court of the US, who early back in 1890s took the view that the law is what a justice decides […] This is a very distinguished tradition […] You might find that interesting, because it is a first statement of legal realism to which a great many American lawyers subscribe. Just as an illustration this school in philosophy of law said "you cannot really understand the law by simply reading it, but you have to find out what a man's prejudices are and do empirical research on the attitudes of judges who are determined by class association". So this was again an empirical approach, similar to the Lazarsfeldian approach to sociology. But again despite the fact that Kaufmann was obviously strongly interested in philosophy of law, he showed no interest, at least to my knowledge, in American legal realism, I don't know whether he really wrote on the philosophy of law after he came to this country.

H.: I don't think so. So what about economics? Would you happen to know about that? Of course in the early German book on methodology, he has a chapter on Kelsen and he has a chapter on economics, marginal utility.

E.N.: Yes that's right. He followed Carl Menger not the son, the mathematician, but his father. It was really a neo-classical approach and again, the work that was done at Columbia by a professor of economics, Wesley Mitchell, here again the approach was strongly empirical, that is, he had very little use for some of these classical theories of economics because he had made out that they didn't correspond to reality. There was a strongly empirically minded stream in this university. And Kaufmann maybe was sympathetic to it but he was outside [...]

H.: One question we didn't discuss at all: what changes if any, took place when Kaufmann came to this country?

E.N.: Well, I recollect very clearly that he said since he has to write in English rather than in German his thought is clearer, but I don't think there was any substantive change in his thinking, because you notice if you compare the *Methodenlehre* and the *Methodology* I don't see any great difference between them [...]

You see Popper's thought in a way is very simple. The one idea that so-called scientific propositions are refutable by experience that's what makes them scientific. Now Kaufmann's thought was not that simple. It was really much more complicated and he had a great admiration for Popper by the way. Tremendous admiration. I remember when I first knew Kaufmann in Vienna, he talked about Popper as a really outstanding person and he had nothing but praise for *Die Logik der Forschung*. Very strange, because Kaufmann's own thought would have made him, at least in part, a critic of Popper [...]

H.: What could one describe as his [Kaufmann's] function within a particular subject? And then again how did he himself look at the task of philosophy?

E.N.: Yes, I think you say that very well. The clarification of ideas, and the reason, I think, why he was such a great admirer of Carnap was that Carnap tried to do that in a very formal way.

H.: Can you think something else in this connection? Something about Kaufmann, something I should say in the book? [...]

E.N.: You know one thing you certainly ought to say is the kind of man Kaufmann was, after all, he wrote on social sciences, he wrote on philosophy of mathematics, he wrote on philosophy of law, he wrote on phenomenology. There was a breadth of interest which is still exceptional. And I think you ought to say that somewhere. Just how much influence he had is awfully hard to say. My guess is that very few people read Kaufmann, even people interested in philosophy of mathematics because in many ways it is outdated and even in sociology, people take Talcott Parsons very seriously, but Kaufmann is not in the picture, he is not a professional among the sociologists, you know […]

The memorial [for Kaufmann] was held at the New School. The person who chaired that meeting was the former president of the New School who is no longer alive. And it was a very large meeting, as I recall […] I know my memorial speech appeared in *Philosophy and Phenomenological Research* [PPR 1950, Vol. X, No. 3, p. 464], when Marvin Farber was editor […]

Paul Rosenstein-Rodan

P.R.: […] Now as a preamble you realize that between 1818 and 1930 Vienna was a place of unusual cultural and intellectual progress […] The synthesis of music, poetry, philosophy is quite extraordinary […] The Austrian School of economics was really the best economics on the continent largely by accident. The Austrian School had this prominent part, because in Germany the so-called Historical School under Schmoller ruined economic theory; it was really economic history and not theory, which is very extraordinary because in the early part, say, 1820 to 1860 or 1870 there was a very high level of economics in Germany. […] They were really victims of a kind of ideological crusade against theory for the Historical School by Schmoller. And from 1870 onwards Schmoller was a kind of dictator of social sciences in Germany. […] Carl Menger created the marginal utility theory in Austria; the home of economic theory, I would say, on the whole continent outside England was in Austria. And then, in physics and the best medicine, one of the best medicines in the world was in Vienna. God knows why: this concentration of intellectual and artistic creativity was quite extraordinary. In certain respects more than Paris […] Immediately after the War, after the First World War, the intellectuals of the whole of the Austrian-Hungarian monarchy somehow remained in Vienna. And, well, Max Weber came from Germany, gave lectures; so it was not only economics although economics, because of marginal utility theory was the kind of leading social science […] Europe as a whole […] was a very strong feeling which unfortunately by the end of the '20s vanished. But in the cultural history, you see, between 1910 and 1930 it's a very important part […] Now Vienna was very much full of intellectuals, and there was a very strong feeling, for international values […] one is an Austrian, one is a German, one is also European. That was very important. Part of it were the young intellectuals who immediately after the War created the club which was called the *Geist Kreis* to discuss general cultural methods, economics is only one among other things [such as] philosophy, and then don't forget that in the history of ideas and ideologies – the fact that it is compromised now is perhaps the greatest tragedy of mankind – socialism is a fantastic and great idea. You see, when the London Times correspondent came to Vienna and interviewed the founder of the Social-Democratic Party, Viktor Adler, and he asked him: "What is the political situation in the Austrian-Hungarian monarchy?" his famous reply was: "In Österreich herrscht Absolutismus gemildert durch Schlamperei". ["In Austria absolutism rules softened by inefficiency.", my translation] So that you had in Vienna kind of three shifts: you had the court, Franz Joseph was very conservative; but somehow the *Schlamperei* allows him many things. Kronprinz Rudolf secretly wrote pro-liberal things for the *Neue Wiener Tagblatt*, was in touch with radicals and liberals; was a kind of pink we would say today. And he would gradually, therefore [tolerate] a kind of center of the Liberals, cultural, but very progressive. They were gradually disappointed, they had some reservations, but only reservations, not dismissing it a priori. The *Geist Kreis* was very much of that part; and people who were very active in the *Geist Kreis* were Felix Kaufmann and Voegelin and Schutz. There was also a very nice historian called Engel-Janosi. And, Haberler's brother in law Herbert von Fürth. And the economists, Hayek and Haberler [...] But economists in the *Geist Kreis* were only half. And then two or three seminars emerged, one a private seminar of Mises, which was held in the Handelskammer [Chamber of Commerce], and there the non-economists were only three or four: the rest were economists. But the economists were a very peculiar brand of economists, twofold [...]: The Austrian marginal utility theory, which is the theory of choice, the theory of rational conduct; and that was very good, incidentally. And there was also under the influence of Mises the ultra *laissez-faire*, saying: any state intervention is wrong. I told you the extreme of it later on was Hayek; and I told you about his book [*The Road to Serfdom*, Hayek 1944].

Now the liaison between economists – marginal utility theory, not the *laissez-fairism* and sociology, philosophy especially, was Felix Kaufmann. To some extent Schutz and Voegelin had great philosophical culture, too. And Felix Kaufmann played a very important role in the way of spreading general culture not only in a kind of blinker's orthodoxy, not only in the Mises Seminar, but among the economists in general. And the following, you see, Max Weber came, gave lectures at the university which were a sensation. And, you know, in methodology of social sciences *Wirtschaft und Gesellschaft* by Max Weber is not only very profound but very important: the continuation of methodology. You see, there is a standard book, a very great book, as great as Max Weber, and that was the second book of Carl Menger [1883] on methodology. The continuation and the synthesis [...] was later Felix Kaufmann. And then it's a very important point: Felix Kaufmann as a typical intellectual was in touch with all the Austrian economists – marginal utility – and unlike other sociologists knew it [...] You see, before 1920, economics was part of the law faculty [...] He was in touch on the one hand with two parts of philosophy in Vienna. One was Schlick, a very important philosopher; and the other, which is a subbranch but separate, he was early in touch and got us all in touch with what later became known as the Vienna Circle; Carnap and Wittgenstein before he left for Cambridge in England. It's on a very high level. I personally consider the *Wiener Kreis* in philosophy a mistake. You see, I personally, but that you should discount. You see, I'm a neo-Kantian. To me the modern philosophy is the *Marburg school* in Germany and Schlick later in Vienna. But Felix Kaufmann, although on balance on the side of Schlick, was very much in touch with Carnap and brought us all in touch because, you know, this is something very different. And then he was very taken by and gave beautiful lectures on Husserl. All that was in the 1920s.

Now he, as every cultured man should, but in Vienna there was a certain weakness among the kind of *Geisteswissenschaftler* not to take notice of mathematics but what is very important is philosophy of mathematics. And he wrote a little essay [*Das Unendliche in der Mathematik und seine Ausschaltung*, 1930] on the notion of the infinite [...] On the other hand he was beautifully cultured [...] He was also in touch with those who consider psychology the basis of social sciences: the law of conduct and especially Franz Brentano. And Brentano played a very important part methodologically in the Austrian school [of economics]. I mean, the two books on methodology after Menger and Max Weber, was on the one hand Brentano, and on the other hand a man called Meinong, who are very much forgotten now, but not in those days in Vienna. They were really very, very alive, Meinong and Brentano. You see. There was this standard methodological problem very much discussed that the basis of economics – unlike classical theory – is that your subjective judgment determines what an amount of importance, of pleasure will derive from one good as against the other: that was called 'utility'. And very early the problem arose which has a long history of literature: utility or my pleasure or *Bedürfnis*; the utility is an intensive magnitude; you cannot say that the pleasure you derive from an orange is 1.23 of the pleasure you derive from an apple. And yet, these utilities, which are only intensive magnitudes determine prices which are quantitative. You see, the problem philosophically was one of quantification. How can intensive magnitudes determine extensive quantities? [...]

H.: Yes. He writes on that in his book, doesn't he?

P.R.: Yes, a lot […] In the 1920s, it was perhaps the central problem in economic theory. And he [Kaufmann] who knew also mathematical philosophy, […] popularized among the economists the problem of definitions […] You see, the foundation of mathematics, is not one, two, three, but the ordinal numbers first, second, third. The cardinal number is only a special case. And you have three types of […] measurement meaning only ranking: first, second, third. You can say about any two things that they are greater, smaller or equal but not by how much greater, by how much smaller. Today we would call them propensities. Second and this is the only thing which is needed for a good marginal utility theory of conduct. But there is a second problem that in some case you cannot only say A and B whether they are greater, smaller or equal, but you can also say where the difference between A and B is greater, smaller or equal than the difference between B and C. Those I call magnitudes. This is a feeling of distance, of interval, not only ranking. And the third thing, where you can say one, two, three, are mathematically or philosophically speaking quantities, you have propensities, magnitudes and quantities. The problem is: are utilities – propensities or magnitudes and so on. And this is, as I said, this was a very important part in Kaufmann's thinking. And via, well, one was published, but via any number of coffeehouses, you see, the intellectual activity in Vienna was in coffeehouses. And, you see, Felix Kaufmann was very prominent among the economists in discussing this problem and putting a kind of order into it […]

H.: I would like to understand how it worked on a concrete level. Kaufmann on a concrete level. Kaufmann did his degree in the, I think, he got the law doctorate in 1920 and the philosophy doctorate in '22, and then he started working with the Anglo-Iranian Oil Company.

P.R.: Both in economics *and* philosophy. That's a very important part to begin. You see, he attended the very stimulating Max Weber lectures; and then he read under the influence of Max Weber a great deal of methodology on (a) the infinitesimal principle, (b) Husserl. To a lesser extent he knew but did not overestimate Max Scheler. But then he was in touch with the whole tradition of marginal utility in Vienna. Wieser was still professor, Böhm-Bawerk and Menger were dead. – And now apart from the *Geist Kreis* and the Mises Seminar there was a third activity by Marxists. But, you see, really the more or less would-be-orthodox Marxist was Max Adler. But there were the left wing Austrian social democrats, you know, not like the German *Sozialdemokratie*, but were independent, largely under Otto Bauer and a man called [Otto] Neurath. He wrote a very important thing about *Wirtschaftsrechnung* [Neurath 1913]: all on utilities, how you can combine the optimum. And although their values were different, they were still in touch with each other. And one of the important intermediaries, the people who maintained the contact with the three schools was Felix Kaufmann. But although a prominent member of his Mises Seminar he never accepted Mises' famous book *Gemeinwirtschaft* in 1920, he had only *laissez-faire*, no interventions [...] It is via Hayek and Machlup that that part of the Austrian school had very great publicity in England, in America [...] It's very important, see, marginal utility theory does not say that any state intervention is wrong. It depends on how much – that's a different story. So it's not true that it's inherent in the marginal utility theory; it's inherent in the *Mises Kreis* and, you see, a very prominent book which I mention here was by a man called Richard Schüller, *Schutzzoll und Freihandel*. The idea that any type of protection is inherently wrong [...] was denied. He was a prominent member both of the *Mises Kreis* and of the *Nationalökonomische Gesellschaft*, and he was a good moderate protectionist. He would not say that free trade is natural, all the rest is the same. Those who follow Hayek think that there is only one way. That's a very important point. And this book on methodology by Felix Kaufmann is in my opinion a very great achievement.

You see, he remained in Vienna. I left Vienna by sheer accident, you see, at the end of 1928. I was in touch by correspondence but that is not quite the same as being there. You see, every week the Mises Seminar, every two weeks the *Nationalökonomische Gesellschaft* which I founded, but, you see, I followed it from Italy, from England and he was still there. Then gradually the crowd was broken up. Haberler left for Geneva, for the League of Nations in 1929/1930. Hayek left for America and then for England in 1931. And, you see, when I arrived in London, you see, I represented the Austrian school: 'non-free trade' and Hayek at the London School, 'free trade' [...]

H.: How did these people – a very practical question – when did Kaufmann do his academic work, since he was working for that company? I don't understand that.

P.R.: Well, you see, first of all, the company wasn't – let me say, three out of seven hours he was reading books there – it was very Viennese [...] I remember talking to Felix and he needed 6 and a half hours of sleep. He always said I've an extra hour and I said I've an extra two. And that was our conversation.

But he attended all the seminars with one exception. Well, we were very great friends and he talked a lot about methodological aspects also [of psychoanalysis] because in all his eighty years of his life Freud only accepted four people who were not doctors or psychoanalysts into his seminar on psychoanalysis. And I was one of the four: on a Wednesday evening in Freud's home and I shudder to think how I behaved. I wouldn't dare to speak to Freud like that today [...] But the methodological aspect of psychoanalysis, the concept of the subconscious and so on is methodologically very interesting; the majority of psychoanalysts are barbarians. They discuss a case, a case; the methodological aspects, philosophical aspects are conspicuously absent [...]

H.: I have to understand this on the basis of my experience: today in German universities, if you belong to one group you don't speak with another group. So here is Kaufmann mixing with three.

P.R.: Yes, the atmosphere of Vienna in the '20s which Paris was in the '30s. And that was not an exception. And it's very peculiar. I mean, a real sociologist should describe why in a society at a certain time there is the cohesion of intellectuals: the coexistence of various views, you could say, a moderate tolerance. Not so far as to give up one's own point of view but to understand that another one's point of view, even wrong, has a right to be formulated. It's very [...] I don't know, you see, the coffeehouse comes in very much, perhaps because there were many different nationalities in the Austrian-Hungarian monarchy. And some of them coexisted very well. You see, the Poles and the Germans were always in the ruling class; the Czechs were always in the opposition and so on. And everyone in Vienna had a Croatian cousin, a Polish aunt and that brings in, subconsciously, it's my explanation, a contest of things. The fact is unmistakable; how to explain this is a different story [...]

H.: I recall that paper 'Soziale Kollektiva' [Kaufmann 1929] because he tried to define certain concepts which would now be considered to be the individualism-holism debate.

P.R.: Yes, it was the very important turn in sociology. Methodological individualism was introduced by Menger in 1887, you see. Max Weber took it up later on; but it was really Menger [...]

H.: Now, Lazarsfeld in one of his autobiographical notes says at the time when they were doing the studies for *Die Arbeitslosen von Marienthal* (1933) there was no empirical sociology in Vienna.

P.R.: That's correct. In my opinion correct. To what extent is very difficult to say. You see [...] there was not a professor of sociology – that came only later. So that had its importance. Well, you see, Lazarsfeld was an empirical sociologist but what appealed very much was [...] what we could call the theory of conduct, psychology. You see, these people who, wrongly in my opinion, opposed what they called 'psychologism' [...] throw the baby away with the bath water [...] psychology is still the basis of all social sciences [...]

There was also the feeling in Vienna in those days that, you see, in economics, you can reduce 100 variables to 3 or 4 constants and therefore can apply the cause-effect method. Now, sociology is hopeless in this respect. I have a partial explanation for it. You see, the progress in method into exactness in economics is, contrary to what is said, fantastically slow ever since Aristotle and the schoolmen […] It took a generation each time and the continuous study of economics is only since the beginning of the 16th century. Sociology came later and has not had this much time […] I wrote it up saying: we are more and more exact; we got less and less relevant results […] And, well, it's a price to be paid, it's a great advantage to have a method; but, you see, economics in general determines the necessary but not the sufficient conditions. That's not as little as it will sound because […] if necessary conditions are not created, this events cannot take place. But when they are created it's not certain that they will […]

Ilse Schutz

H.: Mrs. Schutz, how did your husband get to know Kaufmann?

I.S.: As far as I remember they met through Kelsen, at the university. And – as time went by – they became very close friends indeed. Kaufmann took really an unlimited interest in the development of *Der sinnhafte Aufbau der sozialen Welt* [1932, Engl. tr. 1967] and he in fact discussed every single step of the argument with my husband. He was the one who made my husband direct his attention to Husserl, which later became terribly important for him, as you know. In Vienna it was friendship, not only of the two men, but also of our families and Mrs. Kaufmann. Their son was about the same age as our daughter. Even with the children we got together quite frequently: birthday parties and such like occasions.

H.: How was it possible that both men worked professionally full time and at the same time they did so much intellectual work? I would like to be able to imagine concretely what it was like when they read Husserl together?

I.S.: I don't know if they actually read Husserl together. Felix directed my husband's attention to it, and my husband was able to devote time to his studies only during the night – I think this holds also for Felix – which therefore did not at all interfere with his daily work. Kaufmann's position in his occupation may have been a little more independent, and he could perhaps sometimes occupy himself with [philosophy] during the day – but in general – it was making a living during the day and science by night. This was certainly the case for my husband, but surely to a certain degree also for Kaufmann.

Alan Sweezy

Dec. 12, 1986
Dear Dr. Helling,
I'll start by trying to answer your questions:

1. I first met Felix Kaufmann in Vienna in the early part of 1933. I was in Vienna studying economics on a fellowship from Harvard in the year 1932–33.
2. What was your relationship with him? First, through attendance at seminars in which he was a leading participant. Then, thanks largely to Haberler, I got to know him personally and we became warm friends.
3. What was he like as a person? Certain characteristics stand out in my memory of him. First, strength, both physical and intellectual. One of his friends told me that when they were vacationing in the mountains he would come in from skiing late in the afternoon and go right to work on his book. His physical strength contributed, I am sure, to the pervasive good humor that was so prominent a characteristic. He was sharp as a knife in his analytical thinking and never hesitated to express his disagreements with others. But he was never confrontational. He could argue and disagree without any suggestion of personal hostility. This was one reason, no doubt, Mises accepted him in spite of the difference between them on key methodological issues.
4. Who did he associate with? I can't say beyond the group who attended the seminars I was invited to attend. Of the economists Haberler was, I think, particularly close. Morgenstern and Machlup were also friends, and, I think, admirers. There were others but my memory is rather vague. In any case, you can get that information from Haberler better than from me.
5. What was his position in the Mises Circle? I really can't say. Not only is my memory of our meetings rather vague but I was too much of an outsider to know much about the relations of the various groups in Vienna to each other. I, unfortunately, do not have a clear memory of Mises and Kaufmann together in any of the seminars or other discussions I was invited to take part in.
6. Was he considered a positivist, a phenomenologist, a legal philosopher … I should think all three, although I was not sophisticated enough at that time to make clear distinctions. I remember his talking about Schlick and am fairly sure I once heard Schlick speak. The influence Kaufmann had on my own thinking was, I think, well within the scope of the Schlick positivist ideas. It was only with respect to certain deeper philosophical problems that Kaufmann sought aid in Husserl's phenomenology. Evidence of both Kaufmann's respect for Husserl and of his warm friendship for me can be found in the fact that he arranged for me to have an interview with Husserl when I was in Germany in the summer of 1933. I felt it was an honor to have the opportunity to meet and talk with so eminent a philosopher.

I should like to add something about Kaufmann's influence on my own intellectual development. When I first became acquainted with him, I was struggling to get my bearings in the slippery territory of subjective value theory. I felt there was something wrong but I could not see just what. Subjective value theory starts with some simple psychological observations, such as that the pleasure I get from a second cup of coffee in the morning is less than the pleasure from the first cup. But economists were not satisfied with the role of amateur psychologists. On the other hand, they were not willing to venture on to the slippery ground of professional psychology in studying human motivation. They thought to escape the dilemma by proclaiming universally valid rules of behavior based on logic rather than on empirical observation. You summarize the essential points in Kaufmann's critique: "Mit den logischen Positiven glaubt er, dass in der Philosophie durch die Verwechslung von logischer Wahrheit und faktischer Wahrheit und durch Ansprüche auf notwendige Wahrheit für Tatsachenaussagen viel Verwirrung entstanden ist." Without Kaufmann's guidance I would have found it difficult – probably impossible – to find my way out of this confusion. Many economists, demographers, and other social scientists have not succeeded in doing so.

Essentially the same fallacy as that involved in subjective value theory has been common in the last few decades in the form of the theory of 'revealed preference'. According to this theory it is not necessary to study empirically the processes by which people make choices. Their preferences are revealed in what they actually do. A good example is provided by the debate that has raged in recent years about population growth and population policy. A key question in this debate is why people – especially poor people and people in poor countries – have large families. Until quite recently the dominant view among economists and demographers has been that they have large families because it is in their self interest to have large families. According to the theory of revealed preference it is not necessary to investigate empirically whether this proposition is valid or not. Its validity is established by looking at what they actually do, i.e., have large families. Individuals are the best judges of their own self interest. It follows that if they have large families it must be that having large families is advantageous to them. No further investigation is necessary.

What I had learned from Kaufmann prompted me to examine this argument critically. I discovered, of course, that far from enjoying *a priori* validity it was heavily laden with assumptions that called for empirical investigation. It assumes, for instance, that in poor countries children make a contribution to family income through the work they do. This actually is far from obvious. Eva Mueller, of the University of Michigan, thinks on the basis of what scattered statistical evidence she could find, that the value of what children produce, even in peasant agriculture, is probably less than the value of what they consume. On a more general level the theory assumes that people actually do make choices with respect to family size. This, too, is by no means obvious. An alternative view, suggested by the concept of "natural fertility" has been gaining acceptance among influential demographers [...]

As I have said before, I owe it to Kaufmann that I was alert to the methodological fallacy contained in the theory of revealed preference.

I realize, as I think it over, that I am not able to be of much help to you in your work on Kaufmann. I admired and liked him and was much influenced by his ideas. But I did not know enough about the ins and outs of the intellectual life of Vienna the year I spent there to provide significant information in addition to what you are able to obtain from other sources. I hope you will feel free to use what material there is here in any way you can. You may, of course, quote me if there is anything suitable for that purpose. You may find it better to paraphrase what I have said or to work it in with material from other sources.

I enjoyed your visit very much. It was stimulating and fun to have a chance to talk about Kaufmann and the other Austrians with some one knowledgeable as you. If you do come back to Los Angeles, I hope you will be sure to let me know.

With best wishes,
Sincerely,
Alan Sweezy

Students at the New School

Reuben Abel

H.: [Let us talk] about the way you got to know Kaufmann.

R.A.: Oh, I was his first student [in America]. I remember the meeting in which new students were invited to come to meet the new professors and I introduced myself to him and I remember that I was his first student. And, you know, he was hard of hearing in these later years. I'm not sure what relevance that has right now. We hit it off very well. He was personally a very sympathetic and congenial person of no […] pretence; and I [liked] those courses that I took; perhaps 8 or 10 or 11. I took everything he gave in the course of the years. And I remember urging him to give a course in philosophy of mathematics, you know he was a mathematician, too. What happened was, he began to read the American philosophers, in particular Dewey but others as well; and I know it had an effect on him. Oddly enough, you said that you seem to find no one of the phenomenologists with whom he had contact. I don't think he once mentioned the name of Husserl in class or the word 'phenomenology' except to refer to it. And he had steered away from that. I don't know, of course, that he ever was into that. But he certainly taught nothing of that kind of philosophy, never referred to it.

H.: He didn't teach phenomenology?

R.A.: Not at all. Not at all. Not a word.

H.: Well I can tell you about the early book [the present text, Kaufmann 1936]. In the foreword to the early book he says, even though I greatly admire and owe a lot to the great philosopher Edmund Husserl this book remains outside of transcendental phenomenology. But then in his attempt to show that the logical positivists would not go far enough in their attempt to describe experience; which describes experience; he says the logical positivists cut short an analysis of experience and they settled for simplistic, sensational views and so on [...]

R.A.: That's it, yes. Because everyone was unhappy with all the analysis of sense data which so many positivists and others were engaged in, and just didn't seem to be getting very far. Maybe that's what he had in mind. But really I can't recall anything that he said to that later. I remember that he introduced me to Marvin Farber who was the editor of PPR [*Philosophy and Phenomenological Research*], and was, I suppose, a leading American phenomenologist. But Farber, I have to say, at that time Farber read a paper in which he strongly dissented from Husserl's later supranaturalistic, extra-naturalistic formulations and he thought that no one would go along with them. He wanted to make it as naturalistic a philosophy as could be done. Do you understand it? And the phenomenologist Aron Gurwitsch was very upset about that, a great deal of fluttering in the dove cotes at this heretical, what we call a heretical remark. So I can only say: I just don't know the extent, if any, to which Kaufmann could be called a phenomenologist in this country. I just know nothing about it.

H.: What you are saying is he did not present himself as one?

R.A.: Absolutely not [...] The impression I got was that he [was in neither camp...] He never considered himself, as far as I know, as either a phenomenologist or a logical positivist. I remember distinctly his saying that he had gone to some of the meetings of the Vienna Circle, but was not a member [...] He was eager to become Americanized. He loved America; he found this country wonderful. And I remember two of the things he tried to do: to drink Coca Cola and to read the *Saturday Evening Post.* Do you know the *Saturday Evening Post* which is a good way to observe the culture from the outside? He was a wonderful teacher. His hardness of hearing was a bit of a handicap. But he was a wonderful teacher, did very well in explaining; he was into students' questions, encouraging students. I remember his reading a paper at – what we call – the General Seminar [...] It was an institution which was devised at the New School Graduate Faculty, then called the University in Exile. And, it was – I think – unique in this country in the sense that it involved all five of the disciplines; they weren't sharply separated: philosophy, sociology, economics, political science and psychology. And professors would criticize, and members of the audience would criticize each other's point of view. It was a very fine institution. It had its ups and downs since then and is being revived now more or less. Not quite the same thing. I remember his reading a paper in which he first spoke of the tautological nature of logic and pure mathematics. You know what I mean by that? And the professors didn't get it at all. They just couldn't understand what he was saying. At least some of them. Well, I'm not sure how important that is. These are some of the things I remember.

H.: What did he teach then? Philosophy?

R.A.: Oh, yes, philosophy. He taught courses in the philosophy of the social sciences. He gave a course on Dewey's *Logic* indicating how much he had himself absorbed it and was ready to explain it.

H.: Did you take his course 'Methodology of the Social Sciences'?

R.A: Yes, yes.

H.: What did he do there?

R.A.: Pretty much what's in the book. He was working on the book [*Methodology of the Social Sciences*, 1944] at the time, of course, he was working on the book for many years; and he spoke very much along those lines. […] Although he called this book at his course *Methodology of the Social Sciences*; it would be equally legitimate to call it *Philosophy of the Social Sciences* both for its content and from the way the terms are used in this country.

H.: He does, of course, in the foreword say that people who were interested in the actual method should turn to two textbooks which he then gives a reference of. So his conception of methodology was broader?

R.A.: Not just the method. Not just – the use of tests, a questionnaire and that sort of thing. He was a very learned man. More erudite than it's customarily the case in this country […] The sort of things we discussed […]: philosophy, concepts, analysis of reasoning, use of logic, mathematics, structure of arguments. I suppose – that's a good point – I suppose the fact that he did say methodology rather than philosophy reflects the Husserlian background because the phenomenologists certainly never confused the terms, never think the terms philosophy and methodology are remotely synonymous. But the kinds of things, well, the kinds of things like the analysis of rational action […] what kind of technique would you call that?

H.: I would call that, ideal typical explanation.

R.A.: Would you call that analysis near to phenomenology or to logical positivism?

H.: I would call it nearer to phenomenology simply because in the early days logical positivism did not think that motives were approachable to scientific analysis[…]

R.A.: That then is a phenomenological part of Kaufmann's presentation. Of course, subsequently, as you know, as you just said, the notion that this was a rational thing to do is actually Hempel's phrase […]

R.A.: His American book was widely reviewed. How many reviews, offhand, were there?

H.: Well, there was one by Lundberg, a famous sociologist. There were about five in American journals and then some English ones. And they accused him of a very heavy Teutonic style, etc. [laughter]. It was considered a very difficult book.

R.A.: I wouldn't have thought so.

H.: I think it must be his peculiar mixture of being very basic and detailed at the same time. You know, he always tells you where you are and where you go from here and so on, and maybe this is

R.A.: You know, I read the proof on that book; and we had long sessions in which outside of obvious typographical errors, I felt there were some phrases that should have been changed.

H.: Did he write it in English himself?

R.A.: Yes

H.: He did?

R.A.: Yes. And he made it clear that he was writing it in English; and it was not just a translation of his German book [1936, in this volume] He worked hard on his English and I think he did very well with it […]

For example, here in 'Language, Truth and Knowledge', he gave several paragraphs criticising Kant's interpretation of the analytic and synthetic. And that there is a distinction is, of course, maintained by almost everybody. Nowadays, the line between the analytic and the synthetic is being blurred by people like Quine who maintain that it is a continuum. So does Hempel, by the way. Hempel recently changed his mind, fairly recently changed his mind on that. Nonetheless, that's what the distinction is. Now between Kaufmann and Dewey the […] issue was: these logical structures which we use, which we make use of to draw conclusions and to predict experience and to sort out our data – you understand it? Where do they come from? Dewey thought, rightly or wrongly, that they aren't pre-existing; they aren't given to us in advance but that they emerge when human beings try to solve problems and figure out what is it that produces a successful result in a problem. So that the laws, the simple laws of logic emerge, come out in course of the actual inquiry. Kaufmann thought, on the other hand, that they could not – that that just wouldn't do. That these laws of logic have to be presupposed in any activity. Now there is a sharp difference of opinion and it was never reconciled.

I never read the original German book. Partly because I read German with a great deal of difficulty, partly because Kaufmann didn't indicate that it was warranted. He thought that his English book, his book in English, was the book that he wanted to be put forward as his ideas [...] We were personally compatible because among other things, you know, that he spent the larger part of his life in business. And so did I at that time. I was in business at that time and just studying individual courses in the evening with him and with others – but mainly with him. So he encouraged me, and he hoped that I would become a doctor before I became a millionaire. I'm not sure if he was right about it but I remember his saying that and also, you may be interested to know.

H.: What did you become first then?

R.A.: I'm not a millionaire [laughter]. So I got the doctorate first.

Jules Altmann

H.: How did you met Felix Kaufmann and when?

J.A.: I met Dr. Kaufmann approximately 1938 or 1939. The first time he came to this country [...] I was in the first classes he taught at the New School. My recollection was that he was in the second generation of the Graduate Faculty. He was not in the first group [...] I think he came afterwards. In other words, they had some that may have come as early as 1934 perhaps with Dr. Johnson. That would have included Dr. Lowe, Dr. Wertheimer, perhaps Dr. Staudinger, Frieda Wunderlich [...]

H.: So what classes did you take with him? You are a philosopher yourself, aren't you?

J.A.: Yes. Well, no. I went into Federal Service. By the way, I'm not entirely called doctor. I'm one of those who didn't complete a dissertation. I had all matriculation requirements [...] I took two marvelous seminars which he and Dr. Wertheimer gave. And they were extremely exciting for me. The first was the 'Methodology of Social Sciences'[...]

The procedure would be that at the beginning, I recall, one of the professors would give a paper and then assignments would be made for the rest of the term with the other students delivering other papers. And then the student would speak for approximately half an hour and the remainder of the period would be open to discussion. And the student would be expected to defend the position he developed in his paper against both the professors and the other students.

For me this was a revelation because in terms of my prior experience with American philosophers I had not come in contact with anyone that thought as deeply or as broadly as these men had. Although I had a brief contact with Dr. John Dewey. He did not have the reputation or being a very articulate of systematic expounder of his ideas. Generally, his procedure would be to attempt to develop and amplify his ideas while he would be talking. So there would be a great many pauses in his talk but because of their unfamiliarity with the language, most of the professors there had their lectures written out. And I was so impressed with the quality and the profundity of these lectures that for me it was a very thrilling experience, particularly the General Seminar where only the professors spoke; no student except in a discussion period. So that my recollection will be that each of the years – that will be four years – at the Graduate Faculty of the New School – I took at least one course with Dr. Kaufmann. I particularly remember his seminar on John Dewey's *Logic*. And that generated considerable correspondence between Dewey, Bentley and Dr. Kaufmann (cf. Dewey and Bentley 1964; Helling 1988b). And I have copies of that entire exchange [...]

H.: I'm really very interested in the 'Methodology Seminar', in the one with Wertheimer because I was told – somebody told me that Kaufmann just hated Wertheimer. Is that just gossip going round with no substance to it?

J.A.: Well, let's put it that way. Dr. Kaufmann in my eyes was a very sophisticated poised and diplomatic person: So that if there was any animosity it would be not revealed. But Dr. Kaufmann had tremendous facial control. So that even with regard to his students, whose viewpoints he found either ridiculous or objectionable he found a very polite way of indicating the limitations or his objections to the point of view and rarely would he intend to embarrass anyone. And there was a time when I thought I was intimate with Dr. Kaufmann and I can't recall any comment that would indicate this. Of course, there would be, from a methodological point of view, primarily because of the rigid position Dr. Wertheimer took with regard to *Gestalt* theory, where Kaufmann might feel certain qualifications [...]

H.: I would love to know what he did in that seminar because it is said that his position changed in the United States, you see. In Europe, he described himself as the loyal opposition to the Vienna Circle.

J.A.: I'm sorry, my recollection is that that is not completely accurate. My recollection is that Dr. Kaufmann felt that he was attempting to make a bridge between the Logical Positivists and the Husserl movement. Yes. And, in fact in the Husserl *Festschrift* that was edited by Farber [1940] having a great many articles in English; there is a paper devoted to that theme in which he attempted to formulate certain proposals [...]

H.: This is really the topic I'm mostly interested in: his attempt to build this bridge between phenomenology and logical positivism on the one hand and his later relationship to Dewey and pragmatist philosophy here.

H.: Did he ever talk about phenomenology?

J.A.: Oh yes, there was a talk on phenomenology. And, in fact, he gave a course on phenomenology.

H.: That's what I have to look into, you see. Because Professor Abel did not recall that. He thought that Kaufmann in this country did not ever present himself as having anything to do with phenomenology.

J.A.: But again may I say this partly confirms my suggestion that he was interested not so much in the loyal opposition because being the loyal opposition you would have to champion more phenomenological ideas. But Kaufmann represented almost a renaissance mind to me in terms of the circles where he was accepted as a valid contributor [...]

I vaguely recall his discussing the *Logical Investigations* of Husserl [1922] – and he pointed out that there was a man who subsequently produced translations of other works of Husserl. The name begins with a C. in English [Dorian Cairns]. And it took a great many years but eventually they were published. So that unless there were occasions for him to mention ideas related to Husserl it would not come to my attention because temperamentally I'm not very sympathetic to phenomenology. But this is my own prejudice [...]

H.: What about Kaufmann in his country? Maybe you don't know because as a student one doesn't look in this way at one's teachers, but I want to find out in which circles if in any he moved in this country?

J.A.: It will be more phenomenological circles than American philosophy. Because most of his articles were published in *Social Research* and *Philosophy and Phenomenological Research*. So this is why he did have close contact with some of the members there [...]

H.: So you attended his seminar about Dewey's *Logic*

J.A.: Oh, yes, that was very important for us. I was a Dewey enthusiast. And I attempted in my – let's call it hesitant way – to suggest ways in which his interpretation probably would not be accepted by American pragmatists and attempted to defend that. And this was also the background of this very active correspondence between Dewey, Bentley and Kaufmann [Ratner and Altmann 1964; Helling 1988] and occasionally I would add a letter. It was primarily for me whether or not I could be in the same league as these very superior philosophers because it was a challenging opportunity for me, after all, to read their contributions because of Dr. Kaufmann; and through Bentley I saw both Dr. Dewey's and his own correspondence and so I was able to get copies of them [...]

Interestingly enough I can't recall Dr. Kaufmann referring to Wittgenstein, although Wittgenstein, of course, was a source as well as a worship area for a great deal of Logical Positivism. But remember, one also has to take into consideration the level of the students. There wouldn't be much point discussing Wittgenstein or even Husserl if we didn't have the preparatory background to follow and appreciate that. Whereas to discuss Dewey or rather other aspects of either American or European philosophy we'd be better prepared [...]

H.: Did he ever talk about practical aspects of the social sciences?

J.A.: Oh, yes, very much so. Particularly in his 'Methodology Seminar' and in the General Seminar.

H.: Because that's important for me, since in his books [on methodology] he doesn't. He does concept analysis but rarely touches on methods of data collection and data analysis, etc.

J.A.: But this would be, I would suggest, a rather elementary aspect of methodology, data analysis, because as you yourself suggested he was more interested in conceptual analysis and in foundational analysis. It would seem to me that his attempt to understand the levels of foundations would be what phenomenology contributed to his way of approach. And this, too, was a source of the very strenuous disagreement between him, Dewey and Bentley. Because Dr. Kaufmann believed that there were firm foundation levels, certain things were presupposed by others and they were more or less timeless whereas Dewey and Bentley were more pragmatic in the sense of their functions in terms of a particular problem under investigation. And that their significance and that their function would change in terms of the problems under study. This would be my current recollection of some of the sources of the disagreements. And for me one the most thrilling experiences I had – I met Dewey twice through Dr. Bentley – and one of the times was during the intense correspondence he had; and Dr. Dewey accurately predicted the positions that would be taken by Dr. Kaufmann in that correspondence. And to me, I was just so dazzled, now I'm no longer so because I can see that Dr. Dewey – because of his superior understanding of the assumptions involved in Kaufmann's approach – could see the inescapable conclusions that Dr. Kaufmann would lead himself into.

H.: Did they also meet?

J.A.: Oh yes. Dr. Kaufmann met Dewey a number of times and I can recall how thrilled he was. All I presently recall was his enthusiasm, the honor that he had; the first time he met Dewey he came back and reported that. For him that was a historic occasion.

H.: And he met Bentley in Europe, you said?

J.A.: Definitely. Yes […]

H.: Was Kaufmann generally recognized as a phenomenologist in this country?

J.A.: Yes, more than as a logical positivist. Although I'll be prepared to show in the English edition of the *Methodology of the Social Sciences* [1944] that he adopted quite a few essential positions of the logical positivists.

H.: Yes, you have looked at that?

J.A.: Well, you'll find my name in the prefix. See, Hofstadter, Abel and I were the people Dr. Kaufmann asked to examine the manuscript.

H.: I see. So you had to correct his English?

J.A.: Oh no. We would bring up various questions with regard to the clarity and the coherence. But Dr. Hofstadter was a much greater contributor, a more important one than me. Because I was a comparative novice at that time in philosophic discussion. But I was flattered that he did ask me and in addition he would ask me to examine in manuscript a great many articles that later were published […]

I've got a complete index of all of Bentley's letters to Dr. Kaufmann. In fact, I brought them here. See, I brought my correspondence with Dr. Kaufmann […] Here it is […] Here, the first letter is 1936, when Bentley sent Dr. Kaufmann copies of his two books: *Linguistic Analysis* and *Behavior, Knowledge and Fact*. And here I've got the background. It was a psychologist from Indiana University, who went to Vienna that told Kaufmann about Bentley at that time.

And then apparently, there was an important paper by Kaufmann in 1937: 'Über den Begriff des Formalen in Logik und Mathematik' […] Here, I just came across something: 1940 Kaufmann writing to Bentley something about [his criticism of] Dewey's naturalism. I've got that letter right here […]

H.: [H. dictates] "I have in mind the presuppositions connected with the idea of an objective world, if we keep in mind that we have to define it in terms of experience and that each experience is an experience of a single person at a definite time so that the coherence of all these experiences is not self-evident. This should not be regarded as a revival of the most radical form of solipsism, but only as an approach to analysis which is in some respect similar to Kant's Transcendental Criticism. The aim would be to give a systematic account of the presuppositions implicit in the idea of experience of an objective world." […]

J.A.: This correspondence brings out in what way his conception of methodology differs from Dewey's more restricted use of methodology.

H.: And that has not been published?

J.A.: No, there are others who offered to do that. Yes, but unfortunately not at the present time. This, I think, is of great historic significance, this three-way correspondence.

H.: Yes, you see, in sociology, the struggles are still going on; nothing is solved there, so I think it would be very important. These different schools are kicking each other all the time and there's no agreement whatsoever. So I think – one should look into that.

J.A.: This is a joke. In view of fact that philosophy has not solved anything in 2000 years. Perhaps you are premature in expecting sociologists to do that.

Personally, one of the great contributions that Kaufmann made to my way of thinking was that in value analysis irrespective of the approach one employs, some definition of the nature of 'good' is needed. So that until one tentatively attempts to define how one is using 'good' one will be unable either to make progress in the analysis or unable to obtain a consensus with regard to the significance of a particular problem [...]

Well, I'll tell you something amusing in terms of my own background. When Dr. Kaufmann came here, George Kaufmann was approximately 6 or 7 or so, I believe, and while Dr. Kaufmann was pleased with his intellectual development he had some reservations in regard to the physical coordination; and he thought because the emphasis we give to athletic powers in this country was important for his social acceptance that he improve athleticwise; and he made the following proposition to me: that if I spend three hours improving his son's athletic ability then he would spend three hours discussing philosophy with me – on Sunday. So I accepted it. So that was very important for me; and it was on one Sunday that Dr. Kaufmann said: do you know who's coming for tea today? And I said: "no". And he said: "Bentley". And I said: "not A.F. Bentley". He said: "yes". Would you like to meet him? And I said: "yes, of course". And I read a few articles about him, and it was through that that my acquaintanceship with Bentley began [...] Now, a personal touch. During my oral examination I got into a heated debate with Dr. Wertheimer, the issue about which I can't recall. But what I do recall is that Dr. Kaufmann put his hand on me, on my hand to indicate: "Don't press – he was my adviser – don't get carried away because the purpose is not to be at Dr. Wertheimer in discussion, the purpose is something else." But to me, I was deeply moved by this diplomatic hint from Dr. Kaufmann [...] If he was disappointed by my failure to complete [the dissertation], he certainly did not indicate to me. Which is another aspect, as I said, of his consideration about the feelings of others [...]

Dr. Kaufmann would be outstanding with regard to his diligence in pursuing a particular problem. I can recall Dr. Kaufmann had note-books in which he constantly revised the particular articles and his lecture notes. These were hardcovered note-books, written on one side; the left side would be blank and comments or amplifications or notes to clarify would be put on the lefthand side. And then it would be utilized when he wrote the article. And then this would be the basis of his eventual dictation.

I would suggest that Dr. Kaufmann was not interested in thoroughly covering the literature. He like Nagel had the ability to select what would be the crucial areas upon he would be investigating. And isolate from these, let's call it outstanding papers, significant material to utilize in order to illustrate the methodological issues that he wanted to clarify and analyse. But he would be receptive; particularly suggestions made by his students with regard to anything in American literature that we thought would be significant, relevant, interesting [...]

In addition to all of his educational activities he was an active soccer player. In this country his primary exercise would be walking and he would be one of those who walk up that lengthy hill up to Riverdale. Originally most of the Graduate Faculty lived in Riverdale or then Bronx. And he either could take the bus or walk approximately a mile all of which was uphill [...] I can recall he knew a great deal of the *Meistersinger* by heart. Also *Giovanni*. Yes. He and Bentley could sing the arias because they both knew Giovanni very well [...] I'm not sure but I have a feeling that he dug into poetry, too.

H.: Yes, I know in the Vienna period he wrote these little *Stanzerln* [poems to be sung to the melodies of popular songs].

J.A.: Yes, I was just going to say. And some of them were humorous. Yes. And I think some of them are contained in the correspondence with Bentley which will be the main source [...] He was able to recite a great deal of German poetry, but the German poets I can't recall in the sense that I readily recall his great familiarity with the *Meistersinger*. And it was he, who pointed out to me the mathematical basis of a great deal of Bach's counterpoint [...]

He was very gracious. I can recall how nice he was to my ex-wife. And he did not judge people solely by their intellectual capability. He would recognize that there were other important qualities in men or in humans. But certainly 'character' would be a very important thing. – I have a vague feeling he made some comments about some of the faculty participants in General Seminars. But I can't be more specific than that. I like to think, although this is a vague statement, that of his students I was closest. I'm pretty confident that I saw him more frequently because I used to go every Sunday for tea and also to have these activities with George. And then after his death my relationship with Mrs. Kaufmann continued for quite a while [...] I always remember the morning, you see, when we read that Dr. Kaufmann had died; and my ex-wife and I went up to Riverdale and there was Dr. Kallen and another member of the faculty trying to comfort Mrs. Kaufmann [...]

Martin Dworkin

M.D.: Kaufmann, I think, was either associated with or had as his main client the British-Iranian Oil Company and these people wouldn't believe what was happening in Germany or what the Nazi program would be like and were urging Kaufmann to remain so that he would be able to conduct their business. But then he began to realize, I think he became more and more desperate and here there's the word of the man who was one of the founders and later president of the New School for Social Research, Alvin Johnson, of whom Kaufmann couldn't say well enough. He spoke well of him every time he mentioned him, because he said Johnson saved their lives. And he meant this in the broadest sense because he could come here and work and Johnson put together what may be the greatest graduate faculty in modern history thanks to Hitler because they were all refugees and including some very clever people: Arnold Brecht had been president of the '*Reichstag*' and he was professor at the New School and then there was the economist Adolf Lowe, who was associated with Mannheim and with the London School of Economics and there was Frieda Wunderlich, Fernando de los Rios, who was minister in the Spanish Republic and professor; Leo Strauss, Kurt Riezler [...] And dozens and dozens of other men came together.

Now there was an institution at the New School which afforded an opportunity which is very unusual and should be something which other schools could pick up: it was called the Graduate Faculty Seminar. This was held every Wednesday. What would happen there? Most of the time there would be someone from the faculty; but sometimes an invited speaker of some great prominence, would deliver a paper from the standpoint of his discipline: sociology, economics, whatever position he talked about. I'm remembering one by Eduard Heilmann, the economist. Then, you see, all the other members of the faculty were present in this auditorium and each from his discipline would raise questions. And then the students would raise questions. The idea being not simply a very old notion about what a college is but it was also a notion of what knowledge was, of what social science was all about: [...] that the social sciences had fallen into a terrible state because of the specialists' divisions. That anthropologists couldn't talk to sociologists was an outrage. That the only ones who could talk to them were philosophers and the philosophers were all mixed up among themselves, you see.

Now at the Graduate Faculty one had an opportunity to see someone like Kaufmann respond to historians, respond to a physicist, respond to anyone [...] You saw him in action against a great many other people. And he had a manner that became very unique among the professors. For one thing he was extremely hard of hearing, you know. And he had a very powerful hearing aid on; it also affected his own voice level and how he spoke. And sometimes there were some very, very sharp exchanges that reveal some of the problems that have to come up in the discussion of Kaufmann's work.

At the time, he was very much involved in the *Methodology* (1944), which had been published in English and which was not, however, a translation of the *Methodenlehre* (1936). When he made the arrangements with Oxford University Press to do a translation of the *Methodenlehre*, he started to work and discovered that he had to change it. He said – and this is one of the things that interested me – one of the reasons changing it was his contact with American philosophy. Now there were other professors there; some of them were very great men [...] who used to brag that they knew nothing about American philosophy [...] Kaufmann had encountered Charles Sunders Peirce, the work of Peirce, when he was still in Vienna [...] and it was through Peirce that he discovered the American Pragmatists, but he [...] went really from Peirce to Dewey. He really – if I'm recalling correctly – he really didn't credit James as a serious philosopher which is a mistake, of course. One of the courses I had with him, with Kaufmann, was 'Pragmatism and Logical Positivism' [...] And you see: Kaufmann, maybe this will explain it: Kaufmann saw a relationship between pragmatism as it has developed in America as distinct from the pragmatism in England, which was somewhat different. He saw pragmatism and logical positivism as, if related historically, they had a mutual origin in the problems and the word 'problem' was one of the most used words in Kaufmann's vocabulary. He continually spoke of problems, problems, problems, which is a revealing statement for him. As a matter of fact later on he was saying very proudly that John Dewey was urging him, Kaufmann, to do a book on the relationship between American pragmatism and European Logical Positivism. And he said he would get to that. But he was working on another book first and he wanted to finish that. Of course, he never got to that. He never finished the book. He said that Dewey told him that he was the most qualified to do this, which was a certain position, because there was someone like Charles Morris, for example, and certain other writers who had had contact not simply with the Vienna Circle and with other logical positivists in Europe, but had been trained in American philosophy and you had the connections and the establishment there. And I only felt that it was really tragic that it never worked out.

Kaufmann was a strange and in many ways a magnificent man in these respects. At one point in his course on 'John Dewey's *Logic*' we took that book and we went through every word in a seminar. This was the – I took 1, 2, 3, 4, 5, 6 courses with him and this was the 5th of the 6 courses that I took. At one point, Kaufmann was going on in the class about the correspondence he was having with John Dewey and with Arthur Bentley [Ratner and Altman 1964] and also with Albert Einstein. This was all together and he was making certain references to certain points that Dewey was making that helped to explicate some of the points in Dewey's *Logic*. Principally, the point that I'm talking about, the problem of what constituted indeterminateness for John Dewey. And therefore: what constituted a problem? And then what would constitute a solution or recognition of a solution to this problem? And it sounded more and more to us that he was quoting from letters or at least refined the letters in his mind and we asked Kaufmann why he doesn't publish these letters from Dewey. He said: "I can't do that, I would need Dewey's permission." And so we asked: "Well, write Professor Dewey and Dewey will send you this permission." And Kaufmann, I remember his face fell, and he said: "I would never ask him that kind of question. Because this is the kind of question to which Professor Dewey could only answer 'yes'. He wouldn't have a free choice", he said. He could only in courtesy answer "yes" because if he answered "no", it would be terribly embarrassing. And so here we have one of the reasons why that correspondence (cf. Helling 1988) for so long wasn't brought out – which we thought, I always thought this was an example, one more example of Kaufmann's character. I also thought it was very foolish. I thought that we should have – I told him so as a matter of fact. […]

H.: Maybe you could just say a few sentences about what we talked about downstairs because it's important to tell the reader in which position you knew him.

M.D.: Yes. This is after the War and we were the students who were studying at the New School under the 'GI Bill of Rights' which allowed us to study. It also let me find this very, very wonderful place: the Graduate Faculty of the New School, which was a remarkable collection of scholars, which had been put together mostly because of the efforts of Alvin Johnson, whom Kaufmann so greatly admired […] It was very definitely graded towards the Germans and towards the Germans of the Weimar Republic – politics and philosophy […]

One of Husserl's most profound influences is in jurisprudence, in the theory of law […] And is another reason I think, why Husserl was so decisive in Kaufmann's life is Kaufmann was a lawyer and very much a lawyer, spoke like a lawyer, thought like a lawyer and really behaved like a lawyer – and a lawyer very much in the continental sense [...] It also, I think, bore upon the attraction of Husserl for Schutz, I think that there was such a strong basis in philosophy of law in, well, sociology was constituted so differently for the Europeans with whom I studied from the way it was thought of with the Americans at Columbia University with whom I studied [...]

But I'm trying to get back to Kaufmann and get away from the overall view. Kaufmann, well, as I took a course with him 'History of Modern Philosophy', he did not profess strictly what we would call history of ideas approach. His approach was more a problem approach and if you discussed history of philosophy with him it was a history of problems [...] This is the point where I had one of my greatest disagreements with him. Kaufmann used to say: The history of philosophy would be otherwise if the philosophers admitted that they were wrong, but that problems have persisted in philosophy at this point [...]

I would say: "What you're saying is that there is an universal methodology of thinking according to which problems are recognized, they are stated and then they are concluded. Whether they are solved or not solved you state them in a certain way. Doesn't this presuppose unanimity of conclusion if we all have to subscribe to this methodology?" He said: "Not at all." That is what he called the thought of philosophy [...] he said that this did not presuppose or predetermine a fixed way of looking at problems or of concluding them. It did not determine an outcome [...] Kaufmann, you see, was trying to develop a notion of method which would get some of the problems solved [...] in the same way that you can make progress in the physical sciences [...] He was saying that: certain problem would lend themselves to restatement in a form in which they could be solved. If the philosopher only, let's say all the contenders, if they would only give up their precious terminology and their precious way of seeing things.

H.: I think that's his logical positivist streak.

M.D.: Right. Now he had, however, a concept here, which I have found and lots of others enormously valuable. [...] He spoke of levels of clarification. Now once you accepted Kaufmann's notion of levels of clarification something became possible that first attracted Kaufmann to me or the other way round and that is: it made it possible for you to reach someone with whose presuppositions you utterly disagreed. Because you could see his work as way of dealing with a certain issue and see it that way. This is what made it possible for me to study, say, Thomas Aquinus for 6 years without being a Thomist, without becoming a Thomist [...] Now what Kaufmann would say about that; he would even say this: he patronized what he called mysticism. I'm remembering his pronunciation which was very funny. He would say "müsticism", "müsticism". And he accused some of the members of the faculty of being "müstics", you see. But, you see, for him mysticism was one level of the statement and clarification of the problems that on other levels were stated as philosophical problems. And on that level there were forms of propositions and validation.

And here was another insight of his that I think is one of the most valuable. Matter of fact, it's one that I'm applying today in my own writing. Kaufmann said and I think he makes this clear in his *Methodology* (1944) [...] what happens very often, especially, in the social sciences but also in the physical sciences [...] is that there is an insight or prediction of a solution to a problem. The various steps that have to be done to solve that problem are not yet established, are not yet clear; yet people immediately assume or begin to behave as if the problem is already solved without having settled all the steps in between. [...]

Several mathematicians said that. One of the philosophers of science has a phrase about this, Henri Poincaré says something like this at one point but this is also Gödel whom Kaufmann was continually quoting. Maybe Kaufmann got the idea from the mathematical philosophers because this would be characteristic of mathematical thinkers. [...] Now Kaufmann was continually bringing this up. He used a great many of the problems in philosophy, particularly in discussing the possibility of adapting methods of the physical sciences, primarily mathematical ones to the social sciences which he considered to be one of the principal problems of modern thought. How to make the human sciences more exact without dividing the human sciences? [...]

H.: Now Kaufmann in writing, often refers to phenomenology [as an argument against] atomism because there is a field of experience.

M.D.: Now the reason I smile is because one of the reasons he was able to do this – apart from the richness of phenomenology – is because *Epoché* [...] allowed for suspension of judgment about, if not alternate considerations, the preconsiderations, so that you could certainly proceed to the phenomenal level. But Kaufmann, as we just discussed in talking about his notion that his *Methodology* (1944) was not a practical book, Kaufmann was not interested in the phenomenal level but in the – what shall I say – the cogitator's interpretation of the phenomenal level – the mind's synthesis of the phenomenal level, the mind's thinking about phenomena is what interested him. It was the thoughts [...] It isn't that he refused to look at those things, [...] it has also to do with, I think, the books of Husserl that he considered important. He thought the *Ideen* was more central in Husserl's thought than, say, the *Cartesian Meditations* [...]

But we mustn't proceed for a second if there's a notion here that Kaufmann was not open to new things or not willing to read. Let me tell you a magnificent story [...] the most magnificent statement that

I have ever heard of a graduate professor. He was interested in my work as a writer and he asked me about a lot of the things I had written, I had written on so many subjects. But he was very much struck by a fact that one of my specialties as an undergraduate had been the American Negro Problem, the history of the American Negro Problem and I had written some articles on the Negro Problem. At that time, I still knew the bibliography of that subject thoroughly and oral […] And he wanted me to write from what he considered to be my strength, one of my strengths anyway. So he wanted me to write on the American Negro Problem. Some philosophical study of the American Negro Problem. And I said to him: "Professor Kaufmann, I didn't come to you to study the American Negro Problem. What do you know from the American Negro?" And he said to me: "Mr. Dworkin, it's Friday. You will go home this weekend and you will write down books that are central in the American Negro Problem and you'll bring them to me Monday. I will read those books; now, to make myself competent to judge your study. Then we will establish whether what you want to discuss is a philosophically valid problem and then we will proceed how to do your work". I've never heard this from any other, from any other professor. So he was perfectly willing, you see, to go out into all kinds of things. Also he was much more venturesome than one might derive from – say, seeing him as a strictly, a methodological thinker. He was continually, for example, referring to Wittgenstein. He would say: Wittgenstein is continually talking about logical clarity and all of his examples are from Hölderlin […]

He had a very great sense of literature and of lyric, also another quality, too. He was continually emphasizing the validity of logic as one of the tools that you must use in reasoning. There was one class, one of his logic classes, he was giving, where he was being baited by a student from the back of the room and I remember Kaufmann came into the room, deeper into the room to hear him better, took off his hearing aid and he proceeded to logically destroy this young man who had been questioning him. He just murdered him. It was a beautiful job, you see, and then he turned round to walk back to his desk. But before getting back to the desk, he turned round and shouted: "But never trust a logician" […] This is the kind of man he was, you see. This is why I was so really struck by him.

He had a very profound sense of the problems. Sometimes I felt, in the correspondence that Bentley refers to, Bentley feels he can't get a straight answer, because there is no straight answer. But Bentley thinks there is a straight answer, you see. And Bentley being very much of a pragmatic outlook and so was Dewey. There is that tendency in a good deal of American philosophy that comes from that pragmatic tradition to consider a problem solved if a method that you have adapted works. You know, a lot of working outcomes don't solve the problems. They don't. But they seem to solve the problems. They don't. But they seem to solve the problems. What we are talking about is politics most of the time. It comes down to political structures, to sociological origins of the people involved and how they considered, what they considered to be problematical and what they've done and various outcomes. Very often for Bentley and for a few people in the Graduate Faculty what Kaufmann meant by logic was a metaphysical empiricism of a validity that pre-existed or pre-exists human statements of problems. I don't think this was true. I don't think Kaufmann felt that. I felt that Kaufmann felt that the disciplines of thinking were tools of thinking. That there were different levels of clarification and which these tools will use differently. That one went from level to level by a process of refinement of concepts and definitions.

Kaufmann had this lovely remark about Dewey's writing in the *Logic*. He said: "Dewey's writing is like a witch's mirror in which every man may see his sweetheart or the devil." [...]

I use every opportunity to bring up Kaufmann in my teaching, in my writing I always brought up Kaufmann, spoke of him and cited him [...]

The story that Kaufmann told me, when Kaufmann visited Bertrand Russell in London. And Russell told him, asked him because to Kaufmann Russell was one of the immortals of modern thinking – because of the *Principia*. Everything else he didn't take seriously. But the work with Whitehead was to Kaufmann one of the decisive works: the *Principia Mathematica*, one of the most decisive works in modern history of ideas. And Russell told Kaufmann: "Oh, you're going to New York. You'll be able to meet the greatest American philosopher." And Kaufmann said to him: "Oh, you mean John Dewey." And Russell said: "No, I mean Morris Raphael Cohen" [...]

H.: What did Kaufmann do in this course on phenomenology?

M.D.: Basically, he explicated the *Ideen* [...] Kaufmann used to always talk of two teachers. This may have been decisive. One was Husserl, of course, but the other was Cassirer. And Cassirer is a humanist of the history of ideas and the kind of separation of what Husserl would call methodology: the separation of something called phenomenology is not possible to a man thinking in Cassirer's terms, because there is a spectrum in the history of philosophy and there is a constant as we say in English, an overlapping, an interpenetration so that you cannot really separate it. By the time I encountered him, Kaufmann was very much adjusted in claiming phenomenology as one of his central outlooks.

H.: But some people told me that in this country he did not describe himself as a phenomenologist at all and he didn't want to have anything to do with it.

M.D.: The reason I smile is something that he said once to me that is so marvelous. He said and he was being very quarrelsome at the time, it was almost annoying, you see. He said: "Do you know some people call me a Platonist; some people call me an Aristotelian; some people call me a logical positivist; some people call me a phenomenologist; some people call me a pragmatist." He said: "What amazes me is that all of them think they are calling me names." You see, it didn't bother him really. He fundamentally – I don't think he cared about this. But he knew of Husserl's importance.

Basically, if I can generalize on a very banal level, maybe, a distinction between Schutz and Kaufmann on phenomenology – in one way it should be put that for Kaufmann phenomenology was a very highly refined tool of clarification, very highly refined that allowed you to locate, state but suspend judgment about certain fundamental metaphysical problems. You could shunt them aside because there was no way of verifying them. This was very Kantian. This was a way of doing it, you see. But basically, you then can proceed to the study of what is apprehended in human perception and what is thought about these apprehensions in the working mind. You then can talk about them to yourself and to others because you have suspended judgment on certain basic problems of the '*Ding an sich*', and that kind of thing. For Schutz phenomenology was not – for Schutz it was also a very refined method; but it was a method for coming to grips with human situations, with human perturbations, with the ambiguities of experience and our understandings of experience. Phenomenology allowed us to separate, for example, what is happening to us. You could separate that and determine a relationship, even a logic, between interpretations. In other words, phenomenology was for Schutz something to use in dealing with everyday experience. And here Schutz was closer to Kurt Riezler and to some of the other men there. It was everyday experience to Schutz that was the great problem of human life. Schutz was more, and I had a lot of feeling for this as a writer and as a photographer, Schutz cared about foreground. He cared about what's in the foreground. He was worried about the shape of a doughnut as a fact in your notion. In other words, he had a profound – aestheticians would say aesthetic sense but I would go further than that. Schutz understood the impingement of everything in a field, in a *Gestalt*, on us all at once. But the fact that it comes at us all at once is what creates the problem that phenomenology can be used to help clarifying. Because when we get it all at once we cannot analyze, we cannot think about the various constituent things that are there; but later in an analysis, in thinking about it, we then can clarify simply what is happening to us. Whereas for Kaufmann as I say, at least when I took the course with him and at that time in his life and we are talking about '49, this is when he died; phenomenology was, as I say, it was a very, very highly refined method, very definitely Neo-Kantian in many ways, but if in no other way, decisively a method dealing with the problems of philosophy the Neo-Kantians thought were the problems of philosophy. For many of the problems that Schutz was worried about, Kaufmann didn't think were the problems of philosophy [...]

Well, you see, I can't make that distinction between phenomenology and Neo-Kantianism, maybe it says a good deal about me, surely, because the Neo-Kantians claim that they are not phenomenologists. I think everybody since Kant has to be a phenomenologist.

Perhaps because Kaufmann's discipline, what he professed, was not easily visible as something one could take on as a course, so to speak. He didn't have a following. For example, you couldn't say that about Leo Strauss, who was invited to the most prestigious chair in American Political Science at the University of Chicago, the Charles Merriam chair from the New School. And many of the students followed him there and developed a whole following of students. He's the most influential professor for political science in this country, of what might be called the conservative anti-social science, anti-liberal viewpoint. You see, there is a clear position that students could relate to. There were other men there that had clear positions and then there were men who had followers that were personal followers. But Kaufmann didn't have a personal follower except perhaps for myself and one or two other people [...]

Altmann was more a follower of Dewey and Bentley. But he was a student of Kaufmann and an admirer of Kaufmann, you know. I felt under terrific obligation to bring up Kaufmann's papers. I was so hurt when I discovered, as I said, 9 years after his death that nothing had been done and then for 5 years encountered this cobweb, this mess of cobwebs, of things in the way of bringing anything together [...]

By the way, there's one thing [...] I was really honored by. In the last meeting, that was just the day before he died, Kaufmann said, asked me whether I would read and comment and edit his, the book he was working on for clarity. In other words, he was aware of the problems you are raising about his writing in English. And he wanted to take advantage of the fact that I was a writer, you see. And I said: my God, of course, I'll be honored to do this. But, of course, we never got to that particular thing [*The Pursuit of Clarity*].

H.: What else would you like to be said in a book about Kaufmann?

M.D.: [...] He was a teacher and a teacher in the great tradition. Really – even though I disagreed with him – what we learned from him was an attitude towards thinking, an attitude towards knowledge that made it possible, as I pointed out before, to reach philosophers with whom you profoundly disagreed. But to reach them for their attempts to solve certain problems and, see, here his notion of levels of clarification becomes very, very useful, because you can then see a continuum of trying to deal with certain problems. The person with a religious statement of a problem here and a religious solution for it is not out in space, in another dimension from the man with the rational description and a rational explanation. They are a continuum of trying to contend with a certain problem. The problem is and this was the hope that he had: was there a language and a methodology to be provided on which these two men could meet so that they could talk to one another? And the problems could be finally dealt with [...] Philosophy to me and I think I'm influenced here by Kaufmann, or at least I gravitated to him because I felt this in him. Philosophy is – first of all it's what it is as a subject matter. Yes, there's something in a study, okay, but philosophy basically is philosophizing. It's basically the process of thinking and arguing. It's Socratic; it's dialogue – this arguing with one another. And this is really what I would say he conveyed [...] Kaufmann was no *Guru*. He was not anybody at whose feet you could sit. He was always turning around in class and saying: "never trust a logician", you see or, something like that. You see, in Europe he had been an avocational philosopher. This was a love thing with him. Making his living as a lawyer for an oil firm. In this country he is a professional teacher of philosophy at a Graduate Faculty, of a very unique institution. I think he sensed the difference in role here. Here he was a professional philosopher: this is how he was identified. Felix Kaufmann teaches at the New School in the Graduate Faculty; teaches philosophy at the New School. This is how he sees himself, you know, as his role. Now what does a philosopher do? A philosopher teaches philosophy. A philosopher philosophizes. A philosopher has a philosophy. What is your philosophy? And, you see, I think he didn't see himself that way and resisted this. If he had a philosophy it was his method of philosophizing. I think that is Kaufmann. I think he would have been put off by any prescription because prescription would soon become dogma. He had definite points of view, had this definite point of view that was the way of looking at ideas. This was *the* way. That was my problem with him because he thought that this was *the* way. There was a meta-science of thinking that could be applied here. And where we would use this that many problems which are problems of confusion of thinking, would vanish. But other problems would become clarified so that we could deal with them. Now what the outcomes would be – these would be the functions of your individual life. I would be always asking him: "But does that mean we will all come up with the same answers?" And he would say: "No. It doesn't mean that at all. The answers that come out will depend on your arguments." [...] And this method is related to all the other methods whereby we have achieved knowledge in the history of ideas. Because there is an integral relation here; there are no disjunctions except those of history or sociology. This means we have to fill in blanks with more knowledge that is coming in. There is no vacuum that means we have to enter another dimension, you know. There is a continuum of knowledge, otherwise there is no knowledge. Otherwise there is no methodology which is method of method. Methodology is the study of method for him. He was very attracted to the unified science approach that was attempted at Chicago in which Dewey participated, you know, the universal encyclopedia. And this was to him a restatement of the encyclopaedic urge, you know, of Diderot and all the others. For one thing, a lot of the men like Neurath and some of the others involved in this were men that he respected enormously out of their own disciplines. And the very fact, the enterprise thrilled him that these men were getting together to try to develop something which would amount to them talking to one another. You see, this very much thrilled him apart from the problem. Now the idea that there was a unified science was to him pre-emptive. To support it, however, was what was required now. If there is science there is one science and no other science. The problem is, however, to get there. And there's no use behaving as if the connections were already made because they aren't. And so he says people make statements in biology as if they have already been proved. But they have not. They are predictions, probabilifications to use a Peircean term, and he loved that term – to probabilify is a terrible verb, is a mouthful. But this is the perfect word; there is no other word for it. This is from Peirce, and it also comes into statistics, the statistical language and this was his relationship with Reichenbach and the other statisticians because there – it goes back to Einstein's phrase and he used to quote this also: insofar as it is not [...] – as it becomes about the world, as it begins to deal with the world, as it is applied, it is no longer absolutely true. It can only become statistically true. And we can then only probabilify with certain levels of exactitude which satisfy you on certain levels of clarification, and he was constantly using the example: in Riemannian geometry we arrive at different levels of clarification, Euclidean thinking is adequate for the earth. Truth doesn't figure here: there is the skeptic *Epoché*. Truth, you set aside. Your notion of validity is not truth anymore. Your notion of validity is application [...]

Felix Kaufmann 'Der Nationalökonom im Paradies': A Poem with Translation

Der Nationalökonom im Paradies
Als unser Herr die weite Welt geschaffen,
Die Krokodile, Papageien und die Affen,
Da hat er in die Welt zu guter letzt

Den Wirtschaftswissenschaftler hingesetzt.

Da saß der brave Mann im Paradiese
Mit einem ganz verzweifelten Gefriese,
Weh mir, daß ich kein Material mehr hab,
Es gibt kein Wirtschaften, denn nichts ist knapp.

Mit Gütern wollt ich planvoll disponieren
Und dann mein Handeln streng analysieren,
Und schließlich stolz sein, wenn ich sagen kann:
So handle ich und so tut's jedermann.

Nun muß ich fruchtlos mein Gehirn zerplagen,
Denn gar nichts gibt es hier zu versagen.
Jeder Genuß ist allsogleich parat,
Selbst mit der Zeit man nicht zu sparen hat.

Da sprach der Herr: Du sollst nicht klagen derfen,
Du kannst Dir eine Theorie entwerfen,
Das macht den Menschengeist ja so erlaucht,
Daß er zum Denken nichts zu wissen braucht.

Zwar kannst du niemals einen Satz erproben,
Doch eben drum sollst du mich stündlich loben.
So bleibt die Lehre aufrecht unentwegt,
Wo nichts erprobt wird, wird nichts wiederlegt.

Froh rief der Forscher: Was für ein Tor i,
Von nun an denk ich nur mehr a priori,
Die Empirie, die bleibt mir völlig gleich,
Hier gibt's ja keinen Anwendungsbereich.

Doch seit wir nicht im Paradies mehr wohnen,
Ist scharf zu scheiden zwischen Konventionen
Und Sätzen, deren Sinn darin besteht
Zu sagen, was in Wirklichkeit vorgeht.

The Economic Theorist in Paradise
When the world was first laid out by our Creator
With the monkey, cockatoo and alligator,
God determined, as He viewed His work with pride
To put an economic theorist inside.

So in Paradise there sat this pretty creature.
With perplexity inscribed on every feature:
Alas, what shall I do? he cried distraught,
Economics I can't, - for nothing is short.

I would rationally distribute my ressources
Then discover what the reason for my course is
And finally with pride I would proclaim

That's me, - and every other man's the same.

In vain I put my mental powers to trial
Here is no scope to practice self denial
In profusion pleasures rain upon my lap
And even time is there perpetually on tap.

Then spoke the Lord: My son, thou shouldst not grumble
Wherefore should thereupon thy theories crumble.
The human mind with such nobility is blessed
That without knowledge it can function at its best.
True thou canst not subject laws to confirmation
But that itself should be a cause for jubilation
The theories thus may ever stand unmoved.
What can't be tested cannot ever be disproved.

Cried the student, after listening to this story:
I'm resolved only to think now a priori
For reality, from henceforth, what care I?
There is nought to which my theories should apply.

But, alas, since we've been banished out of Eden,
There's a new distinction we must base our creed on,
And distinguish with precision exact
Conventions from the laws of real fact.
(Kaufmann 1992)

Bibliography of Works Cited in the Introductory Essay

Acham, Karl (1983), *Philosophie der Sozialwissenschaften* (K. Alber, Freiburg/ München).

Anscombe, G.E.M. (1953), *Intention* (B. Blackwell, Oxford.)

Baumann, Zygmund (1978), *Hermeneutics and Social Science. Approaches to Understanding* (Hutchinson, London).

Bentley, Arthur F.(1932), *Linguistic Analysis of Mathematics* (The Principia Press, Bloomington, Indiana).

Bentley, Arthur F.(1935), *Behavior, Knowledge, Fact.* (The Principia Press, Bloomington, Indiana).

Bentley, Arthur F. and Dewey, John (1949), *Knowing and the Known* (The Beacon Press, Oxford).

Berhel, Hal, Eckehardt Köhler, Adolf Hübner (eds.) (1979), *Wittgenstein, der Wiener Kreis und der Kritische Rationalismus. Akten des Dritten Internationalen Wittgenstein Symposiums* (Wien).

Carnap, Rudolf (1935), 'Les concepts psychologiques et les concepts physiques sont ils foncièrement différents?', *Revue de Synthèse 10*, pp. 43–53.

Cicourel, Aaron V. (1964), *Method and Measurement in Sociology* (Free Press, New York).

Dahms, Hans-Joachim (ed.) (1985), *Philosophie, Wissenschaft, Aufklärung: Beiträge zur Geschichte des Wiener Kreises* (de Gruyter, Berlin/New York).

Dewey, John (1915), *German Philosophy and Politics* (Holt, New York).

Dewey, John (1938), *Logic: The Theory of Inquiry* (Holt, Rinehart and Winston, New York).

Dewey, John (1942), *German Philosophy and Politics*, (Putnam, New York).

Dewey, John, and Arthur F. Bentley (1964), *A Philosophical Correspondence*, 1932–1951, Sidney Ratner, Jules Altmann, and James E. Wheeler (eds.) (Rutgers University Press, New Brunswick, N.J.).

Engel-Janosi, Friedrich (1974), *... aber ein stolzer Bettler. Erinnerungen aus einer verlorenen Generation* (Styria, Graz-Wien-Köln).

Ernst Mach (Artur Wolf Verlag, Wien). For Eng. tr. see ('The Scientific Conception of the World: The Vienna Circle'), in (Neurath, 1973), pp. 299–318.

Farber, Marvin, (ed.) (1940), *Philosophical Essays in Memory of Edmund Husserl* (Harvard University Press, Cambridge, Mass.).

Feigl, Herbert (1950), 'Philosophy as Clarification', *12th Street: A Quarterly* (New York: The New School for Social Research) Vol. III, No. 2, pp. 12–13.

Feigl, Herbert (1981), *Inquiries and Provocations*, Robert S. Cohen (ed.), *Vienna Circle Collection*, Vol. 14 (D. Reidel Publ. Co., Dordrecht, Holland and Boston, Mass.).

Feigl, Herbert (1981), 'The Wiener Kreis in America', in (Feigl, 1981), pp. 57–94.

Garfinkel, Harold (1961), 'Aspects of the Problem of Common-Sense Knowledge of Social Structure', in (*Transactions of the Fourth World Congress of Sociology*, 1961), Vol. IV, pp. 51–65.

Gorman, Robert A. (1977), *The Dual Vision. Alfred Schütz and the Myths of Phenomenological Social Science* (Routledge & Kegan Paul, London and Boston).

Grathoff Richard (1978), 'Alltag und Lebenswelt als Gegenstand der phenomemologischen Sozialtheorie', in (Hammerich and Klein,1978), pp 67–85.

Hammerich, Kurt and Klein, Michael (eds.) (1978), *Materialien zur Soziologie des Alltags* (Westdeutscher Verlag, Opladen).

Hayek, Friedrich August von (1944), *The Road to Serfdom* (Routledge, London).

Helling, Ingeborg K. (1979), *Zur Theorie der Konstrukte erster und zweiter Ordnung bei Alfred Schutz: einige Probleme der Explikation und Anwendung (doctoral dissertation, University of Constance*).

Helling, Ingeborg K. (1984), 'A. Schutz and F. Kaufmann: Sociology between Science and Interpretation', *Human Studies 7*, pp. 141–161.

Helling, Ingeborg K. (1985), 'Logischer Positivismus und Phänomenologie: Felix Kaufmanns Methodologie der Sozialwissenschaften', in (Dahms, 1985), pp. 237–256.

Helling, Ingeborg K. (1988a), 'Alfred Schutz, Felix Kaufmann, and the Economists of the Mises Circle: Personal and Methodological Continuities', in (List and Srubar, 1988), pp. 43–68.

Helling, Ingeborg K. (1988b), 'Wirken der Emigration: Felix Kaufmann. Fallstudie einer mißlungenen Annäherung am Beispiel des Briefwechsels zwischen John Dewey, Arthur F. Bentley and Felix Kaufmann', in (Srubar, 1988), pp. 181–205.

Hempel, Carl G. (1935), 'Analyse logique de la psychologie', *Revue de Synthèse 10*, pp. 27–42.

Hempel, Carl G. (1952), 'Fundamentals of Concept Formation in the Empirical Sciences', *International Encyclopadia of Unified Science 2*, No. 7 (University of Chicago Press, Chicago).

Hempel, Carl G. (1963), 'Typological Methods in the Social Sciences', in *Philosophy of the Social Sciences*, Maurice Natanson (ed.), (New York), pp. 210.

Hempel, Carl G. (1972), 'Formen und Grenzen des wissenschaftlichen Verstehens', *Conceptus 6*, No. 1–3, pp. 5–18.

Hempel, Carl G. and Oppenheim, Paul (1936), *Der Typusbegriff im Lichte der neuen Logik. Wissenschaftstheoretische Untersuchungen zur Konstitutionsforschung und Psychologie*, (Sijthoff Leiden).

Husserl, Edmund (1913), *Ideen zu einer reinen Phänomenologie und phänomenologischen Philosophie* (Niemeyer, Halle) (Eng. tr. 1931, *Ideas: General Introduction to Pure Phenomenology*, London and New York).

Husserl, Edmund (1922), *Logische Untersuchungen*, I & II (Niemeyer, Halle) (Eng. tr. 1970, *Logical Investigations*, London).

Husserl, Edmund (1928), *Vorlesungen zur Phänomenologie des inneren Zeitbewußtseins*, (Niemeyer, Halle) (Eng. tr. 1964, *The Phenomenology of Internal Time Consciousness*, Bloomington, Ind.).

Husserl, Edmund (1929), *Formale und transzendentale Logik, Jahrbuch für Phil. u. phänomenolog. Forsch. 10*, Halle, pp. 1–298 (Eng.tr. 1969, *Formal and Transcendental Logic*, The Hague).

Husserl, Edmund (1963), *Cartesianische Meditationen und Pariser Verträge* (Nijhoff, Den Haag) (Eng. tr. 1977, *Cartesian Meditations*, The Hague).

Käsler, Dirk (1985), *Soziologische Abenteuer: Earle Edward Eubank besucht europäische Soziologen im Sommer 1934* (Westdeutscher Verlag, Opladen).

Kaufmann, Felix (1921), *Logik und Rechtswissenschaft* (Tübingen).

Kaufmann Felix (1923), 'Die ökonomischen Grundbegriffe', *Zeitschrift für Volkswirtschaft und Sozialpolitik*.

Kaufmann, Felix (1925), 'Logik und Wirtschaftswissenschaft', *Archiv für Sozialwissenschaft*, Bd. 54.

Kaufmann, Felix (1929/1930), 'Soziale Kollektiva', *Zeitschrift für Nationalökonomie 1*, pp. 294–308.

Kaufmann, Felix (1930), *Das Unendliche im der Mathematik und seine Ausschaltung* (Deuticke, Wien) (Eng. tr. 1978, Dordrecht).

Kaufmann, Felix (1936a), *Methodenlehre der Sozialwissenschaften* (J. Springer, Wien).

Kaufmann, Felix (1936b), 'Remarks on the Methodology of Social Sciences', *Sociological Review 28*, pp. 64–84.

Kaufmann, Felix (1937), 'Über den Begriff des Formalen in Logik und Mathematik', *Travaux du 9ème Congrès International de Philosophie*, Paris, Vol. 6, pp. 128–135.

Kaufmann, Felix (1940), 'Phenomenology and Logical Positivism', in (Farber, 1940), pp. 124–142.

Kaufmann, Felix (1944), *Methodology of the Social Sciences* (Oxford University Press, New York: reprinted 1958).

Kaufmann, Felix (1959), 'John Dewey's Theory of Inquiry', *Journal of Philosophy 56*, pp. 826–836.

Kaufmann, Felix (1992), *Wiener Lieder zu Philosophie und Ökonomie*, Gottfried v. Haberler and Ernst Helmstädter (eds.) (G. Fischer, Stuttgart, Jena, New York).

Kaufmann, Felix (1999), *Methodenlehre der Sozialwissenschaften. Mit einer Einleitung hrs. von Günther Winkler* (Springer, Wien, New York).

Kelsen, Hans (1934), *Reine Rechtslehre. Einleitung in die rechtswissenschaftliche Problematik* (Deuticke, Wien) (Eng. tr. 1992, *Introduction to the Problems of Legal Theory, Oxford New York*).

Knoll, Reinhard, Gerhard Majce, Hilde Weiss, and Georg Wieser (1981), 'Der Österreichische Beitrag zur Soziologie von der Jahrhundertwende bis 1938', in (Lepsius, 1981), pp. 59–101.

Kraft, Viktor (1950), *Der Wiener Kreis. Der Ursprung des Neopositivismus* (J. Springer, Wien).

Lazarsfeld, Paul F. (1959), 'Methodological Problems in Empirical Social Research', in (Transactions of the Fourth World Congress of Sociology, 1959) Vol. II, p. 225–249.

Lazarsfeld, Paul F., Marie Jahoda and Hans Zeisel (1933), *Die Arbeitslosen von Marienthal, Ein Soziographischer Versuch über die Wirkungen langandauernder Arbeitslosigkeit* (S. Hirtel, Leipzig).

Lepsius, M. Rainer (ed.) (1981), *Soziologie in Deutschland und Österreich* (Westdeutscher Verlag, Opladen).

List, Elisabeth and Ilja Srubar (1988) (eds.), *Alfred Schutz, neue Beiträge zur Rezeption seines Werkes* (Rodopi, Amsterdam).

Luckmann, Thomas (1973), 'Philosophy, Science and Everyday Life', in (Natanson, 1973) Vol. 1, pp. 143.

Lundberg, George (1945), 'Review of Methodology of the Social Sciences', by Felix Kaufmann, *the Annals of the American Academy of Political and Social Sciences.*

Menger, Carl (1883), *Untersuchungen über die Methode der Sozialwissenschaften und der politischen Ökonomie insbesondere* (Duncker und Humblodt, Leipzig) (Eng. tr. 1963, *Problems of Economics and Sociology*, Urbana).

Mills, C. Wright (1945), 'Review of Methodology of the Social Sciences', by Felix Kaufmann, *Political Science Quarterly 9*, pp. 470–472.

Mises, Ludwig von (1922), *Die Gemeinwirtschaft: Untersuchungen über den Sozialismus* (Fischer, Jena).

Mises, Ludwig von (1933), *Grundprobleme der Nationalökonomie: Untersuchungen über Verfahren, Aufgaben und Inhalt der Wirtschafts- und Gesellschaftslehre* (G. Fischer Verlag, Jena).

Mises, Ludwig von, *Erinnerungen von Ludwig v. Mises* (1978), (G. Fischer Verlag, Stuttgart and New York).

Mises, Margit (1976), *My Years with Ludwig von Mises* (Arlington House Publishers, New Rochelle, N.Y.).

Mises, Margit (1981), *Ludwig von Mises: Der Mensch und sein Werk* (Philosophia Verlag, München).

Mises, Richard von (1951), *Positivism* (Harvard University Press, Cambridge, Mass.).
Nagel, Ernest (1950), 'Memorial Address for Felix Kaufmann', *Philosophy and Phenomenological Research* Vol. X, No. 3, p. 464.
Natanson, Maurice (ed.) (1963), *Philosophie of the Social Sciences* (New York).
Natanson, Maurice (ed.) (1973), *Phenomenology and the Social Sciences*, Vol. I, (Northwestern University Press, Evanston, Il.)
Neurath, Otto, Hans Hahn, and Rudolf Carnap (1929), *Wissenschaftliche Weltauffassung: Der Wiener Kreis*, Veröffentlichungen der Vereins
Neurath, Otto (1930/31), 'Historische Anmerkungen', in (Neurath, 1981), pp. 389–391.
Neurath, Otto (1936), 'Die Entwicklung des Wiener Kreises und die Zukunft des Logischen Empirismus', in (Neurath, 1981), pp. 673–702.
Neurath, Otto (1973), *Empiricism and Sociology*, Marie Neurath and Robert S. Cohen (eds.), Vienna Circle Collection, Vol. 1 (D. Reidel Publ. Co., Dordrecht, Holland and Boston, Mass.).
Neurath, Otto (1981), *Gesammelte philosophische und methodologische Schriften*, Rudolf Haller and Heiner Rutte (eds.), (Hölder – Pichler – Temsky, Wien).
Neurath, Otto (1982), *Philosophical Papers 1913–1946*, Robert S. Cohen and Marie Neurath (eds.), *Vienna Circle Collection*, Vol. 16 (D. Reidel Publ. Co., Dordrecht, Holland, and Boston, Mass.).
Neurath, Otto (1982), 'Physicalism' in (Neurath 1982), pp. 52–97.
Österreichisches Biographisches Lexikon (1815–1950), Vol. 3
Popper, Karl R. (1935), *Logik der Forschung* (J. Springer, Wien) (Eng. tr. 1959, *The Logic of Scientific Discovery* Basic Books, New York).
Popper, Karl R. (1976), *Unended Quest: An Intellectual Biography* (Long Fontera, Collins).
Quine, William V.O. (1960), *Word and Object* (MIT Press, Cambridge, Mass.).
Quine, William V.O. (1961), *From a Logical Point of View* (Harper, New York).
Ryle, Gilbert (1949), *The Concept of Mind* (Hutchinson, London).
Schorske, Carl E. (1980), *Fin de Siecle Vienna Politics*
Schutz, Alfred (1932), *Der sinnhafte Aufbau der sozialen Welt* (J. Springer, Wien) (Eng.tr.1967 *The Phenomenology of the Social World* Northwestern University Press, Evanston).
Schutz, Alfred (1962), *Collected Papers*, Vol. I, *The Problem of Social Reality*, Maurice Natanson (ed.), (M. Nijhoff, The Hague).
Schutz, Alfred (1964), *Collected Papers*, Vol. II, *Studies in Social Theory*, Arvid Broderson (ed.), (M. Nijhoff, The Hague).
Schutz, Alfred, and Thomas Luckmann (1974), *The Structures of the Life-World* (Heinemann, London).
Srubar, Ilja (ed.) (1988), *Exil, Wissenschaft, Identität* (Suhrkamp, Frankfurt).
Stadler, Friedrich (1979), 'Aspekte des gesellschaftlichen Hintergrundes und Standortes des Wiener Kreises am Beispiel der Universität Wien', in (Berhel, Hübner, Köhler, 1979), pp. 41–59.
Stadler, Friedrich (ed.) (1997), *Phänomenologie und Logischer Empirismus* (Springer, Wien/New York).

Transactions of the Fourth World Congress of Sociology (1959), Vol. II (International Sociological Association, London).
Transactions of the Fourth World Congress of Sociology (1961), Vol. IV (International Sociological Association, Louvain).
Wagner, Helmut R. (1983), *Alfred Schutz. An Intellectual Biography* (The University of Chicago Press, Chicago/London).
Waismann, Friedrich (1930/31), 'Logische Analyse des Wahrscheinlichkeitsbegriffs', *Erkenntnis 4*, pp. 229.
Waismann, Friedrich (1949), 'Verifiability', *Proceedings of the Aristotelian Society*, Suppl. Vol. 19.
Weber, Max (1922), *Wirtschaft und Gesellschaft* (Mohr-Siebeck, Tübingen) (Eng. tr. 1957, *The Theory of Social and Economic Organisation*, Glencoe).
Whitehead, Alfred N. and Bertrand Russell (1910–1913), *Principia Mathematica* (Cambridge).
Zaner, Richard M. (1970), 'The Phenomenology of Epistemic Claims: And its Bearing on the Essence of Philosophy', in (Natanson, 1970), pp. 17–34.
Zilian, Hans-Georg (1990), *Klarheit und Methode: Felix Kaufmanns Wissenschaftstheorie* (Rodopi, Amsterdam).

Theory and Method in the Social Sciences by Felix Kaufmann: An English Translation

Felix Kaufmann

METHODENLEHRE DER SOZIALWISSENSCHAFTEN

VON

FELIX KAUFMANN

DR. JUR. ET PHIL., PRIVATDOZENT AN DER UNIVERSITÄT WIEN

WIEN
VERLAG VON JULIUS SPRINGER
1936

This translation of Felix Kaufmann's *Methodenlehre der Sozalwissenschaften* (1936) is based on a draft by John Viertel, critically reviewed by Carolyn Fawcett, and finally prepared and edited by R.S. Cohen and Ingeborg K. Helling.

F. Kaufmann (deceased)

R.S. Cohen and I.K. Helling (eds.), *Felix Kaufmann's Theory and Method in the Social Sciences*, Boston Studies in the Philosophy and History of Science 303, DOI 10.1007/978-3-319-02845-3_2, © Springer International Publishing Switzerland 2014

Preface

The plan of this work was born of the conviction that many of the most vehement controversies in the social sciences are now ripe for resolution. The work was first conceived more than 15 years ago; but to carry it out, careful preparation was required; the publications cited in the notes will provide some information about this. Aside from the investigative goal of securing the grounds for that conviction, I also had a didactic goal in mind, to convey to the social scientist concerned with more abstract problems, and to the advanced student, an insight into the way problems of social theory have been intertwined with those of the general theory of knowledge and of sciences apparently remote from social theory; together with that, it was also my hope to eradicate a series of deeply rooted prejudices, which stand in the way of understanding the structure and significance [*Sinn*] of social science research, and of its undisturbed development.

Guided by these goals, I have been led to concentrate on the fundamentally significant relationships within the theories of science, to an almost complete exclusion of considerations peripheral to this focus of inquiry, no matter how interesting these may be from the viewpoint of the history of doctrines or the psychology and sociology of knowledge. Consequently, the entire critique in this book is a critique of principles.

This statement appears to require an additional explanation about the philosophical 'viewpoint' from which this critique of principles has been undertaken. But this is not the case: the appearance to the contrary arises through an erroneous conception of philosophical thinking, to be more precisely defined in what follows. I also wish to note that my work should not be called a *phenomenological* theory of the social sciences, although the works of the founder of transcendental phenomenology, the great philosopher Edmund Husserl, have exercised a strong and lasting influence on my thought. For the methodological analyses carried out in this book remain on this side of the problematic of transcendental phenomenology; their goal is a 'formal critique', not a 'transcendental critique', if these two concepts are understood in Husserl's sense.

Nor can this work easily be fit within the framework of any existing school of social theory, economics, or theory of law; yet I cannot refrain from mentioning that I feel especially indebted to two social theorists for the wealth of stimulation I was able to draw from their writings and from many years of discussion: Hans Kelsen and Ludwig Mises.

The reasons which determined the structure of this book can be gathered from the introduction. Here I just wish to point out that the applications to problems of social theory, which are carried out in the second part of this work, on the basis of results of theoretical reflections in the first part, could be multiplied in many ways – which I hope will happen soon through the collaboration of a large number of researchers; therefore I have not placed too much emphasis on a detailed correspondence between the analyses in the two parts of this work. Such a correspondence would have disturbed the straightforward presentation of the general range of problems in the theory of science; this seemed far more essential to me.

The literature listed in the notes is but a small fraction of the writings that have conceptually influenced this work. For the most part I have restricted myself to citing only those works explicitly mentioned in the text.

For the critical editing of the manuscript and of the galleys, I wish to thank most heartily a number of friends in various countries, especially Dr. Karl Bode, presently of St. John's College, Cambridge and Dr. Alfred Schütz of Vienna. Dr. Bode has also taken upon himself the great labor of preparing both indexes.

Vienna, November 1935 F. Kaufmann

Introduction: On the Problematic and Structure of the Book

The main difficulties in the methodological analyses of the social sciences arise from the excessive multiplicity of intersecting and interpenetrating problems. While in investigations in the theory of logic, mathematics and theoretical physics, we are confronted as a rule with a clearly delimited number of questions and can therefore survey the road which the investigation is to take, at least for short stretches, in the theory of the social sciences, on the other hand, the tangle of pathways lures us on into boundlessness. Whoever has followed the struggle over method [hereafter, the *Methodenstreit*] in its various forms and phases, within the most important of these sciences during the last 50 years or so, whoever has let the pros and cons and the cross purposes of the doctrinal positions pass before him, will initially feel almost overwhelmed by the wealth of points of departure, goals and paths of research, and then later come to realize that before everything else it is necessary to create a systematic order among problems. For such an order will make it possible to shed light on the content of the conflicting theses, the meaning of which, insofar as it is relevant for the analyses, is only too often obscured by a terminology burdened with ambitious implicit presuppositions; only then can the true methodological divergences be grasped and evaluated.

Therefore the task is first to attain a clear orientation with respect to the research goals and research procedures of the social sciences. Many methodologists have understood this very well, and accordingly have undertaken either to decree that one method is the only correct one, or else have sought to assign the various methods to certain basic types of personality and/or world views held by the scientists, and in this way to provide for the corresponding multiplicity of 'equally possible' methods.[1] Here a broad field of investigations into the origin of these character types and world views emerges, and thereby an area for psychological, anthropological and sociological research.

But both dogmatism, and the relativism just described, have fulfilled the expectations attached to them only to an inadequate degree. The world-embracing, dogmatically stipulated principles, which were to justify the hegemony of a specific

[1] Along with others, Dilthey, Jaspers and Rothacker would also have to be mentioned.

method, have usually failed when confronted with the requirements of the individual sciences. Subjectively oriented doctrines frequently provide very interesting insight into the motivation for research *ex post facto*, and thus shed light on one or another emotional element as well as the genesis of such elements, which were especially effective in the choice of methods. However, these subjectivists almost always fail to progress beyond the antechamber of the problem, while the internal cognitive relationships of the research, required by 'the matter itself', receive inadequate treatment. Therefore investigations of this kind will be especially important where the uncovering of historical interconnections is at issue: for a methodology that is to serve as an instrument of current research they are of much less significance.

In the face of this we can render, in a first approximation, the conception of methodological problems to be expressed in the analyses below more precisely in the following manner: scientific research is faced with certain questions which it is supposed to answer. Thus certain definite goals are set, and now the ways are to be specified by which we can reach these goals. Therefore what has to be clarified is: what do we want to know, and how can we acquire this knowledge? Which problems are posed, and what procedures for their solution come under consideration? Accordingly, a methodology of the social sciences, as we understand it, has to set as its task a systematic analysis of types of problems and types of procedures. For the present its domain is only delimited in a pre-scientific manner; and the methodologist has to evaluate the capacities of the various procedures in relation to the goals of research which have been set, insofar as this is possible within the current state of knowledge. For this, purely objective criteria for the preference of a method will emerge.

If one approaches the investigation of controversial doctrines involved in the *Methodenstreit* with this aspect in view, then we will very likely have to distinguish between two kinds of statements: namely, first, statements concerning the points at which the doctrines are distinguished from each other with respect to procedure, and second, how the choice of method is justified. In analyses of the second kind, we encounter very frequently the attempt, undertaken with illusory supporting arguments, to designate one specific method as the 'only correct' one; it is one of the most important tasks of the critique of method to examine these claims of validity. Yet in doing so we must take into account that a proof of the falsity of claims to validity of this kind by no means implies a verdict against the method itself, but only against the status claimed for it; not infrequently it turns out that good methods are 'justified philosophically' with bad arguments. In the theory of methodological doctrines, as a rule it is the case that the pseudo-philosophical would-be justifications obscure the true differences in procedures between two contending positions, so that only after removing them can a clear view of two latter be obtained. Then very frequently it turns out, that the apparent incompatibility of the methods, as consequence of which only one is completely correct (adequate) while the others are totally incorrect (inadequate) does not exist at all. This insight is also the determining reason why in the investigations below such extensive space has been accorded to the critique of speculative prejudices. This critique will be mainly turned against rigid methodological dogmatism but it will also seek to set limits to relativism (conventionalism).

We obtain an especially useful guide in carrying out our analysis by directing our attention to the role *abstract natural science* plays in the debate on method. For it will be seen that in almost all the more recent methodological controversies in the social sciences – whether explicitly or implicitly – the (alleged) method of physics is held up either as ideal or as its opposite. Nor is that difficult to understand. The social scientist who wishes to justify the maxims that guide his own research, and sees the immense successes of physics before him, inadvertently arrives at the following conclusion: if the method of physics is at all applicable to the social sciences, then it alone is correct, or at least it is the optimal method. Then according to whether this condition is seen as fulfilled or not fulfilled, the social scientist will fall either into a crass methodical naturalism, or a no less crass antinaturalism. And then in the contemporary philosophy of nature or in the philosophy of mind [*Geistesphilosophie*] he will readily find the 'deeper justification' for the view he has adopted. Thesis and antithesis can be formulated here as follows:

Naturalistic thesis: Research in the so-called social sciences can be called scientific only insofar as it is carried out with the method of the abstract natural sciences; otherwise it is unscientific, or else prescientific. However, to the extent to which these methods can find application in the social sciences, these are natural science disciplines.

Antinaturalistic thesis: The methods of natural science are in principle inapplicable to the problems – or at least to the central problems – of the social sciences, as these are not concerned with the space-time reality of nature (for which alone these methods are tailored) but with a mental [*seelisch-geistige*] reality which is of an altogether different kind. Therefore the social sciences and the natural sciences are *toto coelo* different.

To be sure there is no lack of attempts to bridge this difference but for the most part their persuasiveness is slight because they do not overcome in principle the prejudices which equally underlie naturalism and anti-naturalism. These prejudices are rooted above all in mistaken conceptions concerning the nature of reality and truth, as well as the validity of the laws of nature; for this in turn, the misconstrual of the meaning of mathematical propositions and of their application to the world of experience, is to a large extent responsible. Now the state of affairs today is such, that the problems just described can be considered as finally solved at the level of methodology, so that it has become possible to radically remove the relevant errors. But with that, the problems raised in the *Methodenstreit* appear to be quite different. In place of formulations framed in terms *aut-aut* [either-or], we now can develop formulations which, on the basis of careful analysis, seek to establish the extent of similarity or difference between the methods in question. After removing the superimposed speculative conceptions, the structures of the specific procedures are revealed, and their description is possible without great difficulty. This in turn is the most important precondition for a systematic classification of the problems of the social sciences, and beyond that, for a theory of induction in social science – a goal toward which to be sure, the present work only clears a path which it is not able to pursue.

If we have just advanced the thesis, that the relation of the social sciences to the natural sciences is the dominant element in the main variants of the *Methodenstreit*, then it is by no means intended thereby to maintain the – totally inappropriate – claim, that this relation forms the central theme of all these controversies or even of the majority of them. What we mean is merely that in the arguments and counterarguments certain conceptions concerning this relation have come to be effective conceptual motives, – especially the mistaken views concerning the nature of mathematical and physical lawfulness.

Accordingly, in order to acquire a foundation for decisions in the *Methodenstreit*, we will first subject the determining ideas about the essential quality of our knowledge of nature to a critical analysis. The preconditions for this are formed by general epistemological considerations concerning the truth criteria of experiential judgments, and of *a priori* judgments. Among the latter, the propositions of logic and mathematics require a more precise analysis, which then leads directly to overcoming the most dangerous prejudices concerning the nature of natural laws, thereby opening up the way to fundamental statements about the relation between fact and law.

With that, a foundation in the true sense has been laid for those high-level problems of a general theory of science, the treatment of which is indispensable for radical methodological investigations in the social sciences. These problems are above all the problem of vitalism, the psycho-physical problem, and the problem of value judgments.

These considerations determine the structure of the first part of our work, in which the questions mentioned, will be examined in the order listed above. This part will be concluded by an examination of the nature of inquiry in the theory of science and its delimitation with respect to metaphysical speculation, as well as by the proposal of a *universal methodological schema*, intended to make possible a classification of scientific procedures.

The second part contains the application of these results which characterize science in general to the theory of the social sciences and takes into consideration the fact that the definition of fields of research and of procedures in the social sciences follows mainly from a four-sided delimitation, namely, with the respect to the natural sciences, to psychology, to axiology and to history. In the *Methodenstreit* bitter controversies are fought over these boundaries.

It is clear that the analyses appropriate here are all contributions to the question of the *independent character* of social science inquiry and therefore represent a suitable precondition for the analysis of the nature of laws and of concept formation in the social sciences; this investigation is carried out in the two sections that follow. The conclusion of the book then consists of the evaluation of the results which these reflections have provided for judging a series of important methodological controversies: the next to last section is devoted to problems in economic theory, and the last to those in the theory of law.

As can be seen from this outline, the work proceeds from more general considerations to more particular ones. Although I am by no means unaware of the difficulties which this path presents to the philosophical untrained reader, I have selected it because, in my opinion, only in this way is it possible to arrive at a classification, according to objectively based principles, of the questions, which, in traditional formulations and treatments of these problems, would, for the most part, lie helter-skelter in confusion.

It is these principles then that form the guideline for a systematic critique of method. In order to be convincing, such a critique must reach down into the most general strata of thought, for it is shown again and again that most fatal speculative errors are rarely bound to the more subtle specific differences of the material investigated, but instead most frequently involve very formal traits of thinking, which appear within the special fields as if they were peculiar to these fields. We can obtain a clear view of this cognitive fact, in a first approximation, by looking at the relation of a specific [abstract] scheme of equations to a corresponding set of examples or so-called 'applications of these equations'.

It is also immediately evident that, aside from their critical results, such investigations open up rich possibilities for valid *analogies* between different domains of knowledge. Since the beginning, extensive – often too extensive – use has been made of the cognitive tool of analogy in science and in the theory of science: but the results frequently represented false hypotheses or also pseudo-philosophical speculations, because the point of application for the analogy, the *fundamentum analogiae*, had been wrongly taken. For the formation of analogies only offers assurance of true gains in knowledge when it has been clearly established just what the two correlated domains have in common. Therefore a consistently pursued attempt to make knowledge within one domain of knowledge useful for another leads to a stepwise order of domains of knowledge. Only through this ordering does it become possible to determine how great an impact the insights gained in one domain will have, and in many cases this ordering alone also permits us to reach a conclusive resolution of methodological controversies. Therefore, in the theory of science, one must not shy away from 'detours' in achieving our goals.

Part One. Elements of the General Theory of Science

1. Basic Philosophical Considerations

From its origins, philosophical thought was guided by the idea of arriving at *ultimate truths*, and its vicissitudes can to a considerable extent be described by relating how these ultimate truths were conceived.

In its first stage, in which philosophy was still barely distinguishable from myth, these truths were conceived as insights into the origins of the world, including those of human beings. Philosophy is cosmogony, in which the systematic quality of seeking unity frequently is expressed by conceiving the world as emerging from an original, 'primal substance', [*Urstoff*] such as water, earth, fire or air, or from a small number of such substances, without any more being said about this process of development, than that it resulted from an act of will of a divine being.

In contrast to this, truly immense progress is achieved by the conception of ultimate truths as *sources of knowledge.* For here an analysis of thinking is already carried out, while in the initial stage, basically, only everyday experience of the transformation of substances into other substances, is uncritically interpreted and arbitrarily declared to be the ultimate truths. In the second stage, however, the most influential conceptual theme is the following: our entire knowledge consists in part of fundamental insights which do not have to be grounded upon other sources of knowledge, and in part of such further insights as are derived from the fundamental ones. A complete system of knowledge would thus be created if (1) all the sources of knowledge could be indicated, and (2) the rules of derivation could be shown. With this the two groups of epistemological tasks have been identified, which have dominated Western philosophy since Plato.

As far as the first point is concerned, we can classify philosophical schools (to be sure, quite one-sidedly and by no means adequately) according to their ranking of the importance of the various sources of knowledge. Sensualistic empiricism assigns the dominant place to the evidence given by sense data, spiritualism to the evidence of inner experience, and rationalism to the evidence of rational truths. Of course, actual historical systems most frequently appear as mixed forms, which differ from each other especially with regard to their judgment as to what degree the laws of physical and psycho-physical events are accessible to finite human understanding.

The problem presented by the *process of derivation* seemed to have found its solution in principle in the axiomatic method, as it was created by *Euclid* for geometry; and thus argumentation *more geometrico* became a philosophical instrument, which was to guarantee the complete rigor of deductions.

The most famous example of this is Spinoza's *Ethics.* But though the rigor of the procedure is a necessary condition for the truth of statements of a deductive system, it is not a sufficient condition. It depends also on the truth of the premises, which, in keeping with the presuppositions, is to rest in these premises themselves. Thus the question of the criterion – or in some cases criteria – of *original* truth is of decisive importance.

Now this *criterion* was most frequently regarded as a feeling of *inner evidence*, which was supposed to be the infallible sign of truth; here *three types of evidence* were distinguished:

1. the evidence of sense perception, which provides knowledge of the existence of things and events of the external world;
2. the evidence of inner experience, which provides infallible knowledge of the facts of one's own consciousness;
3. the evidence of the laws of reason, which contain truths valid in all places and at all times, and therefore occupy a higher rank than changeable factual truths.

First a few words have to be said here about the *rational truths*. These are supposed to embrace the laws to which all being is subject. In theological guise these are the ideas on which God's plan of the creation of the world is based.

Geometric knowledge appears as their prime example, brought to its highest flowering especially by the Pythagorean School, and then placed in the service of astronomy. Here the subjection of Being to laws is clearly manifested, just as it appears in simple forms in everyday experience. For 2,000 years of philosophical thought, Plato made geometric knowledge, and together with it logical and arithmetic knowledge, the *nervus probandi* for asserting the existence of fundamental truths that are independent of experience and yet govern it, thus for the basic position of rationalist metaphysics.

This phase of thought is definitively marked by the fundamental assumption of a *polarity* of the knowing *subject* and the *object* known or to be known. The world with its laws is what it is, independent of whether acts of thought are directed toward it or not; on the other hand, the task of knowledge consists in grasping the self-existing pre-stabilized world as it is, and mastering it intellectually in the most complete way possible. To be sure there is by no means unanimity of opinion as to what degree this task is achievable by human beings, as has already been pointed out.

But the other conception, which I called '*naïve realism*', bears within it the seeds of its own supersession. For as soon as we *reflect* on the process of knowledge and see in it a task that can or cannot be solved, a task in which the examination of both the errors blocking the correct solution as well as the source of these errors becomes pressing; as soon as we arrive at the differentiation of mere appearance from true being, which must prove itself by certain criteria, then an essential step has already been taken away from the idea of the world that has being in itself, towards its gradual replacement by the insight that all being is *consciousness-related.* For the criteria of Being are themselves facts of consciousness, and the distinguishing of Being and Appearance can accordingly only take place in such a way, that certain facts of consciousness (lived experiences), or certain groups of such facts, are distinguished as sources of knowledge from other experiences, to which no objective Being is supposed to correspond. With that, obviously the separation of being and appearance is transferred into the sphere of lived experience, and it becomes a question on what grounds can we assert for lived experience and therefore also for Being itself any significance beyond that sphere? The appeal to evidence can be of little help here, for the feeling of evidence too is a 'subjective' (lived) experience, and it remains an open problem by what means it can be legitimated as criterion of truth [*criterium veritatis*]. As is well known, this is the question which Descartes decided – or cut off – by appeal to the truthfulness of God [*veracitas dei*].

Such considerations very frequently lead to the thesis that the sole true reality is the reality of the phenomena of one's own consciousness, and thus leads to the replacement of naïve realism by so-called *phenomenalism.* The road to this doctrine is marked particularly by the experiences of the deception of the senses, and indeed of the 'subjectivity of sense perception' in general. For from these experiences arises the question, whether any properties at all can be regarded as belonging to the things themselves, and the denial of this then leads precisely to phenomenalism.

A little deeper reflection shows, however, that this doctrine, which seeks to surmount subject-object dualism by the formula *esse est percipi* itself still remains in the *ante-chamber of reflective insight* and, compared to naïve realism, merely represents *"naiveté on a higher level"* (Husserl), as it operates with an uncritical, confused conception about lived experiences of consciousness. Because of this it does not become aware of the dualistic tension that inheres in consciousness as *"consciousness of something"*. But at least it contains the fundamental recognition of the consciousness-relatedness of all Being, and from this it is not too long a road to grasping the fundamental task of philosophy, to comprehend by means of rigorous analysis of the phenomena of consciousness, and especially of predicative thinking, how the world is constituted in consciousness. This task – first comprehended by Kant with full clarity in his 'Copernican revolution' – has dominated the problematic of philosophical criticism to no less a degree than that of the more refined positivism and pragmatism; it was advanced greatly by Franz Brentano through his emphasis on the intentional character of psychic phenomena, and received its most profound treatment to date in Husserl's phenomenology.[2]

As we shall show immediately, these analyses of the phenomena of consciousness contain the radical overcoming of naïve realism as well as naïve phenomenalism, which, to be sure, does not prevent these two views from continuing to be influential in uncritical and semicritical thought, and thus they continue to form the focal point of the methodological dispute within the natural as well as the social sciences.

We wish to characterize these analyses, in a first approximation, by stating that in them the question of the *essential nature of being [Wesen des Seins]* appears transformed into that of the *truth-conditions of judgments*, or more precisely, is revealed to be that question. For the manner in which the data of a verification series cohere determines the 'nature' of the object about which the judgment predicates something.

A simple example may serve to make this clearer: Let us look at the judgment based on an optical impression: "Here is a house." This assertion contains not merely an optical impression, but beyond that, a series of anticipations; for example if the person making the statement closes his eyes for a time and then, without changing his place or the position of his head, opens them again, he will have a conforming optical impression – unless in the meantime something completely extraordinary, say an earthquake, has taken place. Furthermore he assumes that upon changing the position of his head, and in the case of certain changes of his entire body (for instance, walking around the house), he will have a continual sequence of optical impressions of a kind which are predictable to a certain degree. Similar anticipations exist about the tactile and kinesthetic impressions resulting from touching the house or passing one's hand over a part of its contours; finally, it is assumed that a fellow human being with normal sensory equipment, behaving in the same way in the same place, will receive corresponding impressions.

Thus one will recognize that a perceptual judgment concerning the reality of things in the outside world, is by no means to be conceived as the simple ascertaining of sense impressions – *receptive* [passive] *findings* but that it contains an *open series of assumptions* directed towards intertemporal, intersensual and interpersonal (intersubjective) validation. This insight forms the core of both the formulation

[2] See also below, Part One, section "Life and Consciousness"

(of critical philosophy), that what has Being is an infinite idea, and that it is *not given* to thought, but *set as a task for it*, as well as Husserl's dictum, according to which things are "*open horizons of possibilities*". It removes the vague conception of an absolute transcendence of the part of whatever has Being in the world, as it confronts consciousness, the conception that characterizes naïve realism: it removes as well the phenomenalistic doctrine of the immanence of the world in consciousness. And it replaces these by the result of reflection, that while the world is 'constituted' in thought, and therefore to be described by means of concepts of the facts of consciousness, and only by these, it is by no means, therefore, 'contained' in the acts of consciousness in any manner that could be conceived as real or logical inclusion. Thus we can well call this complex of ideas 'immanent transcendence'.

Accordingly the formula '*esse est percipi*' is, in a first approximation, to be replaced by the formula '*esse est percipi posse*', but the problem of the essential nature of being is by no means solved thereby, but merely outlined; for the true task now consists in grasping the character of that potentiality, which is merely indicated by the '*posse*', i.e., in making explicit the *implicit presuppositions* which are contained in empirical judgments, thereby revealing their true conditions, the 'criteria of true Being'. As anticipations of future being are among these criteria, it follows that – in principle – at no point is a judgment of experience to be regarded as 'final', 'uncontrovertible', or 'absolutely' secure: rather it always remains conceivable that newly occurring events will lead to a revision of the judgment, and indeed this is true – contrary to widely-held position – not only for judgments of external experience but also for the judgments of inner experience.

Also the results of these reflections already lead to important consequences for methodology in the narrower sense. They show that assertions about the reality of things, the presence of properties, the existence of relations, contain other elements, or contain other elements as well as the data of passive lived experiences. Furthermore, they direct our attention toward the specific ways in which those syntheses of past and anticipated or imaginatively projected perceptions function, which are contained in empirical judgments. Thus they lead to the removal of deeply-rooted prejudices, which have been a great constraint upon scientific research and are still so today.

But above all, owing to this, the doctrine of the 'feeling' of evidence as a source of truth is superseded. For as soon as it is clearly seen that even statements of fact, directed toward the present moment, contain assumptions about the future, upon the fulfillment of which their truth is dependent, we can no longer hope that through an immediate 'possession' of reality or 'participation' in it (which allegedly displays itself in the 'feeling' of evidence) we can conclusively assure ourselves of that reality. As a judgment of fact contains more than the plain registering of the isolated life situation of the person making the judgment, it can never be verified definitively by any single finding, no matter what texture of feeling may accompany it. This is also true of our judgments concerning our own lived experiences, and thus of the facts of inner experience; for by declaring such a fact to be a fact of a certain kind, for instance, of doubt, purpose, grief – and we must do that insofar as we make any statement about it – we place it in relation to other facts of a similar kind. Whether this relation really holds, i.e. whether the experience in question really was an experience of doubt, purpose, grief, cannot be decided by a directed 'act of looking' [*Schau*] alone (with accompanying feeling of evidence) that is confined to the experience. Rather, here one must refer to the knowledge of what doubt, purpose, grief is, in order to ascertain whether the experience to be characterized has the relevant attributes. But every such assertion must be regarded in principle as something that can be tested and revised.

Thus it becomes clear that in judgmental (predicative) thought the 'matter' of sense perception, which is supposed to represent the correlate of pure receptivity, cannot be considered in isolation, but instead enters sense perception as already formed, i.e., situated within a context of experience. Indeed those implicit presuppositions of which we spoke are nothing else than the incorporation into the current store of acquired and habitual experience, which is composed largely of merely perceived, and not apperceived date. Therefore we can say that *predicative experience* has its foundation in *pre-predicative experience.*[3]

With this insight, the idea that the whole of human knowledge can be constituted by combinations of a number of immediately evident truths can be regarded as superseded. This idea already began to appear in the conception of the axiomatic method in antiquity, and it exercised an enduring influence on late scholasticism and later on Spinoza and Leibniz, and most recently found its most significant formulation in Wittgenstein's doctrine of elementary sentences.[4] The same holds true for the idea of a *science or philosophy without presuppositions* (epistemology). For all predicative thought presupposes a foundation, one which, to be sure, is itself subject to change with the continued progress of experience; without such foundation, the truth-conditions, and therewith the meaning of the judgments themselves, could not be determined. Science and philosophy are not the *spontaneous generation of knowledge out of nothing*, but the augmentation, restructuring and clarification of previously gained knowledge.

This insight has been obscured above all by two considerations, which in themselves are appropriate and very important, but from which misunderstood, mistaken conclusions have been drawn. The first of these concerns the '*privileged status*' of judgments based on perceptions, thus, those which contain findings about sense data as opposed to judgments in which this is not the case. This consideration is indeed of extraordinary significance; for it allows us to understand of what the difference between reality and mere fantasy consists, and it directs attention towards the foundational order which exists among primary (originating) perceptions, and secondary memories that refer to them, as well as (tertiary) reproductions; and it clearly establishes that perceptual judgments as findings about 'originally constitutive' [*urstiftende*] experience,[5] form modes in the network of experiential knowledge. But from the insight that judgments based on sense perception contain the material of sensation that confers on them their privileged status, we may by no means conclude

[3] See Husserl, *Formale und transzendentale Logik* [cited hereafter as *Logik*], p. 185 ff.

[4] *Tractatus Logico-Philosophicus*. With an introduction by Bertrand Russell. London 1922.

[5] Husserl, *Logik*, p. 278.

that this 'material' can be isolated and that it contains the objective content of all knowledge, as sensualism [sense-data empiricism] wants to maintain.

The second of these considerations has led to the sharp separation between *descriptive analysis* of objects of thought and the investigation of their *genesis*; carrying out this separation is the great achievement of Kant's transcendental philosophy, which has also been so path breaking for modern descriptive psychology. For it is easy to see that the question "What is the meaning of acts of thought?"and the question "How did these acts of thought originate?" differ from each other, and that we stumble onto false paths, if we combine them in 'problems of origins' as has been done by all varieties of empiricism. Pointing out this methodological syncretism forms the core of the critique of psychologism in logic, and in psychology, and the various cultural sciences [*Geisteswissenschaften*]. We will examine this more closely below.

Establishing this does not, however, say anything against the insight that the descriptive analysis of the meaning of judgmental acts will reveal 'sedimented strata of meaning',[6] which point to the previous acquisition of experience, as is the case in every incorporation of perceptions into a context of experience; therefore this is also the case in every determination of the kind (categorization) of whatever is 'given' at the time of immediate perception. It only signifies that the temporal stratification of meaning implications does not become *thematic* – as temporal – in the descriptive analysis. The presuppositions of knowledge contained in a judgment, and the anticipations linked to them, which form the truth-conditions of the judgment, present themselves, so to speak, in a two-dimensional projection of the stratified structure, in which the time dimension no longer appears.

In carrying out descriptive analyses of scientific judgments presented in linguistic form, it will frequently occur that no *unambiguous meaning* can be assigned, since the individual terms appearing in the sentence have no precisely defined meaning. From this, then, arises the task, first, of separating the various meanings from each other, and then of assigning to each of them its 'systematic place', i.e., its place within the conceptual structure of science; thereby a certain number of the meanings might be eliminated as not belonging to the system. However, it must not be thought, as so often happens, that with the *elimination of equivocation*, the error introduced into thought by language is eliminated; for ambiguity of language is the *expression* of muddled thinking, and linguistic precision lies in the elimination of this confusion.

As is customary, we have called the operations of thought characterized above 'descriptive analysis'. But this designation is not fully appropriate insofar as these operations are not limited to pure descriptions of what is intended (the content of though), but also include constructive elements since, taking into consideration the goal of knowledge, they emphasize what is 'essentially intended'. Therefore the designation '*rational reconstruction*' – also employed frequently – is to be preferred. But we have to take into consideration that the constructive elements play a secondary role here, and do so as regulative principles in the selection and systematic

[6] *Ibid.*, p. 279.

grouping of meanings, which were combined in confused thought, and which is now being subjected to analysis; this aspect is well expressed by the word *re*construction. That this has not received proper consideration has contributed greatly to the confusion of conceptions of the nature of philosophical investigations in general, and of methodological investigations in particular.

To designate, at least by indication, the stratificational structure of experience is required, especially in the present work, because a considerable number of points at issue in methodological controversy arise from the fact that problems of the formal analysis of *predicative* thought, as completely clear idealized thought, are continually being confused with problems of the 'analysis of origins', which project into the *pre-predicative* strata of thought. These controversies become pointless as soon as this confusion has been removed.

One of the most important questions that must be discussed here, and a problem requiring clarification in various directions, is what kind of connection exists between the *meaning of judgment* and its *truth-conditions*. On the one side we have the thesis that one must first know the meaning of the judgment in order to derive truth-conditions from it, and on the other side the counter-thesis, that the meaning of a judgment coincides with its truth-conditions. Deeper reflection shows that the thesis of the priority of meaning over truth-criteria holds for confused (unexplicated) thought, while for clear thought, although the counter-thesis itself does not apply, a more cautious formulation related to it does hold.

In order to understand this, we have to distinguish above all between what is *presupposed* in a judgment and what is *posited* (asserted) in it. For if, as is the case in unexplicated thought, one conceives of these presuppositions as assumptions about absolute reality, then the meaning, as far as it is determined by them, appears as prior to that which is posited, i.e., to the additional assertions made about that previously characterized Being which must prove to be valid in order to be designated as true. So, for example, we will say that in order to give the truth-conditions of the statement "Aniline boils at 184 °C", we must know the meanings of the concepts 'aniline', 'boils', and '184 °C', while the inverse relation does not hold. On the other hand, in the idea of clear thought, the descriptive analysis of the concepts that occur in the statement is conceived to have been already accomplished; thus, to stay with our example, it is assumed here that the constitutive operations for the concepts 'aniline', 'boil', '184 °C' and their syntactic connections are clearly evident to anyone making that judgment. But thereby, so the argument runs, the truth-conditions for the judgments are also given, and the determination of these therefore requires no separate establishment, logically dependent upon comprehension of meaning. Accordingly, that the meaning of a judgment was 'prior' with regard to its truth-conditions, means no more than that *unexplicated* thought, is prior to *explicated thought* and with that an essential element of our controversy has indeed been clarified.

But apparently the problem becomes more complicated, as soon as we turn our attention to the *positing* of the judgment and thereby go back to our previous consideration about the *open-endedness* of experience. Earlier we said that, included in the conception of being as a 'horizon of open possibilities', are not only well-defined access points, but also *vacant positions*, which are open to further, and yet undetermined experience and which will, so it is anticipated, be congruent [*einstimmig*] with the context of the modes of comprehension already characterized. In this way Being 'transcends' any fixed number not only of individual lived experiences, but also types of lived experience. And in the same way the meaning of judgments transcends any fixed number of truth-conditions. But we can understand without difficulty that this 'transcendence' must not be misconstrued as a supersession of the correlation, i.e., not as an absolute transcendence. This will happen, for example, if we were to make exaggerated use of the notion of causality, which conceives of knowledge as the knower being affected by a 'thing-in-itself'; for each determination of possibilities that as yet remain open is in turn the result of the specification of a mode of comprehension, of a path of empirical access.

But against the above exposition, a fundamental objection can be raised, which must be taken seriously; this we will now have to discuss. This objection maintains that the resolution of the meaning of judgments into truth criteria leads to an *infinite regress*. For – so it is argued – if a judgment, according to the thesis to be examined, cannot be confronted with a Being-in-itself, but is to be made congruent with experiences, which are to be specified, then the truth of a judgment is made dependent on its congruence with other judgments, which are presupposed to be true. A simple example: A makes the assertion to B that a signpost is 500 paces from here in their line of march. Finding that they have arrived at the signpost after 500 paces, A and B observe that they see it, that therefore the signpost 'really' is there, and hence that the assertion of A seems to be confirmed. Thereby the truth of A's assertion has been traced back to the truth of these perceptual judgments, but these in turn now require confirmation; and so on *ad infinitum*.

The reply to this argument is obvious: indeed, infinite regression cannot be avoided, if one always wants to trace the truth of a judgment back to its congruence with other true judgments; but such an attempt is in itself misguided. The ultimate (originating) truth criteria of a judgment do not lie in its congruence with other judgments; but such an attempt is in itself misguided. The ultimate (originating) truth criteria of a judgment do not lie in its congruence with other judgments, but in its congruence with the *given*.

This reply, however, which represents a widely held conception, itself cannot stand up to closer examination. For if we ask what the criteria of this congruence are, then we must recognize that the existence or non-existence of such a congruence can only be asserted if the 'given' has been generally characterized, since only through this can the possibility of a comparison between thought and Being be justified. If this is not the case, the 'in what' [*Worin*] of the congruence remains completely open, and thus the idea of a congruence devoid of content. Such a characterization, however, is an interpretation, a coordination within general contexts, and therefore its explicit execution corresponds to the act of judgment. For example, that a given is blue – whereby the judgment, which maintains that in a certain location there is something blue, appears to be confirmed – cannot be inferred from a sensation conceived of as isolated, as has already been stated above, but only from a *sensation interpreted as a judgment*.

The objection we have just examined, against the claim that an infinite regress is implicit in this truth concept, thus proves not to be valid; and indeed, strictly speaking, this assertion cannot be refuted at all. For understood precisely, this means nothing but the realization of *the horizon-character of all that has Being*, a two-sided open endedness of all experience, as a consequence of which no judgment of experience can be separated on the one hand from previously gained knowledge, on the other, from knowledge to be gained in the future. Because of the relation just stated, every judgment of experience is fundamentally something provisional; it has merely the validity of an assertion.

However, the insight into the fundamental *infinity of the confirmation series* is confronted with the fact that in every case the prescientific and scientific tests of assertions find their termination at a certain point; it is only that these points are not fixed unequivocally for all time.

Therefore we can say that the horizon character of experience corresponds to an *indefinite regress* in the verification of judgments; it is indefinite, because there is no point within the regression at which its termination would impose itself logically, but it is not infinite, because *de facto* it is always broken off at some point, and has to be. For anyone who has overcome the erroneous conception of knowledge without presuppositions, this state of affairs will no longer seem paradoxical.

Now we can formulate the answer to the question of the relation between the meaning of empirical judgments and their truth-conditions as follows:

1. The meaning of a judgment is 'prior' with respect to its truth-conditions, just as confused thinking is prior with respect to clear thinking.
2. In clear thinking, the meaning of a judgment must never be conceived as a ready-made, closed unit, which coincides with a number of truth-conditions. Instead, within it there are empty places for an indefinite number of further (unequivocal) truth-conditions. Insofar as the meaning is 'more' than the mere substance of rigidly delimited, closed truth-conditions, however, it does not contain any element unrelated to the process of verification. The assertion of an, in principle, *unknowable*, transintelligible Being is a *contradictio in adjecto*.

With the insight just gained, the difficulties connected with infinite regress at other points of epistemology and methodology also disappear. The result of the reflections which have removed these difficulties, we will call the *principle of finite formulation*. It states that the concept of infinity which arises in empirical statements is nothing other than an expression of the absence of an ultimate, immovable limitation; the *infinite* thus reveals itself to rational reconstruction as the *indefinite*.[7] An important application of this insight will appear in the analysis of the concept of [scientific] law.

[7] Ernest Nagel came to similar results in his study, 'Verifiability, Truth, and Verification'. *Journal of Philosophy*, vol. XXI, (1934), pp. 141–148.

The finding that no judgment within a process of verification imposes itself as the terminal point of this process must not be misinterpreted; however, to mean that every judgment is actually accorded *equal weight* in prescientific verification procedure. This is by no means the case; rather, as we have already mentioned, judgments of a certain kind, especially the judgments of perception, the judgments about 'inner experience' which has just taken place, function as typical concluding points, as nodes within a series of experiences. What is to be rejected is only the *exaggerated interpretation* of this epistemological fact, according to which these judgments are supposed to be self-contained sources of truth that reveal themselves through specific evidence. In particular, such a misinterpretation has the fatal consequence for methodology, that attention is distracted from the implicit presuppositions with which such 'evident' judgments are burdened. Recent analyses, carried out by noted physicists, motivated by the immediate needs of their science, and also by philosophies of nature,[8] have brought to light how much 'theory', what a wealth of general assumptions, is involved in the use of measuring instruments in physics (e.g., the telescope, theodolite, microscope and ultramicroscope). It is to be hoped that the persuasive results of these analyses, which *mutatis mutandis* can be extended to all empirical knowledge, will in the end also help to bring about a breakthrough, leading circles of social scientists and social philosophers to the fundamental insight that there is an indissoluble linkage between fact and theory, observation and general assumptions. Rigorously thinking this insight through will lead to the dissolution or reformulation of a considerable number of problems in the *Methodenstreit,* especially the controversies about the relation of history and theory, and of statistics and theory; the extreme antitheses in these spring from epistemological prejudices. One of the most disastrous of these prejudices finds its expression in the postulate of common-sense philosophy, to keep to the bare facts. As a battle cry against unconstrained speculation in its time, it had a beneficial influence on the development of research, but sooner or later it must become a serious inhibition. For a lack of clarity concerning the general presuppositions contained in assertions of fact must also bring with it a lack of clarity concerning the possibilities and limits in the evaluation of observations.

In a far-reaching analogy with the problem of the relation of the meaning and the truth-criteria of judgments stands that of the relation of the *subject matter* and *method* of a science. Is it the subject matter, we ask, which determines the method, or on the contrary, the method which determines the subject matter? It is obvious that naïve realism will decide this question in favor of the first alternative. For at this level the world appears as a given, prior to all knowledge, and the correct method will appear to be that which is 'best adapted' to this given. But with the critical dissolution of naïve realism, which includes an analysis of the verification process, the operative aspects appear. Now the world no longer appears as a given, prior to thought, but instead as constituted in thought; only when spontaneous thinking carries out its syntheses are the objects of knowledge 'generated'. However, these

[8] See especially P.W.Bridgman, *Die Logik der heutigen Physik* (*The Logic of Modern Physics*, N.Y., 1927) with an introduction to the German edition by H. Dingler, Munich 1932.

syntheses have various levels; upon the pre-scientific syntheses, which ‘generate’ the objects of everyday experiences, are based the scientific syntheses, which, according to pre-designed principles, aim at ordering the pre-scientific ‘material’, and it is these principles which determine the method, and thus constitute the objects of science.

With respect to this, we can make the following remarks: that the critique of naïve realism contained in the above argument is justified, we have already noted, and there is indeed no fundamental objection that can be raised against it. It is, however, subject to the danger of going too far in its rejection of the over-estimation of the receptive elements of knowledge, according to which these contain all of knowledge *in nuce*, and thereby it tries to *dissolve all of receptivity in an infinite process of spontaneity*, and in keeping with this, determine immanently, within a complete system, all the concepts belonging to this system.

Now here the results of our reflections find obvious application. It may well be that the *receptive* (passive) elements of knowledge *cannot be isolated*, but they also *cannot be eliminated*. The ‘infinite process’ of the dissolution of matter into form cannot be completed, but this is not due to the inadequacy of the human mind, but, rather, we may say, the assertion of this interminable character is but one – not very fortunate – formulation of the irresolvable dualism of the receptive and the spontaneous, of the passive and the active elements of knowledge. From this the relation between the subject matter and the method of science follows: every method of knowledge is based not only in the (genetic) sense on the prior givenness of an object, i.e., that this offers the ‘stimulus’ for the development of the activity of knowledge, but the prior given also enters into scientific procedure as a presupposition in the logical sense: however, it is always only a ‘relatively last’ datum, which, when more profound foundations are sought, is subject to further analysis. Furthermore, in scientific procedure, a selection of the ‘essential moments’ within the prior given takes place, which are then declared to be thematic moments and incorporated within the system of the science. Through a synthesis of these thematic moments, the scientific object of knowledge is then constituted. This we can say that the original ‘*object of experience*’ is prior with respect to method, which for its part is prior with respect to the ‘*object of knowledge*’.[9] But in this formulation, we must not forget that the methodological analysis is not concluded with this distinction, but rather that its further task consists in revealing in each case the relation of ‘object of experience’ and ‘object of knowledge’, and thereby rationally reconstructing the transformation that leads from the one to the other.

This analysis also keeps us from assuming that a scientific system is conceivable, which is closed in such a manner that its concepts ‘implicitly define’ each other. For such an assumption cannot withstand the deeper reflection, that the meaning of the fundamental concepts of an empirical science can never be comprehended, without reference to *pre-systematic experiential knowledge*. In relation to this reference, the open-ended character of experience is of crucial importance for

[9]This excellent terminological comparison can be found in A. Amonn, *Objekt and Grundbegriffe der theoretischen Nationalökonomie*, (2nd ed., Vienna 1927).

the *Methodenstreit*: as a consequence of this empirical judgments not only point toward verification by an indefinite number of experiences of one type, but also by the experiences of other types, that for the present still remain unspecified. For it is above all the manner of evaluating this epistemological fact for the interpretation of scientific thought which distinguishes modern *Realism* from modern *Positivism.*

We wish to clarify position and counter-position by again resorting to an example from physics where, as is so frequently the case, the two fronts are delimited most distinctly. It is a debatable question, whether reality is to be attributed to the atom, or whether the atom is only to be conceived as a rational construction, serving certain goals of knowledge, and therefore capable of being relinquished as soon as it is no longer confirmed. Positivism advocates this thesis, and in doing so can point to the confusion which arises due to the confounding of vague conceptions of realism with the concept of the atom. For example, we combine color qualities with real objects of the external world, but to attribute color to the atom would mean to place ourselves in contradiction to the whole of optical theory. Thus we have to make clear that the concept of the atom is defined by a fixed number of well-determined physical operations. Accordingly, the conception that the atom possesses still further 'properties' would be just as unreasonable as the conception that to disregard the definition "a roan is a brownish horse" would be to assume characteristics for 'roan' that are not included in the meaning either of the concept 'horse', or of the concept 'brownish color'.

Now to this the Realist replies: the conception of the atomic structure of the spatio-temporal world marked a breakthrough in the course of chemical research; meanwhile it has been seen that with the aid of this concept, a wealth of problems in apparently remote fields could be solved, and that therefore the concept of the atom brings with it a large number of 'connotations'.[10] As a consequence of his doctrine, the Positivist would have had to limit the concept of the atom, formed in reference to Avogadro's Principle, and the law of multiple proportions, to the relevant operations, and would thereby have blocked himself off from the path to important physical discoveries. This path, however, stands open to those who see more in the atom than a hypothesis correlated to specific well-determined experiences, namely that the atom is a reality transcending those experiences. Even the Positivist – if he is at the same time a successful experimental physicist – does not himself comply with his doctrine in carrying out his physical research, but trusts completely in the experience of a coherent reality. This Reality provides the possibility of continuous progress for the cognitive activity directed toward it. The rational reconstruction of research procedure as offered by positivism is inadequate.

The reply of the Positivists to this objection will go as follows: In case the anticipations which the Realist wants to include in the concept of the atom have been determined, nothing stands in the way of formulating this concept in such a manner that they will find their place in it; and the experiments undertaken in the direction

[10] (John Stuart Mill) *System der deduktiven und induktiven Logik*, translated (into German) by S. Schiel, 3rd ed., I., 1st vol., ch. 2, § 5. [A System of Logic Ratiocinative and Inductive, (1843) 9th ed., London 1975].

thereby delineated will then have to show whether this formulation was appropriate. But with the empty expectation that new, completely undetermined elements, will insert themselves into a given experiential complex called 'atom', science can do nothing; for this indicates no direction for verification.

The resolution of this quarrel is brought about by the insight that the positivist conception cuts a *temporal cross-section* through scientific research; according to it, the *present state* of assured research results is to be fixed is a rigid, closed conceptual system, and every extension of empirical knowledge signifies a fundamental restructuring of this system. Realism, on the other hand, operates with a *system with an indefinite number of empty positions*, which are to be filled by future research, progressively, but never conclusively.

Therefore, the fundamental assumption of realism is that the store of congruent experiences and types of experience can be expanded without limit, and the conception, resulting from this assumption, with respect to the meaning of concepts, distinguishes the methodology of realism from that of positivism. This state of affairs is obscured by the fact that the realist side operates almost entirely with a vague pre-critical idea of an 'absolute' reality, i.e., one not related to consciousness. That is why in the preceding discussion, the substantive core of the doctrine had to be extracted from its speculative shell.

Now the character of the confrontation just discussed, of '*rigid concepts*' with '*expandable concepts*', has hardly ever been understood in methodological investigations, and the most intricate pseudo-problems and controversies have resulted, from them. In what follows, we wish to make a terminological distinction between expandable concepts as 'empirical concepts' and rigid concepts as 'system concepts'.

The results of reflection up to this point furnish the suppositions for a more profound analysis of the concept of reality, in that they make it possible for us to understand the connection between the two main meanings which appear to be united in this concept. These two meanings will emerge clearly through the confrontation of their opposites in the concept pairs: '*real objects – objects of imagination*' and '*real objects – ideal objects*'.

Let us clarify for ourselves the meaning of the first pair by an example of a real material object. The judgment that a thing of a specific kind, say, a wooden cube, is real, is verified when a thing of this kind is localized spatially and temporally in congruent – and in some cases also intersubjective – experiences. In this sense each assertion of reality contains an individuation, a specification of place, – which, to be sure, can confine itself to the demarcation of an arbitrarily wide space-time framework. In the pertinent verification series, acts must occur which, as we have already established, include among them, receptive moments (observations). If the assertion does not stand up under verification, then we say that no reality corresponds to it, or respectively, to the conception underlying it. The same thing holds true, when no locale at all is asserted for the occurrence of an object of the kind designated, and therefore no verification series is intended, when we have merely 'thought up' something, have fantasized it. The real world is thus distinguished from the world of fantasy by the fact that statements with respect to it can be verified or falsified in a specific way. The basic form of a judgment that states something about the world, of an empirical statement, in the sense we have established, thus is: "At a certain place there is something of a certain kind." If this place is space-time determined, we speak of the physical world (external world); if it is personally-temporally determined, we speak of the psycho-physical world.

Let us now turn to the analysis of the second concept-pair 'real objects – ideal objects'.

The open-ended character of the world that we have discussed is reflected in the fact that basically we can continually discover *things to fill* the predesignated places, and we can fix these in a judgment such as occurs in the statement, "the white cube-shaped object which is now located there is sweet." The positing contained in the judgment – which can prove true or false in a verification process – consists in the correlation between places and qualities, as a consequence of which certain places are 'filled' by certain qualities. Accordingly, the qualities and places which are asserted to be linked in such correlation must also be ascertainable independently of their coordination, for otherwise the assertion would not be testable. But this 'lying ready in the mind' (Kant) is not to be interpreted as though there were places and qualities existing 'by themselves' [*für sich*]; their independence from each other is not to be conceived as *absolute isolatability* but only as *independent variability*, by which we want to say that basically any arbitrary place can be filled by any arbitrary quality. (The surface there is red now, but it could also be blue.) Thus the qualities are non-independent objects. But the abstraction by means of which these qualities are obtained must not be interpreted, as so often happens, as if, in this way, those moments, from which the abstraction is made, were eliminated, for such a conception would bring the activity of thinking into a false analogy with the external activity of acting, by conceiving the former as changing the world; instead they are *left open*, i.e., they are set as variable.

Therefore the qualities are not independent entities 'beside' or 'above' the real things in which they occur: they do not form a world for themselves [*für sich*] beside the real world, but they are *dependent* moments in the construction of this real world. If we designate these abstracta (Universals), keeping to Plato's terminology, as '*ideal* objects', then consequently we must be most careful to avoid being seduced into an extreme dualistic conception with respect to the relation between reality and idea, such as was put forth by Plato, and already criticized by Aristotle. As is well-known, the question of the manner of existence, of what is general, then became one of the basic problems of scholastic philosophy, under the name of the 'controversy over universals'[11] and it continues to play a role today, especially in the controversies about *a priori* knowledge.

The three positions taken in the *controversy over the problem of universals – 'universalia ante rem', 'universalia post rem', 'universalia in re'* – or, rather, the arguments raised in support of these, already contain the essential elements for the aporetic puzzle of questions and objections to this problem: we want to present

[11] The controversy proceeded from a passage of the Isagoge of Porphyry in the translation of Boethius.

them here, divested of all nonessential features. From the viewpoint of epistemology the thesis of the 'reality of concepts' (Platonic), *'universalia ante rem'* is based upon the argument that in the localization of any thing, and thus in the establishment of a space-time localized foundational context in which qualities are based, these qualities are already presupposed[12]; in order to establish that at a certain place there is something of a certain kind, one must first know this kind.

The nominalist (empiricist) thesis, *'universalia post rem'* on the other hand, is based upon considerations which concern the 'origin' of the universals. Here the question is asked: What is the source of universals? And the answer to this question must be the experience of 'concreta' [concrete objects], i.e., of physical things and psycho-physical persons. If, for example, we want to show what a 'color' is, then we must point to the perception of a colored thing (and to a blind person we cannot make this comprehensible); the answer can only be gained from a complex total experience through abstraction 'after the fact'.

Finally, the third thesis, *universalia in re'* is guided by the correlation of place and quality in an empirical judgment. In every proposition which states the existence of a thing, it is judged (as has also been established above), that at a certain place there is something of a certain kind; in it a 'now (once, then), here (there) thus' is asserted; the idea of the general cannot be dissociated from the idea of the thing, nor the idea of the thing from the idea of the general.

We can see that each of the three theses presented has in fact objective foundations and that all of them arise from valid insights into the position of the general within the complex of experience. Therefore it is an unavoidable task to ascertain in what respect these theses are compatible with each other. Now the key to ascertaining this lies in the reflections undertaken above about the relation of pre-predicative and predicative experience.

By means of these, first of all, a distinction between *genetic analysis* and *analysis of origins* becomes possible; the confusion between these has played an important role in the prevailing lack of clarity on these issues. By means of the genetic analysis of a judgment of perception – for instance, one in which the filling of a certain position by a certain color is asserted – the act of judgment in question is integrated into a causal interrelationship, and its coming-into-existence is investigated in a manner analogous to that employed in the case of facts of the external world with respect to their causal explanations. If this point of view were maintained consistently, then we would almost always arrive at the result, illustrated in particular by the investigations undertaken by *Gestalt* psychologists during the last few decades, that the comprehension of a complex phenomenon (in our example, a color at a certain place) is genetically prior to that of its elements, which only emerge independently into consciousness upon reflective analysis. Thus if we proceed from a specific act of perception – i.e., performed by a specific person at a specific point in time – then we can say that in it the grasp of the synthetic unity of the elements (moments) precedes the comprehension of the isolated elements.

[12] See Husserl, *Logische Untersuchungen*, 3rd ed., [Eng. tr. J.N. Findlay (1970)], (*Logical Investigations*) 3rd ed., Halle a.d.S., 1922, vol. 2, part I, p. 261 ff.

The *genetic priority of what is connected*, however, is confronted by the *logical priority of the elements*, according to which what is connected can only be described with the aid of the elements; and in this description it becomes apparent that in a certain sense the elements also have temporal priority over their synthesis in the judgment of perception in question. In the light of what has just been established, let us consider an example: although a specific perception of blue must not be interpreted as if it actually came about by means of the completion of a linking of the element 'blue' with a 'specific location in space-time', we can still say, on the other hand, that by this act of perception, the person who perceives blue not only learns what 'blue' is, but also possesses this knowledge independently of this act. It is, therefore, justifiable to ask from which sources knowledge of the 'nature of blue' originates; and we will have to answer this question too 'from perception'. But this answer is not intended to mean that a specific act, which can be fixed temporally, has caused this knowledge, but rather that the knowledge of what 'blue' is points back to perceptions as 'constitutive experiences' [Husserl]. It belongs to 'sedimented experience' in the sense characterized above and as such forms an element within the framework of the knowledge of facts that can be isolated in thought.

These findings seem to show that the empiricist doctrine *'universalia post rem'* is right to the greatest extent, for after all, even 'sedimented' – thus not individually distinguishable – acts of perception are still acts of perception, and empiricism maintains only that the universals are derived by abstraction from such acts. But it characteristically overlooks or misinterprets above all a fundamentally important point, namely the circumstance that this sediment of perception is not pure receptivity, but as descriptive analysis shows, presents an already complicated synthetic structure. From this it follows directly that the subsumption of the individual under the general (e.g., this there is blue) can by no means be conceived as a mere harmonization – both present and past, but contained in memories – of sense impressions, but rather contains within it elements of the "spontaneity of thought" (Kant).

The error of empiricism, committed by the various kinds of histories of ideas under consideration here does not consist at all in maintaining that all, even the most abstract knowledge, is linked to experience. Indeed, in this assertion, which, to be sure, requires more careful interpretation, it is entirely right. Instead it conceives experience as an *aggregate of pure receptions*, and therefore never even catches sight of the central question of the theory of knowledge: what is the nature of the structure of experience.

The fundamental error of the thesis of the reality of concepts *'universalia ante rem'*, on the other hand, lies in insisting on the *logical priority* of the elements of knowledge (*Being-thus [So-Sein]*) over the synthesis of the now, namely, a priority which exists only for *one* level of experience, and is thus valid only in terms of one particular cross-section of total experience, rendering this experience absolute as *ontological* priority, and so definitive for total experience. By this, the 'origin' of the universals is obscured, and the structure of total experience cannot be adequately comprehended. A typical consequence of the realist doctrine of concepts lies in the attempt to determine the relation between the general and the individual with the aid of the category of causality, and this is the point at which it comes into conflict with inductive science.

There remain only a few words to be said about the formula mentioned last, '*universalia in re*'. This fits in best with the results of our reflections, for the interrelation just described between the experience of individual phenomena and the 'experience' of the general can most readily be described as 'within each other'. Only of course it must remain clear that this formula only furnishes the point of departure for the task of the *analysis of this 'within each other'*, a task to which [this doctrine of] conceptualism has done justice only to a very modest degree, and toward the accomplishment of which the present investigation represents only a step.

Let us now briefly summarize the results of our analysis of the concept of reality, with a few supplementary remarks. In order to comprehend the 'nature of reality', we first had to determine the basic form of statements about what is real [*Realsätze*], that is of empirical judgments: and in so doing we recognized that these proved to be assertions of a certain 'occupation of filling' of given 'places'. We say of such a place that its occupation is 'predicated' and therefore we can designate it as the 'logical subject', and this kind of occupation as the 'logical predicate', which should be noted in view of the attempts to create an aprioristic grammar.[13] Now if such an assertion is consonant or congruent with the context of experience, i.e., with the complex of anticipations and the verifications pertaining to them, then we say that it proves correct (is true), or also, that the state of affairs asserted by it is real (belongs to reality, exists). On the basis of this fundamental assumption, which has generally proven right but is not capable of an ultimate 'justification', that every congruent complex of experience can be extended congruently, especially by incorporation of new series of verifications (types of experience) we can deal with the tasks contained in the reality concept and the connotations linked to it. What is involved here is that in certain places, or rather (more correctly expressed) in the more closely designated environment of these places, new congruent experiences will be attainable, experiences that can be more or less clearly determined with respect to their kind. The inprinciple limitless expandability of knowledge pertaining to the course of reality is therefore linked most closely with the *principium individuationis*, and with the interpenetration of place and what occupies it; for this reason, it is clear that knowledge about the essence (the structure) of these occupations cannot be expandable in the same sense. For example, the experience of blue objects can be expanded by establishing that in the 'environment' of such objects, electrical waves of a certain length occur, but in this finding the knowledge of what 'blue' is already presupposed, and this latter knowledge is not enriched by that assertion.

This consideration is especially important for methodology because, as already mentioned, the *quasi-expansion of knowledge* of ideal objects obtained by reflection on 'what was really meant' and by *rational reconstruction,* is confused again and again with the *authentic expansion of the knowledge* of facts by means of additional experience. Linguistically this confusion is displayed by the fact that ideal objects are allowed to occupy the subject position in a sentence, as 'substantivized properties', or 'substantivized relations', in the same manner as real objects. If we compare, say, the two sentences: "the wooden cube here on the table is blue" and "every

[13] *Ibid.*, Husserl, p. 294 ff. and *Logik*, p. 259 ff.

blue has a certain brightness", then the temptation to assume that the relation between subject and predicate in both cases is of the same kind structurally is only too great; but actually this is by no means the case. For the first sentence is of the form: This-there, which has the properties $P_1, P_2, \ldots P_i$, also has the property P_n; thus a new occupation (filling) for a specific place is indicated. However, in the second sentence, this sort of interpretation is excluded, as the ideal object 'blue' is the product of an abstraction from every place whatsoever. Thus we state (if we wish to designate the form of the empirical judgment as 'statement') nothing about the color blue in our sentence; for in the concept 'blue' the idea of a brightness amplitude is already contained (intended with it), and instead we are only *clarifying* for ourselves what we really mean by 'blue'. When, however, we consider the sentence "This blue (localized in a specific manner) has a certain brightness", which obviously contains an authentic statement, we recognize that the statement is not about the color as such, but rather that it states that in a specific place a color of a specific kind is to be found. The meaning of the sentence is thus: This place is occupied by a specific kind of blue and indeed a blue specified according to its brightness. But it is obvious that the kind of statement, 'blue of a specific brightness', cannot be predicated of the genetic concept 'blue', for the meaning of the abstraction, which leads from the species concept to the generic concept, consists just in this, that in the latter the specific differences remain undetermined (variable). Therefore in the two sentences "This blue has a specific brightness", and "Every blue has a specific brightness", the term 'blue' has a different meaning. In the first sentence it refers to blue places, in the second, to the concept 'blue', thus to that which is understood as such by 'blue'.

Accordingly, the logician can state that this is a highly dangerous type of equivocation, and seek to pursue the errors in the reasoning produced by it with all possible vigor. He will find a rich harvest here. Above all, the *iteration of the concept of property* must be mentioned here; 'properties of properties', 'properties of properties of properties', etc., which lead to the semblance that a 'property' of a property is the same thing as the property of a thing. That this is not so, has just been shown. In any case, from the viewpoint of the psychology of knowledge this mistake is comprehensible, owing to the fact that 'properties' of properties are just as much products of the process of abstraction as the properties of things. But with the radical difference that in the first case the abstraction from place is never included, but in the second case it always is. The basic error is thus that an insufficient distinction is made between the grounds of determination that concern a place, and those grounds of determination that concern the occupation of a place. Though people like to deride the scholastic concept of '*haecceitas*', the conception which is conveyed by it by no means appears to be radically superseded. In the next section we will have to refer to the fact that the fusion of the concepts of 'set' [*Menge*] and 'property' arises from this error as well. Furthermore this lack of clarity is also expressed in the ambiguity of the term 'relation', for on the one hand sameness of properties and difference of properties – similarity and dissimilarity – are designated as relations; while on the other hand, sameness and difference of position, space-time proximity, (coincidence and succession), and also real incompatibility, are also designated in this way. In this connection, then, the remarks are made about 'properties of properties' can obviously be transferred to 'properties of relations', 'relations of properties', and 'relations of relations'.[14]

Since the preceding results, in which 'authentic', empirical judgments, a linking of places with qualities, of being-there with being-thus is carried out, are distinguished from inauthentic judgments in which the subject is merely analyzed, they are of great significance for a theory of the social sciences: for they furnish the basis for a radical analysis of the concepts of the highly complicated ideal objects, which are under consideration in those fields, and together with that make possible – and this is especially important – a critique of the *a priori* in the social sciences.

It will be no secret to the reader trained in philosophy that a great part of the discussion in this section is most closely related to the problems of *a priori* knowledge, and we have not pointed to this connection earlier only because we wished to hold at bay as long as possible the many and varied accompanying conceptions linked to this *term, which is so much in need of clarification*. Now we have to define this relation with all possible clarity, and in doing so to examine to what extent these conceptions are objectively founded and are compatible with each other. For this purpose we want to raise to clear awareness the most important 'properties', which are linked by 'a priorists' to the idea of the *a priori.*

If, to begin with, we ask ourselves to what extent the designation *a priori* is referred, then the answer is (a) to judgments (knowledge [*Erkenntnisse*]), (b) to concepts. Thus, on the one hand, the propositions of pure mathematics as *a priori* judgments are confronted with the judgments *a posteriori* (empirical judgments) of the empirical sciences; on the other hand, the concepts of pure mathematics (e.g., the concept of number) are also distinguished as *a priori* concepts from empirical concepts (e.g., the concept of a table). Whether in this all concepts of ideal objects, and thus also concepts with material content (e.g., blue) or only the formal concepts, are to be designated an *a priori*, depends on whether in the sphere of judgments, as *a priori* with material content would be recognized along with the formal *a priori*. Since, generally speaking, the problems of concepts *a priori* and those of judgments *a priori* are correlated, in what follows we can confine ourselves to an analysis of the latter.

A second question concerns the relation of *a priori* to experience. It is most often specified by the following two theses: (1) the *a priori* is independent of experience; (2) experience is dependent on the *a priori.*

A third question concerns the validity of *a priori* judgments. Here, as a rule, the answer is: judgments *a priori* have apodictic validity; they convey necessary truth. This answer then finally leads to the two further question concerning the criterion

[14] See Kaufmann, *Das Unendliche in der Mathematik und seine Ausschaltung* (cited subsequently as *Das Unendliche*), (*The Infinite in Mathematics and its Elimination*), p. 18 f. and "Bemerkungen zum Grundlagenstreit in Logik und Mathematik" (subsequently "Bemerkungen"), (Remarks on the Controversy about the Foundations of Logic and Mathematics), in *Erkenntnis*, vol. 2, pp. 262–290, (1930–31) [Eng. tr. *The Infinite in Mathematics,* Vienna Circle Collection, vol. 9, Dordrecht and Boston, 1978, p. 23 ff].

and the origin of this validity. The apriorists' answer to the first is most frequently that the truth of *a priori* propositions mainly manifests itself in specific evidence; the origin of this validity, however, is 'explained', depending on whether the 'objectivist' or 'subjectivist' variant of apriorism is involved, either by the thesis of the participation of thought in being, or else by the thesis of the generation of the objects of experience in thought.

The conceptions very briefly sketched here are confronted most directly with the *empiricist* doctrine, according to which there can be no talk at all of specific *a priori* knowledge. According to this, what is called *a priori* knowledge are only greatly *generalized experiences*. Evidence as a 'source of truth' is nothing other than an expectation developed by habit, which, as is generally the case with habits of experience, frequently proves correct, but also may lead to disappointments. And so, a sharp line of demarcation between propositions generally designated as empirical and so-called *a priori* propositions, including the propositions of logic and arithmetic, cannot be drawn at all. The latter also derive from experience and can be falsified by experience, and what is true for propositions, is consequently also true of concepts.[15]

Beside these two doctrines, which for their basic theories can, with some justification, call upon the authority of Plato, or Aristotle, stands a doctrine, which, in spite of a number of attempts in nominalistic philosophy, only received rigorous elaboration during recent decades, namely conventionalism. According to this, the so-called *a priori* propositions are *conventions*, i.e., regulations, and more precisely, regulations on the use of certain terms. For example, for any numbers a and b, the proposition '$a \times b = b \times a$' is valid, but this is neither knowledge *sui generis* (a synthetic judgment *a priori*, as Kant thought), nor a generalized experience (as John Stuart Mill thought), but this epistemological fact is posited together with the operational definition of the multiplication operator, and therefore included in it analytically. That this is indeed the case emerges clearly from the fact that this operational definition can also be changed: for example, it can be set up as '$a \times b = -b \times a$', which actually does occur in the theory of quaternions, a calculus that is applied particularly in vector analysis. The most famous examples, and also those that are the most important for the development of conventionalist theory, can be given with respect to geometry, or as one must now say, more correctly, to geometries: we will deal with this in more detail in the next section. The conventionalist conception thus points out that the secret which lies at the root of *a priori* knowledge is the same fundamentally as that which is ridiculed in the verse:

Das ist doch sonderbar bestellt,
Sprach Hänschen schlau zu Vetter Fritzen,

[15] When we speak of 'empiricism', we have in view the older empiricism as represented above all in the nineteenth century by John Stuart Mill. Modern logical positivism of the Vienna Circle (Carnap, P. Frank, H. Hahn, Neurath, Schlick) does not conceive of logical and mathematical sentences as empirical judgments, but (correctly) as analytical judgments, where at times (following Wittgenstein) a distinction is made between the character of logical sentences (tautologies) and of mathematical sentences (equations).

Dass nur die Reichen in der Welt
Das meiste Geld besitzen.[16]

That is most curiously arranged,
Said little Hans to his cousin Tess,
That only the rich in this world
The most money do possess.

According to this doctrine, apriorism is right as against empiricism, first, in that the conventions called '*a priori* propositions' are not subject to refutation by experience – for they are resolutions and not judgments – and secondly, in that they actually do form elements of experience in a quite specific sense. For total experience – but especially scientific experience – is permeated by numerous and varied convention, and it is extremely important to distinguish these from the 'material of experience' (in the narrower sense) in order to establish which elements fall within the testing of scientific judgments. On the other hand, empiricism, to which, incidentally, conventionalism is much closer in its basic intellectual attitude, opposes, with complete justification, the aprioristic doctrine that *a priori* insights offer knowledge flowing from sources of a distinctive kind. It holds that one cannot even talk about such knowledge: on the contrary, every proposition which is not empirical is analytic. That in the conventions (definitions), which form the source of analytic judgments, some content of true knowledge is mistakenly supposed to be contained, can be traced back, above all, to this: that the conventions are chosen with respect to their efficacy for gaining knowledge, that they are thereby oriented toward experience.

Let us now look at these three doctrines with respect to their interpretations of the 'a priori with material content', again taking as our basis the example of the sentence: "Every color has a certain hue, a certain brightness and a certain degree of saturation." For the apriorist of strict observance, the truth of this statement is apodictic, comprehensible by its evidence in the intuition of essence [Wesensschau]. It forms a precondition for the experience of color and cannot be refuted by any experience. In opposition to this, the empiricist declares the knowledge that color can be ordered according to these three variable dimensions to be a general experience with respect to color, to be gained inductively and in principle falsifiable by a contradicting induction, even if from a practical epistemological viewpoint one would hardly take this possibility into consideration.

Finally, the conventionalist argues: the term 'color' has been so *defined,* that this *triad of variable dimensions* is included in the term; it therefore follows logically from that definition, and insofar as one retains this definition, it is neither capable of, nor requires confirmation by experience. This assertion remains undisturbed by the fact that the decision to make a choice, or else a modification of a certain definition will be typically motivated by taking empirical findings and historically preestablished goals of knowledge into considerations.

In the evaluation of these three doctrinal opinions we can directly invoke the considerations we have presented concerning pre-predicative experience and ideal

[16] Cited in Somlo *Juristische Grundlehre*, Leipzig 1917.

objects. Pursuing this argument rigorously will lead to the following results: *a priori* propositions can be interpreted in two ways that is either as propositions about the world, to which a specific immediately characterizable position in the complex of experience is attributed, or as conventions (definitions).

Let us now take a closer look at the first of these interpretations. The position within the complex of experience under discussion here is – implicitly or explicitly – that of a *fundamental assumption*, i.e., a judgment, the truth of which is presupposed without further confirmation within the complex of inquiry concerned, and which serves as basis for pertinent argumentation. For our example of the variable dimensions of color, this interpretation will be as follows: It is assumed, without further examination, that for each given color three series of colors can be constructed (in a manner to be more precisely specified), in such a way that in each series one of the elements, and a different one within each of the series (i.e. hue, brightness, and saturation), will vary, and the two remaining elements will remain constant. This assumption will form a basis for further investigations of color.

According to the interpretation just described, *a priori* propositions can be designated as implicit or explicit hypotheses; the testing and possible refutation of which has been *suspended.* It is in this *uncontested character* alone then that their 'necessary validity independent of all experience' consists, and their function as basic assumptions declares them to be presuppositions for all pertinent experiences. But obviously the choice of such basic assumptions does not result from free and arbitrary whim, but in most cases virtually forces itself on thought. We can readily see that this interpretation is very closely related to the empiricist conception of the *a priori*: it is distinguished from it only in one point, that empiricism has not properly recognized the aspect of the *suspension of testing* of basic assumptions, and consequently equates them with hypotheses in the ordinary sense.

The second of the interpretations mentioned, on the other hand, corresponds completely with the conventionalist conception. According to this, *a priori* statements are *not at all assertions* about the world but *regulations fixing* the use of a specific term. Thus our statement concerning the variable dimensions of color does not assert anything about what exists or goes on in the world, but contains instead a regulation about the manner in which the term 'color' is to be used. Such regulations are usually made by taking experience into consideration, and with concern for traditional language usage; but for the acquisition of validity, this circumstance is irrelevant: the 'validity' of our propositions – insofar as one can use this expression here at all – is '*validity due to definition*' and as such is independent of experience; but therefore also no experiential judgments can be derived from it. The confusion arises only because the two irreconcilable interpretations are continually confounded; thus through interpreting the meaning of concepts one hopes to arrive at knowledge of the real world. This confounding, however – and here lies the decisive point – can be understood as arising from the *stratified structure of experience*. It is the sedimented experiences associated with traditional language usage which, clarified in rational reconstruction, are codified in the pertinent definitions; then this process of rational reconstruction is illegitimately interpreted as at once experiential knowledge, on the one hand, and insight gained from the 'essence of the concept', on the other; from this arises the contradiction *in adjecto*, 'apodictic propositions about the world.'

So, for instance, in our example, the rational reconstruction of the concept of color will lead to the result that each individual color is conceived as an element of a color system, that thus the three variable dimensions in question, hue, brightness and saturation, are a 'connotative meaning' of the concept. But from this we cannot deduce that the phenomena, whose real interrelationship, according to the definition, is the criterion for the occurrence of color, will always and everywhere 'necessarily', appear in the combination in question. However, if such is the case, it will not be permissible to speak of the 'occurrence of color phenomena' in the sense of the definition.

But aside from this, – and related most closely to the above definition from the viewpoint of the psychology of knowledge – is the experiential statement that these phenomena *actually do appear in combination*. This assertion is, in principle, subject to confirmation, if confirmation does not appear suspended. Consequently, the fundamental error of the empiricist doctrine with regard to the *a priori* lies in the fact that the conventional moment is not taken into consideration; while the fundamental error of conventionalist doctrine is that it does not consider the empirical aspect which is contained in *a priori* conceptions, and which is to be revealed in rational reconstruction. Apriorism, finally, errs in that it does not distinguish these two viewpoints from each other: they are merged in confused pre-predicative experience, but in clear thought are irreconcilable, and hence apriorism mistakes an 'unquestionable nature', assured only by the lack of deeper analysis, for evidence guaranteeing truth. Nevertheless, the development of the basic conceptions of the *a priori* by Plato represents one of the most important turning points in the history of philosophy, for here for the first time the problem of the structure of experience was raised in its profundity, and the untenable character of the sensualist mimetic theory [*Abbildtheorie*], according to which experience is pure receptivity, was made visible. A further decisive turning point is the Kantian variant of apriorism, with its union between the idea of the *a priori* and the idea of the formal. Considering this doctrine, which, to be sure, is not wholly coherent internally, and is, from the viewpoint of philosophical insight achieved today, superseded at various points, we may perhaps say that its most important achievement consisted in directing attention to the special place which must be accorded to the formal in the stratified structure of experience.

This thesis is now to be discussed briefly. We have already shown that rational reconstruction leads to a separation of the receptive and spontaneous elements in experience, although the receptive 'moments', the unformed material of experience (prime material) cannot be isolated. We have further recognized that this formed character of the material consists in the lived experiences, with material content appearing as already embedded within a *horizon of lived experience*, and that with these, certain anticipations are posited. But now the question emerges: are the basic functions of spontaneity that make such synthesis possible, i.e., which can be shown to be essential elements in these syntheses? The reflection that begins here leads first to the recognition that in each such synthesis, the comparability of objects is presupposed on the one hand in terms of their qualities, on the other, in terms of their place. For indeed every anticipation consists in the assumption, that under certain conditions, at a certain place, something of a certain kind will exist, or will happen. Comparability, however, includes the ability to identify and to distinguish. Therefore, – expressed in subjectively colored terminology – the cognitive functions of identifying and distinguishing are presupposed in every experience. The exceedingly difficult, subtle and profound analyses of pure descriptive psychology devoted to the constitution of these formal *basic functions* of cognition, in the temporal course of the stream of consciousness, cannot be indicated by us here, even in outline, and we have confined ourselves to referring to the pertinent fundamental investigations of Husserl.[17] One finding, however, which should be understood without difficulty in the context of our discussion, cannot be omitted, namely, that even these basic functions of spontaneity cannot manifest themselves in isolation. For every comparison, every identification or distinction is bound to the material upon which it is carried out, and the criteria that in each instance are decisive for identification or distinction, are in turn determined by sedimented experience, which proves, upon rational reconstruction, to be structured in a highly complex manner.[18] Thus the decision as to whether the same thing is being perceived in two perceptions, separated to a greater or lesser extent temporally, is based on implicit assumptions about objective changes and about the movement of the objects; a distinction can result with respect to both the dissimilarity of the two perceptual images, as well as the discrepancy of the localizations.

Therefore also logic and pure mathematics, which involve these formal basic concepts (as we shall show in the next section), are not independent of experience in the sense that to understand their propositions no knowledge of the world is required. *To be sure, no statement about the world is made in them, but they presuppose the world.*[19]

In conclusion we have to say a few more words to clarify the problem of the *schemata of pure possibilities*, as projected in imagination. For it is not only by the combination of known elements that the concrete formations of the imagination are 'thought out', for which no examples can be found anywhere in the world (e.g., centaurs), but also unknown qualities (more correctly: not completely known) can be 'thought out', e.g., colors which lie between known colors. In this reference to 'anticipation' of experience in free thought, i.e., *thought far removed from observations*, lies the strongest argument for apriorism, the one most difficult to oppose,

We by no means want to presume to solve this far-reaching problem here, in passing as it were, but only to indicate the path which, according to our view, can lead to a solution. It might well turn out that the freedom of imagination, which is confined within given limits, can be traced back to the structure of sedimented experience, in which both narrower and more extensive *spheres of similarity* are set up, and which also contains knowledge concerning the interrelationship of these as

[17] [Husserl], *Vorlesungen zur Phänomenologie des inneren Zeitbewusstseins,* [Eng.tr. I. Churchill, *The Phenomenology of Inner Time Consciousness*, (Bloomington 1964)], *Jahrbuch für Philosophie und phänomenologische Forschung*, vol. IX, Halle, 1928.

[18] For this, see the investigations in the next section.

[19] Husserl comes to the same conclusion in his *Logik*.

'higher level sedimented experience.' According to this, free imagination would only be distinguished from the mere reproduction of constitutive experiences by the fact that it is based on *sedimented experience of various levels.* Owing to this, the borderline between free imagination and reproduction is assumed to be fluid; 'simple' reproduction is to be conceived solely as an idealized limiting case.

With this clarification of the *a priori* completed, we close this section; it is certainly not a radical clarification, but should be sufficient to serve as a basis for the investigation that follows. The character of the results gained in it will emerge still more clearly in their application to the various partial domains of scientific thought.

2. Logical-Mathematical Thought

The reflections in the previous section have made it clear that it is the question of the nature of logic and mathematics which aims at showing the systematic place occupied by logical-mathematical thought within the complex of experience, for we have established that this sphere of knowledge cannot be conceived as a domain isolated from experience. Moreover, the conclusions just reached can be applied immediately to the question of the meaning of logical and mathematical concepts, the character of the validity of logical and mathematical statements, the singularity of logical and mathematical procedures. For it is through them that the foundations have been laid for understanding both the abstractions that lead to the concepts of logic and of pure mathematics, and the meaning of logical and mathematical 'propositions'. We can understand this meaning by using the distinction between what is 'posited' and what is 'presupposed'.

Let us begin our investigation with the basic concepts of logic: 'truth', 'judgment', 'concept', and let us examine *the constitutive invariants* for their thematic isolation within the framework of the complex of experience.[20]

For the concept of the truth of judgments – and for the time being let us consider only empirical judgments – we arrive at the following result:"A certain judgment is true" means "whoever makes this judgment, wherever and whenever, judges *correctly* (in accordance with the facts); he asserts what is the case". We have already said what we needed to about the criteria of 'correctness' in the previous section. Thus as far as the concept of the truth of judgments is concerned, the *invariance of their correctness* is expressed in contrast to the *variations of the facts of judgment*, such as the person making the judgment, and the place and time at which the judgment is made.

Now the state of affairs just described has been reinterpreted as if, aside from what exists, there were also a *'realm of truth'* and as if a judgment were correct only if it fell within that realm. But to insert something like that is not justified; there is no realm of valid truths which unites thinking and existence [*Sein*]. 'Judgments as such' and 'truth as such' do not stand 'beyond' or 'above' the facts of judgment: but

[20] See Kaufmann, *Das Unendliche*, p. 23 ff. [Eng.tr. p. 26 ff.]

by means of these terms, invariance, on the one hand, as opposed to the variations of judging persons and of the space-time data of the act of judgment and the mode of clarity of thought, on the other, are to be characterized.

Proceeding from this insight, the meaning of the expression 'judgment in the logical sense' can be comprehended directly. An act of judgment is thinking of a state of affairs as existent or nonexistent; now, if only the state of affairs present in thought is taken into consideration while abstracting from occasional data, i.e. who judges, when, and where, and if *completely clear thought is presupposed*, then we obtain a 'judgment in the logical sense', with its 'truth value'. Accordingly, in fully formulated propositions about the truth of judgments, a judgment in the logical sense is not the subject, and 'truth' is not the predicate, for such propositions must read: "Every arbitrarily selected person – whenever and wherever – judges correctly when he judges in the manner indicated."

Analogous to the process of abstraction that leads to the 'judgment in the logical sense' is the process that has, as its result, the 'concept'. The *concept* of an object (state of affairs) is the *clear thought* (intention) of this object (state of affairs). *Whoever* thinks this thought, and *where* and *when* it is thought is left open. To be sure, these findings, which seem to be obtained so simply, no longer seem so as soon as we realize that in the 'idealization' contained in the presupposition of clear thought, the profound problematic of clarification (to be discussed briefly below) is also contained, as well as the relation between predicative and pre-predicative thought. Yet, it is very important, especially from the methodological viewpoint, to grasp that the essential difficulty lies here, *between* the strata, so to speak, and not in the problem of the relation of the abstract to the concrete *within* the stratum of clear thought.

For this stratum, to which all investigations of formal logic refer, one of the most significant consequences of the reflections of the previous section is the recognition of the independence of the abstract from the *number* of existing concreta to which it applies. Locke, Berkeley and Hume, with their sensualist theory of abstraction, erred in their failure to recognize this independence, and from the nineteenth century to our own day this failure has continued to have fatal consequences for the logical calculus by making 'property' the equivalent of 'set'. This is the error of thinking that Husserl has called the fusion of *individual generality* and *specific generality* in his critique of the empiricist theory of abstraction.[21] I have discussed the consequences for the problems in the foundations of mathematics more exhaustively elsewhere, and in the present content I must confine myself to a reference to those investigations.[22] Here I can only mention the critical insight gained from that critique: it is completely erroneous to interpret the comparison of the extension of logical concepts as if each concept had a specific extension determined by the number of the individuals covered by it.

A further question of great importance that must be discussed is what, after all, is it that is actually set as *invariant* in those abstractions which lead to 'truth',

[21] *Logische Untersuchungen*, vol. 2 p. 110 f. [Eng.tr. p. 340 ff.]

[22] Kaufmann, *op.cit.*, p. 30 ff. [Eng.tr. p. 30 ff.]

'judgment', and 'concept'? This question can also be answered immediately on the basis of the preceding considerations. It is those conditions which are declared to be *truth-criteria*. Therefore if we only know these conditions, in examining the validity of a judgment, we do not have to be concerned with the *occasional data* of the judgment at all. So the same process of abstraction leads to 'truth' and to 'judgment in the logical sense': they are correlative concepts. There is also a correlation between 'judgment' and 'concept', for on the one hand, judgment is 'composed' of concepts, but, on the other hand, the 'meaning of the concept' is determined by the judgments that specify what falls under this meaning and what does not. Accordingly, for the abstraction 'concept' as well, the uniformity of confirmation is the constitutive invariance. Thus it is clear that within the concepts 'truth', 'judgments' and 'concept' there is the presupposition, as implicit assumption, of the possibility of an intertemporally and intersubjectively consonant experience, and corresponding to this assumption is the 'objectivity' of truth.

At the root of the erroneous doctrine of 'logical *psychologism*' lies the failure to realize that the process of abstraction that leads to logical concepts, disregards the occasional data of thought. The basic thesis of this psychologism is as follows: concepts, judgments, inferences are all products of thought processes, thus of psychic acts, and therefore are subject to the laws of that science whose task consists in the investigation of such acts, i.e., psychology. The noblest tasks of logic consist in ascertaining the 'natural laws of thought' and especially in interpreting logical principles experientially. Thus the law of contradiction is 'explained' by an inner compulsion to think in a non-contradictory manner.

That such an interpretation is untenable can already be seen from the experiential fact that completely normal people often enough think in a contradictory manner. But the principal objection against this interpretation is that it does not even make evident what is to be explained. For if the meaning of the law of contradiction consisted in an inner compulsion to bring about thought free of contradiction, then the validity of this law would be dependent of an empirical investigation of the nature and intensity of this compulsion; this obviously is not the case. Thus the critique of the psychologistic theory is the *critique of an erroneous rational reconstruction*.

The critique of psychologism that began prior to Husserl's *Logical Investigations* stressed the assertion that the validity of logical propositions is not a 'must' like that of natural laws, but as 'ought' and this distinction then also became of great importance for the theory of the social sciences as it led to a contrast between the *sciences of Being* and *normative sciences*. But, as Husserl showed in the first volume of his *Logical Investigations*, although this distinction constituted great progress in its rejection of erroneous psychologistic conceptions, still it could not in any way be regarded as a real clarification of the problem. For the norms of logic are not the heteronymous norms of an authority that requires no further justification for its commands; rather, they are supposed to have their origin in principles of reason, and thus normative logic leads back to those principles of reason on which the norms are based. A pure logic must be presupposed as foundation of normative logic; to shed light on the epistemological character of the latter constitutes the real problem.[23]

[23] *Ibid.*

With that we come to the questions of the meaning of *logical principles* and *logical operations*. In treating these we must of course confine ourselves to a concise formulation of their guiding conceptions.

To begin with, as far as logical principles are concerned – and by these are usually understood the law of *identity*, the law of *contradiction*, and the law of the *excluded middle* – we want to proceed from the distinction, developed in the preceding section, between what is *posited* in an empirical judgment and the *presuppositions* contained in it. This is because the following holds true for *logical principles*: They contain *nothing at all that is posited*; their knowledge content, insofar as one can speak of such a thing at all, instead consists exclusively of the *implicit* presuppositions contained in them.

Let us exemplify this thesis first in terms of the law of identity. It is usually formulated thus: "Every object is identical with itself." What this quite obscure sentence means only becomes clear, when we ask *how it is applied*: in short, the conception current in logic – the one expressed in most presentations of the subject – to the effect that the manner of application of a logical law is something more or less different from its meaning, is completely off the mark and dangerous. – Now the application of our law obviously lies in the *indication* of the possibility of identifying a real or ideal object of knowledge, either by the same person at different times, or by different persons; the presupposition implicitly contained in the concept of identity, that there is a *capacity for identification,* which points back to acts of *retention* and *protention*, thus forms the real content of the law of identity.[24]

Just as the law of identity presupposes the cognitive functions of identifying, so the law of contradiction – a statement and its negation cannot both be true (are incompatible) – presupposes the cognitive function of *distinguishing*. But as this law contains not only the concept of *negation*, which corresponds to this cognitive function, but also the concept of *incompatibility*, its content extends beyond this presupposition. What can be demonstrated in an especially illuminating manner in this law is the nature of the connection between the formal functions of thought and the material content of the world; this is a connection that has been treated frequently in the history of philosophical thought under the heading 'intuition and thought' [*Anschauung und Denken*]. For it shows that negation always points back to the presence of incompatibility in material content; to be sure, here the kind of incompatibility is left open – and this is what constitutes its formal character.

A simple example will make this clear. Of the two sentences: "D (a specific thing) is red" and "D is spherical", the one will not be understood as a negation of the other; although, without doubt, 'red' is not 'spherical'. While the sentences "D is red" and "D is blue" – at the same point of time and in the same place – negate each other, i.e., they cannot both be true.

[24]A thorough treatment of the problem of validity of logical principles is contained in my essay, 'Über den Begriff des Formalen in Logik und Mathematik', [published in Travaux du 9ième Congrés Internationale de Philosophie, (Paris 1937), vol. 6, pp. 128–135 – Ed.]

Now it is clear that the root of this *incompatibility* is to be sought in the 'essential laws' of the material sphere of 'color',[25] according to which there cannot be things of different colors in the same place at the same time. Thus the principle of incompatibility and the law of contradiction are, generally speaking, also based on the fundamental assumption of the unambiguous determinateness of what exists, that is on *coherent unity of experience.*

In conclusion we want to clarify our thesis still further by a brief analysis of the principle of inference: "What is valid for the general is also valid for the particular."

There is *no 'gain' in knowledge* if every determination valid for the general is also valid for the particular – if the meaning of generals and particulars are presupposed as given; for this is precisely what characterizes the pair of correlated concepts. Still this principle expresses a fundamental insight: the *presupposed possibility of knowledge of the general.* The concept of the general derives its meaning from one feature of knowledge, namely that it is possible in the thought process to extricate the being-thus [*So-Sein*] of something – say, a certain color – from the context of reality in which it appears and to consider it in isolation, varying it according to certain aspects (hue, brightness, saturation). It is especially evident here that the meaning of 'inauthentic' logical propositions – and thus not only the meaning of 'authentic' empirical propositions – can only be clearly grasped by going back to the *sources of our knowledge.* Thus, it is trivial to assert that the particular is 'contained' in the general; however, the knowledge of *what the general is* that is presupposed in it holds insight of fundamental importance.

This insight leads us directly to the understanding of logical inference, of *deduction,* and teaches us in particular to recognize the most dangerous prejudice contained in the typical conception of logical inference. This is the assumption that there is a positing of reality in logical inference – even if only a hypothetical one.

Let us leave aside the question, which is not important in this connection, of whether all logical inferences can be brought under the form of syllogism; instead we shall use a syllogism *(Barbara)* to clarify the above-mentioned erroneous conception. The meaning of the inference: "all men are mortal; Socrates is a man; therefore Socrates is mortal", is interpreted to mean that with the truth of the major and minor premises, the truth of the conclusion is necessarily given. 'Therefore' it could best be expressed by the hypothetical formulation:"If all men are mortal, and if Socrates is a man, then Socrates is mortal". Now in order to recognize that this formulation is faulty, we want to compare it with a hypothetical judgment, where the truth of the assertion also appears to be linked to two 'conditions', namely the statement: "If hydrogen is pumped into a balloon, and the hydrogen is ignited, then the balloon explodes". Here an assertion is advanced that can be tested empirically, for its validity; our statement therefore contains something that is 'posited' with respect to the complex of events in the world, and obviously this is also true of its transformation. If the judgments "Hydrogen is pumped into balloon" and "This hydrogen is ignited" are true, then the judgment "This balloon explodes" is also

[25] Concerning the origin of such essential laws in sedimented experience see above. Part One, section "Basic Philosophical Considerations".

true. But this last judgment is *not logically contained* in the first two; what we have here is a real relation between facts, not a logical relation between propositions. Just the opposite is the case with respect to the link between the three judgments in the syllogism. The truth-condition is contained in the premises under all circumstances (its meaning is 'intended together' with the premises), and this says – under the presupposition of clear thinking that is constitutive of logic – that its truth-conditions form an (authentic or inauthentic) part of the truth-conditions of the premises. Thus, to be sure, the non-existence of the state of affairs asserted in the conclusion is incompatible with the existence of the state of affairs asserted in the premises, but only because the former state of affairs contains nothing different from the latter state of affairs. The principle of the logical syllogism thus does not contain any assertion about the world; it only presupposes a world in which identifications and distinctions are possible.

The failure to recognize this epistemological fact, which is also completely valid for pure mathematics, has led to a series of inconsistencies in the logical calculus, where initially no proper distinction between the two kinds of 'if-then' procedures between *logical entailment* and *empirical implication* was made.[26] In order to remove these inconsistencies a *modal logic* was created,[27] in which necessary connections (strict implications) were distinguished from merely factual connections. Although this distinction is impeccable from the standpoint of logical technique, it is nevertheless unfortunate insofar as it does not allow us to see clearly the core of the difference between *real relations* on the one hand and *logical relations* on the other, which contain no positing concerning the real world and result solely from the analysis of judgments.

The main reason for the confounding of these two spheres is to be found, however, in the deep psychological link between deductive and *inductive* thought, with which we dealt fully in the last section – although we did not use the term 'induction'. In the following investigation, based on the findings made there, we wish to use as point of departure John Stuart Mill's critique of logical deduction, in which he asserts the primacy of induction over deduction.

In the syllogism, as Mill explains,[28] the conclusion is only apparently derived from the major premise; actually the conclusion is already presupposed in it. Thus, for example, we must already know that Socrates is mortal, in order to be able to assert that all men are mortal. *De facto* the assumption that a man now living will die is based on the experience of the deaths of many other men. The general statement only expresses the fact that we consider a generalization permissible on the basis of available experience. The actual process of (extending) knowledge thus lies in establishing the major premise, i.e., in induction; the deduction that follows is

---

[26] Compare below Part Two, section "The Way to Overcome the *Methodenstreit*", furthermore, Kaufmann [*Das Unendliche*], p. 36 ff. and Carnap, *Logische Syntax der Sprache* [*Logical Syntax of Language*], *Schriften zur wissenschaftlichen Weltauffassung,* vol. 8, Vienna 1934.

[27] See for instance Lewis and Langford, *Symbolic Logic*, New York and London 1932.

[28] Mill, *System of Logic*, vol. II ch. 3.

nothing but an interpretation of a notice that we, because of our experience, consider ourselves justified in assuming the conformity of further cases.

Sigwart[29] attempts to refute Mill's argument by pointing out that the "universal major premise should not be understood as the statement of this *numerical generality*, it is the statement of the necessity of connecting the predicate with the subject. This necessity cannot be attained, even by a complete enumeration; indeed it can never be known at all in a directly empirical manner."

But on closer examination this objection does not prove to be sound. For a 'necessary link' of subject and predicate is only preset when the predicate has been specified by definition as property of the subject, in which case the major premise is an analytic judgment. The major premise of our example would then state: the property 'mortality' is contained (thought together with) in the concept 'man'. Now this interpretation is obviously not what Mill meant; rather he wanted to have the major premise understood as an empirical judgment. Instead, what characterizes the weak point of his presentation is that, like almost all empiricists, he was not sufficiently aware of the stratified structure of experience, nor of the cognitive function of sedimented experience, and for this reason the significance of rational reconstruction, which is posited with logic as the doctrine of *clear* thinking, remained hidden from him. Thus he was not able to do justice to the cognitive achievement that lies in the *analysis of judgments*. Sigwart's objection doubtlessly also points in this direction, but he misses the mark when he fails to realize, that the transition from the concept of a 'real connection' to the concept of a 'necessary connection' presents itself as *μεταβασιS ειS αλλό γενοS* [change to another kind].

Now we can characterize the *relation between deduction and induction* in the following manner: in deduction it is shown that certain propositions (conclusions) are implicitly contained in other propositions (premises). The truth of the premises is not involved in this at all; it is posited neither categorically nor hypothetically. Consequently, under certain circumstances, true propositions can be deduced from false propositions just as true elements can be asserted in false assertions. On the other hand, in induction, general assumptions based on less than general basic assumptions (especially individual observations) are established, and the *rules* of induction state under which conditions this 'progression from the particular to the general' can take place.

The fundamental error in the conception of induction – which recurs repeatedly despite the fact that Hume[30] pointed it out as clearly as possible is the assumption that the inductive method can be justified ultimately by reason, or by experience. To support the latter assumption, it is argued that induction had proven valid in such an exceedingly great number of cases that it could be counted on as a fact that it would continue to prove to be valid. In this argument, however – as Hume has shown – the supreme principle of induction is already presupposed, namely the basic assumption, essential for induction, that there is a certain *uniformity in the events of this*

[29] *Logik*, 3rd ed., vol. I p. 477 ff.

[30] See above all, his main philosophical work, the *Treatise of Human Nature*.

world. Thus to seek the foundations of induction in experience is a *petitio principii*, as all experience is 'inductive'.

Still less can induction be justified by 'pure reason'. The thesis that represents it as a necessity of thought, because without it experience would not be possible, states only that induction forms a constitutive element of experience. This amounts to stating that we cannot experience (in our sense) a world in which no induction can be made. But it is a fact *that* we live in a world in which predictions gained inductively prove to a great extent to be right – or more correctly, that for a long time they have to a great extent proven to be right – and every attempt to justify this fact leads to a circular argument, as the fact must be presupposed in the justification. Nor in this instance may one appeal to Kant's often misused dictum that human understanding prescribes its laws to nature, for if this statement is to stand up to further scrutiny, it cannot be interpreted as if experience were *pure* spontaneity: rather experience *also* contains spontaneous elements and is not, as the mimetic theory of naïve realism has assumed, pure receptivity. Even if Kant did not adequately describe the role played by mathematics in the framework of the investigation of nature, and therefore arrived at a doctrine of the aprioristic foundations of the empirical sciences that requires revision in many respects, he did not overlook that the laws of nature have to be validated by observations, and thus by partially receptive acts.

Thus, as has been recognized with increasing clarity, especially by the leading philosophers of contemporary natural sciences,[31] a *theory of induction* cannot consist of a justification of this procedure, but only of a *rational reconstruction* and systematic ordering of the inductions actually carried out in prescientific and scientific thought. Thus such a theory has above all to ask, what kind of 'interferences from the particular to the general' actually have been carried out – up to now successfully – i.e., what constellations of facts have proven necessary and sufficient for this or that anticipation. In doing this it becomes evident that such anticipations are by no means bound exclusively (or even predominantly) to the occurrence of a certain minimum number of similar observations – as happens solely in the elementary case of *inductio per enumerationem simplicem* – but that the significance or 'weight' of various kinds of individual observations is evaluated quite differently. For their evaluation depends on how they fit into that section of the total system of experience that is considered *relevant* in this context, on the basis of an induction that is generally carried out pre-predicatively.

The first extensive attempt to set up rules of induction was carried out by Francis Bacon,[32] compared to his work, John Stuart Mill's well-known principles of induction already represent a significant advance, but modern philosophy of natural science has progressed far beyond Mill. In spite of this, there still remains a great deal to be done today in the theory of induction in the natural sciences; and for the social sciences, the primary area of concern in the present work, the theory of induction is still taking the tiny steps appropriate to little feet.

---

[31] Compare for instance, the writings of Schlick, Reichenbach, Frank, Feigl, Hempel, Popper.

[32] Compare especially his *Novum Organum Scientiarum*.

It will already be seen from what has been said that within the empirical sciences the deductive method and the inductive are *irrevocably interrelated*. For on the one hand, the general propositions (principles, laws, hypotheses) from which deductive procedures take their departure are the results of induction; on the other hand, however, general propositions, i.e., the results of induction, are tested by ascertaining the correspondence or non-correspondence of certain inferences, deductively derived from these propositions, with observations. The idea of a purely deductive empirical science, which has played a by no means insignificant role in the social sciences, also goes back to the erroneous assumption that here are immediately evident propositions that do not require further corroboration and which, in spite of this, contain knowledge about reality.[33]

Of course the material connection of deduction and induction by no means rules out the possibility that in the course of scientific collaboration, one group of researchers, the 'theorists', is occupied mainly with the derivation of inferences from general assumptions, while another group, the 'empirical investigators', is devoted more to the observation of facts. As we shall have to show in more detail below, the quarrel about method between the 'theorists' and the 'empirical investigators' arises, however, from disagreement about the 'cognitive value', correctness, fruitfulness, status and sphere of application of the 'theorists' 'general assumptions'. The weak points of the 'theorists' argument lies in their frequent failure to recognize the empirical character of their basic assumptions i.e. their testability and refutability – as well as in a false conception of their basic assumptions – i.e. their testability and refutability – as well as in a false conception of the relationship between analytic judgments and factual judgments. The 'empirical investigators', for their part, usually overlook the 'theoretical content' of facts, i.e., the general assumptions implicit and contained in factual judgments; therefore they do not evaluate correctly the 'theorists' achievement, which lies in the rational reconstruction of implied presuppositions. This is especially evident in the empiricist's objection that deduction does not lead to *new* knowledge. What is most important about deduction and induction in the present context has probably been said; yet we shall have to deal with the latter concept in the next section.

The inadequate distinction between empirical propositions on the one hand, and propositions in which the result of rational reconstruction (analysis of concepts) finds its expression, on the other, leads not only to erroneous opinion about the character of ideal objects, *a priori* knowledge, and the procedure of deduction but also to a lack of clarity concerning the conception of *definition*. But here, as we shall see right away, a further element of confusion is added.

Every definition contains a determination of the use of a certain term which, as such, obviously contains no knowledge. Thus, for example, the definition 'a rhombus is a parallelogram with all sides equal' states: "we want to call a parallelogram with all sides equal a 'rhombus'". Here 'parallelogram with all sides equal' is called a '*definiens*' and 'rhombus' a '*definiendum*'. " Now we can recognize at once that the introduction of the word 'rhombus' could be dispensed with,

[33] *Tractatus Logico-Philosophicus, p. 96 ff.*

without the propositions of geometry being changed in the slightest, as far as their content is concerned; we would simply have to replace 'rhombus' by 'parallelogram with all sides equal' in all of them. However, there are two aspects that make it appear as though this imposition of a name signifies an enrichment of knowledge. One of these aspects relates to the new introduction of terms, the other to making existing terms more precise.

As far as the first case is concerned, the 'creation of concepts', i.e., the introduction of new terms, often marks the conclusion of a process of gaining knowledge, and this gain appears to be captured as it were, in that concept. Thus in the concept of thermal equivalents, there is the knowledge (discovered by Robert Mayer) of the possibility of comparing the energy of quantities of heat and quantities of mechanical work. As analysis of this state of affairs leads to the following distinctions: (1) the idea (imaginative projection) of searching for a constellation of facts, in reference to which a given quantity of heat can be replaced by a precisely specifiable quantity of mechanical work; (2) the possibility of the empirical realization of this idea; (3) the terminological fixing by means of the words 'mechanical equivalent of heat'.

It is understood that our term could also be used in an imaginary context, that is, the possibility of its empirical aspect is not a necessary condition for a term's meaningful use. However, the empirical aspect is taken *de facto* as part of the meaning of our concept, and because of this, we have the false impression that defining, or more precisely giving a name, contains knowledge. Here again we encounter the dangerous confusion of 'positing' and 'presupposition', which, as we have previously established, also stands in the way of an understanding of logical principles and logical inference. The confrontation of '*nominal definition*' with '*real definition*', proposed by scholastic logic since antiquity, does indeed aim at the distinction we have just made, but it does not provide a clear expression for it, as if fails to indicate that every definition, qua definition, is a nominal definition. This insight can lead directly to the understanding of the second major source of prejudice with respect to the epistemological character of definitions, insofar as they are linked to existing pre-scientific and scientific language use. As we have already anticipated, it consists in misconstruing the procedure of concept analysis (rational reconstruction), so that the term that appears as the 'object' of the analysis is not distinguished from the concept symbolized by it (the thought content). Hence comes that fetishism of symbols which has created so much mischief, even in logic and mathematics. Later on, we shall have to examine the place of this fetishism in the social sciences.

This contrast between 'deduction' and 'definition' also lets us avoid the misunderstandings that are linked to the concept of '*tautology*' as understood by Wittgenstein.[34] We cannot discuss Wittgenstein's treatment in detail, which is closely linked to his theory of atomic sentences, but shall limit ourselves to the assertion that from here on 'tautology' means essentially the same thing as 'analytic judgment'. A paradigm example of tautology is the sentence: "It rains or it does not

[34] Compare below, Part Two, section "Remarks on the Methodological Controversy *[Methodenstreit]* over the Theory of Marginal Utility".

rain." Such a sentence only apparently contains a statement about the world, but actually it is nothing other than an explication of the meanings that are given to certain words – in our example, the words 'or' and 'not' – or certain combinations of words. Therefore there is a close affinity between tautology and definition. In our section on economic theory[35] we have to show that derived propositions in a deductive system were mistakenly equated as 'tautologies,' so that there is the false impression that all propositions of a science constructed as a hypothetical deductive system (axiomatized) are tautologies.

Now before we turn to our findings about mathematical thought, we still have to say a few words about the *formal* character of logic. We have already shown that the thought functions of identifying and distinguishing correspond to 'formal objects'. Now it can be shown that the logical concepts, that is, those concepts on which logical inference depends (not, and, or, follows, all, some) can be defined by means of two concept-pairs:' same – different' and 'general – particular', both of which can be traced back to those two thought functions.[36] The formal aspect of logic, not to be confused with the analytic character of logical statements, lies precisely in this abstraction from any material content found in logical concepts. Here we want to say *'absolutely formal'*, since the concept 'formal', depending on the degree of material content, frequently has different *levels of the formal*. Thus for every thematically coherent science, a *formal theory* can be constructed, i.e., an ordered survey can be given of those concepts whose content is contained in the content of the fundamental concepts.

The absolutely formal concepts determine the content not only of logical propositions, but also of the propositions of pure mathematics, and here we have the reason why *logic* and *mathematics* in principle form a *unity*, as has been shown beyond question in the more recent studies of their foundations. However, in mathematics, where a multiplicity of complicated symbol connections must be investigated, the interrelationships are much more difficulty to penetrate than in logic. We must limit ourselves to indicating some of the most important results, ignorance of which has also led to serious misunderstandings in the numerous analogies drawn between the concepts and methods of the social sciences and those of mathematics.[37]

A. All the concepts of pure mathematics can be traced back to those of the *theory of natural numbers* 1,2,3, . . . i.e., they can be defined in terms of these 'natural numbers'. Therefore the so-called extensions of the number concept – negative numbers, fractions, irrational numbers, imaginary and complex numbers – are not extensions in the true sense, as every statement about such numbers can be *translated back* into a statement about natural numbers.
B. The entire theory of natural numbers can be derived from the following description of the number series: (1) there is an initial element; (2) for every element there is precisely one immediately succeeding elements; (3) for every element,

[35] With respect to this point to the reader is referred to my essay mentioned in footnote 24.

[36] See Kaufmann, *Das Unendliche*, p. 76 ff. [Eng.tr. pp. 68 ff.]

[37] *Ibid.*, p. 84 f. [Eng. tr. p. 72].

with the exception of the first, there is precisely one immediately preceding element; (4) by means of the definitions (1)–(3) the object of thought 'number series is completely defined.[38]

These definitions of the number series, which correspond in essence to Peano's five classical axioms, also allow us to derive all of the propositions of so-called 'higher mathematics', for, generally speaking, the line of demarcation that has been frequently drawn at the infinitesimal calculus, between lower and higher mathematics, for teaching purposes, does not correspond to any fundamental differentiation of subject matter.

C. With regard to the *infinitely small*, which allegedly appears in the so-called infinitesimal calculus, what we have here is a mistaken interpretation of a specific symbolism. As has been clear to mathematicians, but not to philosophers, for more than a century, we understood the actual sense of the derivative, and, thereby, of the entire infinitesimal calculus, only by translating it back into finite language.[39] It can be easily shown that all of the concepts of the infinitesimal calculus can be derived from those of the series of natural numbers, and all of its propositions from Peano's axioms.
D. We cannot settle the matter of the *infinitely great*, the *actual infinite* of Cantor's set theory, as we did that of the infinitely small. But here too, the breakthrough of the *finitist* conception, which, in my opinion, is the only legitimate one, has become more and more decisive in the last few years. It is my conviction that within a short time the assumption that there is yet another kind of infinity, besides that of the infinite progression of the number series, accounted for by the expression 'there are infinitely many number', will be finally rejected.[40]
E. But now, what *is* the meaning of the concept of natural number and what is the *nature of the validity* of mathematical laws? In the analysis of logical laws we have established that their validity is based not on act of positing, but on what they presuppose. The same is true with the laws of numbers. For it will be seen that the abstractions that lead to the concept of natural number already contain within them the principles which are then formulated in the laws of *arithmetic*. The arithmetic laws are by no means about numbers in the strict sense, but laws about certain *operations*, which regarded from that viewpoint of abstraction, are to be considered equivalent or as different, in a manner that will now be precisely indicated.[41] We wish to briefly clarify this:

In order to describe how thinking proceeds from experience, in a narrow sense, to numbers, it is best to take as a point of departure the *process of counting*, investigate what implicit presuppositions are contained in it, and then state how one arrives at the concept of number from the process of counting. Then, guided by a simple

[38] This was already demonstrated in the nineteenth century, especially in the analyses of Cauchy and Weierstrass.

[39] See Kaufmann *op. cit.*, p. 135 ff. [Eng.tr., pp. 14 ff.]

[40] See footnote 35.

[41] See footnote 35.

example, we should ask ourselves what assertions are made in the process of counting. Let us count the people found in a certain room at a certain time. The first assertion is: "Here is a person; let us call him 'A'". To this corresponds the act of counting '1'. To '2' corresponds the assertion: "Here is a person who is different from the person 'A'; let us call him 'B'". Thus the two sentences: On the one hand, "In this room is one person A and in this (same) room is a person different from him, 'B'", and on the other, "In this room are *two* persons A and B" are completely equivalent in meaning. The assertion that leads to '3' then states: "In the room there is a person who is different from the person B and from the person A, who is different from the person B", and so on, where the limitless progression of this 'and so on' expresses the insight that any specific assertion of difference is *independent* of the assertions of the difference previously made.[42]

What, therefore, is presupposed in the act of counting? First, the knowledge of those *features* that are decisive for whether something is included in the count or not, further, a specific *domain* – in our example, the room – within which things of this kind are to be counted; and finally the ascertainability of difference within this materially and space-time delimited sphere. In principle, it is of secondary significance that the act of counting proceeds in this manner; that a certain *symbol* is assigned to each epitome [core meaning] of the various assertions of difference with respect to the same object just characterized; that objects are thus designed as 'first', 'second', 'third' – for this assignment is nothing more than the marking of these assertions. Now as the two assertions: "A is different from B" and "B is different from A" have the same meaning, so the *order* of counting has no bearing on the result of counting, as long as *what* is to be counted has been established. For a change in order of counting is nothing other than the displacement of the succession of the individual assertions of difference. Thus, as soon as it has been established unequivocally *what* is to be counted, then the *last thing* counted – if, for instance, in any series of counting it is the eighth – then in every other possible series it will also have to be the eighth. The *cardinal number* 8, for example, expresses the invariance of the symbolic mark (ordinal number) of the last counted element, as opposed to the variations in the order of counting. To say that there are eight light bulbs in a room means nothing other than that the last-counted light bulb will be eighth, no matter in what order they are counted.

The mathematical operations of *addition* and *multiplication*, which are governed by the *law of commutation*, the *law of association* and the two *laws of distribution*[43] as well as the mathematical laws derived from them, by no means refer to the numbers themselves, as the concepts of the numbers are already abstractions of differences that are expressed in these laws. Instead, the laws are nothing more than (statements about*) the invariance of the result of the computation* (of the concept of a number) with respect to certain *operational* variations. Thus, the sentence '3+5=5+3' says nothing about the numbers 3 and 5 as such, but rather about the

---

[42] See Hölder, *Die Arithmetik in strenger Begründung*, 2nd ed., Berlin 1929.

[43] More emphatically than in the first edition of the *Critique of Pure Reason* which appeared in 1781, this was done in the *Prolegomena* (published two years later).

operation of counting, by asserting the invariance of the result of counting with respect to temporal variation [*sic*] within the counting process. Accordingly, the *'arithmetic a priori'* is the totality of those *presuppositions,* which are contained in the concept of the series of *natural numbers*. In the sense of what has just been said, it embraces the entirety of *pure mathematics*.

But what is the situation with regard to *geometric* knowledge? As was previously mentioned, geometric knowledge, especially, in the axiomatic from given it by Euclid, was regarded as a model of knowledge that was exact and yet empirically fruitful. To begin with, we wish to deal not with the axiomatic mode of presentation, but with the problem of the knowledge contained in geometric propositions. The decisive question here is whether the propositions of geometry, or more precisely, the propositions of Euclidean geometry, while not derived from experience, contain truths about the real world. As is well know, Kant who saw in geometry the paradigm for synthetic propositions *a priori*, and based his 'transcendental investigation' of space on the fact of geometric knowledge, believed that this indeed was the case.[44] But this Kantian doctrine, in which geometry was identified simply with Euclidean geometry, was already shaky when the so-called non-Euclidean geometries – i.e., those for which the Euclidean axiom of parallel lines does not hold – was shown to be logically free of contradiction. However, Poincaré's conventionalist interpretation of geometry and finally Einstein's General Theory of Relativity, which showed that Riemannian non-Euclidean geometry was better suited to the description of the physical world than Euclidean geometry, dealt this doctrine its death blow. The relation between geometry and the physical world was characterized by Einstein himself in the famous words: "As far as the laws of mathematics refer to reality, they are not certain; and as far as they are certain, they do not refer to reality".[45] That is: in geometry – or, more correctly, in the geometries – one must distinguish carefully between their internal formal character and their application to reality; pure mathematics is concerned exclusively with the former.

Such a system of formal relations does not become a *theory of space* until we use it to describe the *relations of positions of objects in the external world.* Therefore the relations contained in a geometric system are beyond truth and falsity, they are not propositions, but only rudiments of propositions, or, as it is formulated in logistics, *propositional functions* (sentential functions).[46]

Using this concept, we can characterize the following state of affairs: as we have seen, in order to comprehend clearly the meaning of an empirical proposition, we must ask for its truth conditions, and thus ask under which conditions it would be designated as true, under which as false. However, for relations that are conceived without reference to the data of experience, verification is not conceivable, hence the formulations of these relations are not to be designated as propositions, and no

---

[44] *Geometrie und Erfahrung.* (*Geometry and Experience*), Berlin 1921, p. 3 [Eng. tr., 1922, p. 28].

[45] For this, see for example, Carnap, *Abriss der Logistik* (*Outline of Logistic*), *Schriften zur wissenschaftlichen Weltauffassung*, vol. 2, Berlin 1929 [See also revised Eng. tr. *Introduction to Symbolic Logic and its Applications,* 1958 – Ed.].

[46] Carnap, especially, pointed this out.

clear meaning can be accorded to the terms contained in them. Thus, in the axiom of pure geometry, for example: "Two straight lines intersect at a point", the concept 'straight' and 'point' have no well-defined meaning if there is no specification of how one is to ascertain whether something is straight or a point. Accordingly, one can conceive of both concepts as *variables*, which are only accorded a fixed meaning by having assigned to them an empirical significance. By means of such 'assignment', then the *propositional functions* become propositions. The formal internal relations of the system, however, remain invariant vis-à-vis these differences of meaning and therefore can be treated in isolation, without the need to link any of the 'assignments' to them conceptually.

Thus the conception of geometric axioms has undergone a radical change. Whereas formerly they were conceived as absolutely valid judgments about the external world, it has now been recognized that their 'irrefutability by facts' is possible because they do not assert anything at all about facts, but only represent a general schema which permits empirical 'assignments' of various kinds.[47]

Only the argument of empirical suitability – not those of logical necessity – can determine whether, on the basis of a convention, we use this or that system of propositional functions to characterize the relations between the positions of objects in the external world, whether we describe these relations in the 'language' of one geometry or of another. Here and in similar cases we want to speak of '*materially based conventions*'. For the physics of small velocities (as compared with the velocity of light). Euclidean geometry is particularly suitable and therefore imposes itself on the physics of everyday life as virtually 'self-evident', but for high velocity physics, Riemannian geometry is simpler. Therefore we may say that Riemannian geometry is the 'correct' geometry; but in saying this we must recognize clearly that such an assertion is not a statement about *a priori* status but about empirical suitability. It is also entirely possible to formulate high velocity physics in the language of Euclidean geometry; it is just more complicated. There is an analogous relation between the two assertions: 'The earth moves around the sun' and 'The sun moves around the earth'. Neither of the two is absolutely true; a geocentric astronomy that does justice to the facts is quite readily conceivable; but the heliocentric astronomy is essentially *simpler*.

Now a few more words about the *axiomatic method*, which was brought to high perfection in the various geometries. Even today the use of the word 'axiom' is not uniform, although the ancient, classical conception that axioms are self-evident truths, and Kant's conception, which is in essential agreement with this, in that he designates these axioms as '*synthetic principles a priori*', have become outdated in the meantime. For at times, by 'axioms' we understand an epitome of those propositions that stand at the head of a deductive system, and in keeping with this, the term 'axiomatic system' is used synonymously with the term 'deductive system'. This is usually the case where the postulate of 'axiomatization' of an empirical science has been set up. But as has already been noted, the axioms of geometry are not

[47] In logistic, one speaks of the circumstance that every judgment has one of the two truth values, 'true' or 'false'. This terminology has nothing to do with axiological considerations.

propositions but propositional functions, and therefore have no truth-value at all.[48] Thus Hilbert, in his *Grundlagen der Geometrie*, the classical work of modern geometrical axiomatization, characterizes his axiomatic system as follows:

> Consider three distinct sets of objects. Let the objects of the first set be called *points* and be denoted by A, B, C. …; let the objects of the second set be called *lines* and be denoted by a, b, c,…; let the objects of the third set be called *planes* and be denoted α, β, χ. . . .The points are also called the *elements of line geometry*; the points and the lines are called the *elements of plane geometry;* and the points, the lines, and the planes are called the *elements of space geometry* or the *elements of space.* The points, lines and planes are considered to have certain mutual relations and these relations are denoted by words like '*lie*', '*between*', '*parallel*', '*congruent*', '*continous*'. The precise and mathematically complete description of these relations follows from the *axioms of geometry*.[49]

Formerly it was said in this context that the axioms are '*implicit definitions*' of the basic concepts contained in them, and from the erroneous conceptions linked to this term, far-reaching conclusions with respect to the theory of science were frequently drawn. By means of the insight that the axioms are propositional functions and the basic concepts are variables, the ground has been cut out from under such speculations. In addition to the absolute requirement of freedom from contradiction, the most important requirements for axiomatic systems are the logical independence of the individual axioms from one another, and their completeness. The methodological significance accorded to the axiomatization of a science or of a scientific discipline lies in the clarification of implicit presuppositions contained in the fundamental assumptions, as well as in the individual steps of inference. As far as the latter point is concerned, the resolution of the deductive procedure into elementary operation was carried out in the logical calculus[50] and elaborated to perfection in Hilbert's theory of proof,[51] with the axiomatization of logic and mathematics. With regard to the presuppositions, the extent of the difficulties confronting their clarification can be measured by the fact that it was only at the end of the nineteenth century that an axiom of Euclidean geometry was discovered (and named after its discoverer, Pasch),[52] which had been used for 2,000 years without anyone being aware of it.

---

[48] See Kaufmann, *Das Unendliche*, p. 23 ff. [Eng. tr. p. 26 ff.]

[49] Hilbert, *Grundlagen der Geometrie* (*Foundations of Geometry*). [Eng. tr. p. 3]

[50] Main work: A. N. Whitehead and B. Russell, *Principia Mathematica*, vol. I, 1910, (New ed. 1925), vol II 1912 (new ed. Unchanged, 1927) vol III, 1913 (also 1927).

[51] See Hilbert, "Die Grundlagen der Mathematik", (*op. cit.*) (with discussion by Hermann Weyl and an addition (*Zusatz*) by Paul Bernays), *Abh. Math. Seminar der Hamburger Universität*, vol. 5, pp. 65–92, 1928.

[52] It states: let A, B, C be three points not lying in a straight line and *a* a straight line in the plane ABC which does not intersect any of the points A, B, C; if the straight line *a* intersects any of the points of the line AB, then with certainty it will intersect either a point on the line AC or a point on the line BC.

Finally one more fundamental remark with respect to the *probability calculus,* which, like geometry, was thought to offer *a priori* insights about events in the world. In fact, the probability calculus is the mathematical theory of combinatorics; it is this arithmetic in the broad sense, and can no more furnish empirical knowledge than can the multiplication table. Here it is especially the fusion of the Bernoulli-Poisson *theorem*[53]- one of the laws of combinatorics with the *law of large numbers*, which has created much confusion. But this is, in my view, a *materially based convention*.[54]

The preceding discussion is especially intended to delimit the power of the mathematical method in the empirical sciences. In this we see that neither logic nor mathematics produces *materially new* knowledge, but can only serve to clarify knowledge gained elsewhere, and to present it in a more comprehensive and systematic form.

The analyses in the sections to follow will enable us to make a series of fundamental assertions about the character, in a narrower sense, of *empirical* knowledge.

3. Fact and Law

While we have recognized that the concept 'fact' by no means designates an 'absolutely ultimate' given – indeed it seems to involve further difficult problems of constitution – for the scientific endeavour of extracting laws, facts are the *'relatively ultimate'* given, i.e., the starting point for research. Therefore, in the following analyses of the concept of law, we shall assume that the nature of fact is unequivocally determined, but without losing sight of its multi-layered character.

First we must determine how judgments that state the existence of a fact are distinguished from those that assert the existence of a law. The difference is that every judgment of fact contains an *absolute localization*, while every judgment of law, on the other hand, contains only a *relative localization*. Propositions of the first kind are always concerned with the occupation of specific places – i.e., of places ultimately fixed in relation to the body of the person making the judgment; in propositions of the second kind, on the other hand, this fixedness is lacking – in this case, only the relative position of phenomena is stated.

From this it follows that lawful relations do not exist between individual facts or individual groups of facts as such, but between *classes* of facts (groups of facts), and thus between any *arbitrary* facts of the kind *E, F, G,*...on the one hand, and any *arbitrary* facts of the kind *M, N, P,* ...on the other. Every empirical law can thus be put into the form: if facts of the kind *E, F, G,* ...appear in one given

[53] Compare here also Karl Popper, *Logik der Forschung* [*The Logic of Scientific Discovery*, tr. 1959], Schriften zur wissenschaftlichen Weltauffassung, vol. 9, Berlin 1935.

[54] See above, Part One, section "Basic Philosophical Considerations".

domain, then facts of the kind *M, N, P, …* will appear in a determinate environment of the first-named facts.

Every law asserts the *isolatability* of generally designated phenomena or 'aspects' of phenomena within the totality of events. Whatever else may happen – so the law goes – that relation between facts *E, F, G, …* on the one hand and *M, N, P, …* on the other, remains valid. But, as far as it goes, this formulation is too broad, because 'disturbances' of the lawful course of events owing to the existence of 'abnormal' conditions (for example, an unsupported iron ball is kept from falling because of a nearby magnet) must not be considered incompatible with the idea of the law.

The '*ceteris paribus*' clause allows for this epistemological fact in that it embodies the condition of *freedom from disturbance* in the law itself. But the formulation of this clause, which plays an important role in the *Methodenstreit* in the social sciences, is very unclear. *All* the surrounding circumstances of two *different* events cannot be the same, because their occurrence in different space-time or personal-temporal places is absolutely nothing, but their occurrence in different surrounding circumstances. Strict fulfillment of the clause would thus eliminate any difference between the events in question.

In fact, however, the sense of the *ceteris paribus* clause is that certain, more or less sharply delineated classes of '*neighboring phenomena*' adjacent to the facts under consideration, are to remain *unchanged*, while others are considered to be *irrelevant* to the casual relation from the outset. This can be seen especially clearly in natural science experiments, designed to discover laws, in which the experimenter tries to keep a certain small number of magnitudes of the environment *constant,* for instance, temperature when measuring lengths, while the remaining phenomena are disregarded as they can be expected to be *irrelevant*. However, should the results of his experiment conflict with the secured state of theory in his field, then one of the experimenter's most important considerations will be to find out whether he failed to account for certain 'sources of error'. Here he may very well be induced to assume that the constancy of some other factor, disregarded until then, may be essential for the validity of the law in question.

Accordingly, analysis yields the following meaning for the *ceteris paribus* clause: it does not demand the congruence of the total environment of the phenomena under consideration, which is unrealizable in principle, but merely a similarity of certain traits of that environment. However, in view of the expected modifying experiences no rigid finite limit on the number of these elements under consideration can be specified. The sense of the – *prima facie* – transfinite formulation, i.e., the idea of infinite magnitudes expressed by the word *cetera*, is, accordingly, *indefiniteness*. Thus here we have a case for the application of the *principle of finite formulation*, which we set up in our first section.

There is a further application of this principle in the *domain of validity* of laws. We have determined that the relation between events, or groups of events, as stated in the traditional formulation of laws, refers not to individual events as such, but to any arbitrary event of a certain kind; in this lies the general validity of laws. However, it is a misunderstanding to say that a law is valid always and everywhere. No matter how much one's powers of observation are increased, such a statement is not verifiable, and therefore completely lacks empirical significance, although, of course, it can be refuted by conflicting facts to the extent that it is confronted with facts. In order that the meaning actually accorded to a law in the empirical sciences emerges in the formulation, a *transfinite manner of speaking* must be *replaced* by an *indefinite* one, by operating, say, with the concept, '*within the limits of experience'*.

A counterpart to the formulations expressed in the terms of the concept 'always and everywhere' are those that rely on the concept 'somewhere and sometime', included, for example, in the completely indeterminate link between cause and effect, as soon as one asserts nothing more than that events of type *U* will follow those of type *I*. While in the one case, verification was impossible, here refutation appears to be ruled out, as there is still the possibility that somewhere, at some time, event *I* will occur. Therefore, the framework within which the causal events as well as those caused must take place, can be extended arbitrarily, yet still must have some limit. In the cognitive practice of the natural sciences, these limits are drawn very rigorously at most points, i.e., the time and place of the occurrence of the event – relative to the time and place of the occurrence of the cause – can be computed with considerable precision.

The meaning of the assertion of the '*hypothetical character'* of laws requires clarification no less than the meaning of the assertion of the 'universal validity' of the laws. This last term contains a mish-mash of various facts concerning knowledge. By distinguishing among the following three sets of questions, we can disentangle these: (1) of what kind are the assertions that form the content of the law? (2) what supports these assertions? (3) what are the criteria for verifying or refuting the assertions? We can characterize these questions as being directed at the meaning, the origin and the validity of the law.

To be sure, these three questions, if taken in pairs, are not independent of one another, because for the ideal case of completely clear thought, the assertion of the meaning includes the assertion of the domain of validity. But even if we disregard this, our grouping will prove advantageous as a means of presentation, because *de facto* the formulation of laws is hardly ever carried out with complete clarity.

We can already see from the formulation of laws that the occurrence or non-occurrence of events of a certain type is asserted only on the condition of the occurrence or non-occurrence of certain other events. Therefore, a law, of and in itself, does not allow any predications; moreover, an *assertion about the state of the data* is required. Even for the fictional Laplace's Demon, knowledge of the laws would not suffice to determine future or past events; rather, he had to possess knowledge of the state of the data at some point in time in the course of world events. In characterizing this state of affairs, it would be appropriate to speak of the '*conditional*', instead of the 'hypothetical' character of laws.

We get a different concept of the 'hypothetical' when we turn to the second point mentioned and ask about the *origin* of laws. By 'origin' we do not mean here those psychic processes that lead to the formulation of laws, but rather those facts on which they are based or, in other words, from which they are derived inductively. As we have already outlined the principal characteristics of induction in the preceding section, we can confine ourselves here to pointing out that the second meaning of the 'hypothetical character of laws' consists in the fact that the laws are based on general assumptions which are themselves not reasonable *a priori* and therefore they can be designated as '*mere hypotheses*'.

Now for the third set of questions differentiated above: those concerned with the criteria of verification or falsification of the laws. From these we will arrive at a third meaning of 'hypothetical', which is that the validity of an empirical law must not only be regarded as dependent on facts with respect to their origin, but also even once it has been accepted, it is still the subject of continual confrontation with the facts, and will be discarded if it cannot be reconciled with them. Every empirical law, then, is a 'mere hypothesis' in the sense that eventually it may have to be discarded, and the distinction frequently made between 'hypothesis' and 'theory', in which a definitive validity is attributed to the second, but not to the first, is, when scientific procedure is examined in the long-run, one not of principle but of degree.

However – and this assertion leads us to an important new point – even if the conception of absolutely valid laws is untenable, still the outstanding significance to be attributed to the gradual differentiation of validity must not be underestimated. It is extremely important in research, especially as the acceptance of those observations that do not agree with the hypotheses depends on it. Thus Michelson's experiment, which led to results that contradicted Maxwell's well-founded theory of light, was repeated for decades, because, in spite of all the precautionary measures taken in conducting the experiment, one was more inclined to assume that there was an error or gap in observations than that Maxwell's well-founded theory of light needed correcting. Other hypotheses, however, have been given up at once when observations contradicted them.

Under certain circumstances, trust in the validity of a law can be so great that at a certain stage of research *every* observation that does not correspond to it will be regarded as false or incomplete. In such a case we can say that the possibility of refuting a law is *suspended*: if an observation cannot be made to agree with such a law, but its correctness is completely above suspicion, then additional factors are introduced as 'disturbances' or 'data changes' to eliminate the inconsistency. This creates the following cognitive situation which can easily lead to a misunderstanding: the given general proposition, i.e., the law, is in the form of an empirical statement, because, first of all, it links facts with one another, and secondly, in testing its origin we find that it is based on observation. Nevertheless it seems irrefutable, as every inconsistency between the law and observations of facts is interpreted as a result of an error in, or incompleteness of, these observations. Thus we get the illusion that there are absolutely valid empirical laws, whereas in truth what we have are materially, *factually founded conventions*. The history of the concept of the principle of [the conservation of] energy during the nineteenth century offers a striking example of this.[55]

[55] Compare Philipp Frank, *Das Kausalgesetz und seine Grenzen,* (*The Law of Causality and its Limits*), *Schriften zur wissenschaftlichen Weltauffassung,* vol. 6, 1932, p. 136 ff. [Eng. tr. M. Neurath and R.S.Cohen, Vienna Circle Collection, Dordrecht and Boston, p. 138 ff, 1998 – Ed.].

As for the epistemological significance of assuming absolutely valid laws, which then turn up frequently in procedural rules or heuristic postulates, we must consider that such assumptions are based on the conviction that the structure of the world is uniform and relatively simple, and thus on a conviction that has often been brilliantly justified in the history of scientific thought. In this connection we have only to point out to Galileo's admirable hypothesis, that the motions of the heavenly bodies and those in his physics laboratory obey the same laws. Or his still more general assumption that the course of all events in the external world can be traced back to uniform principles.

But there is the danger that convictions of this kind can degenerate into dogmatism, and that all observations that could modify an original formulation are either wholly ignored or, at the least, will not be given their due. This danger grows still greater because many metaphysicians see very fertile soil for their speculations in such general principles. Generally speaking, scientific dogmatism, whether the apparent consequence of metaphysical doctrine or not, is capable of imparting a strong impulse to scientific research in that it decrees, with considerable emphasis, that one specific path of research is the *sole* correct one. In other kinds of cognitive situations, however, where a derivation from prevailing methods seems *de rigueur*, the same attitude can be a serious hindrance to the progress of knowledge.[56]

From what we have established up to now certain methodological consequences follow which are of great importance for the *Methodenstreit* in the social sciences. Insofar as they are more closely linked to these sciences, we shall have to deal with them in the second part of this work; here only a few general results will be stressed.

First of all it follows from our deliberations that the sharp *break* between strict *laws* and mere *rules or tendencies* – a break seen as marking a fundamental difference between method in the natural sciences and in the 'sciences of the mind' – cannot be maintained for as has just been established, the laws of nature can only be considered as 'inviolable' insofar as they are fixed by convention, i.e., no experience will be recognized as a counter-example. However, once we realize the untenable character of this fundamental separation, we are faced with the task of analyzing the natural-scientific laws and the social-scientific laws in terms of their *specific* differences, for, once they are *no longer totally different* entities, they can no longer be barred from all comparison with one another. And such an analysis (which we shall outline in the second part of this book) will yield important insights into the procedural distinctiveness of both groups of science. In particular it will prove that certain conceptions about the significance of mathematical method are wrong, especially the one that attributes the rigor of natural-scientific knowledge to the applicability of mathematical method in the natural sciences, said applicability – and this, too, is a mistaken view – being based, supposedly, on the spatial structure of the external world as opposed to the psychic-cultural [*seelisch-geistigen*] sphere. Our reflections, however, have made it clear that the alleged epistemological fact of the

[56] Thus, since the circular orbit was considered the most perfect and therefore the only suitable orbit for the courses of the stars, it was only with great difficulty that Kepler struggled to teach the knowledge that the planets moved in elliptical orbits around the sun.

absolute rigor of natural laws, supposedly justified by recourse to mathematics, does not exist at all.

Now we can deal successfully with a further question, which in its time played a big role in the natural sciences and today, after having undergone necessary clarifications there, still arouses considerable confusion in the methodological controversies surrounding the social sciences, namely the question of the relation between *causal laws and functional laws*, or, the replacement of the concept of cause by that of function in the formulation of laws.

Thus Mach's definition:

> The laws of nature are equations between the measurable elements α β γ λ μ ν of phenomena … if we know *all* the values of α β γ by which, for example, the values of λ μ ν … are given, we may call the group of α β … the cause and the group λ μ ν …the effect.[57]

In order to understand the meaning and direction of the struggle against the concept of causality, we must remember the *intellectual-historical roots* of the concept of 'law' on the one hand, and of 'cause' on the other. To begin with, as far as the concept of 'law' is concerned, its origin does not lie in the natural, but rather in the mental sphere. The archetype of the law is the *norm* and every observed regularity in natural events is traced back to such a norm.[58] Thus a famous fragment of Heraclitus states: "The sun will not transgress his measures. If he does, the Furies, ministers of Justice, will find him out." From this also comes the idea of the absolute and perfectly exact validity of laws, which is nothing but a correlate of the perfect precision of the plan of a Supreme Being and of unlimited power in the realization of this plan. This explains the great significance that the discussion about the nature of miracles, as the violation of natural law, acquired for the philosophy of nature.

It is understandable, then, that with the secularization of natural science, the idea of necessity, being linked to the concept of law, was subjected to critical analysis. Its most radical critic was the father of modern empiricism, David Hume. However, since 'law' and 'causal law' were almost always identified with each other, Hume's critique of the concept of law found expression in his critique of the traditional conception of cause.[59] We will now have to say a few words about the history of this concept.

Just like the concept of law, which is most closely linked to it, the concept of causality is of *anthropomorphic* origin. The archetype of causality is the effect brought about by human force directed by will, and, at a somewhat later stage of thought, also the relation between will and bodily motion. As far as the explanation, of nature is concerned, this produced the dictum to interpret the regularities of natural events in such a way that forces working as *qualitates occultae* [hidden qualities] were attributed to bodies and formed the center of causality. The natural science and

---

[57] Die Mechanik in ihrer Entwicklung historisch-kritisch dargestellt, 4th ed. 1901, p. 356. [*The Science of Mechanics.* 5th English ed., tr. T.J. McCormack, pp. 605–60, Chicago 1942].

[58] This has been pointed out on numerous occasions, among social scientists, for example, by Eulenburg and Kelsen.

[59] Compare especially his main epistemological work, the *Treatise on Human Nature.*

natural philosophy that were developing in the seventeenth and eighteenth centuries most decisively turned against such *qualitates occultae*. Thus Newton, in his *Opticks*,[60] writes:

> To tell us that every Species of Thing is endowed with an occult specific Quality by which it acts and produces manifest Effects, is to tell us nothing: But to derive two or three general Principles of Motion from Phaenomena, and afterwards to tell us how the Properties and Actions of all corporeal Things follow from those manifest Principles, would be a very great step in Philosophy, though the Causes of those Principles were not yet discovered…

David Hume sets forth the critique of the principles underlying the scholastic conception of causality when he shows that using the concept 'necessity' or 'impossibility' (as the case may be) to define the concept of cause does not furnish any empirically specifiable meaning, for as we only have one world, we can only ascertain in each instance that, of two events or groups of events, E_1 and E_2, E_2 did actually follow E_1. Thus the *'post hoc'* we can establish directly; it is, however, in the conclusion *'post hoc ergo propter hoc'* that the problem lies. When we consider this, we see the decisive turning away *from* naive Realism, where the world with its laws is asserted unequivocally, to be simply accepted, *toward* a critical approach to knowledge, which characterizes Hume (next to Leibniz) as the most important predecessor of Kant.

Incidentally, Hume's critical results were by no means refuted by Kant's analyses of the concept of causality, as has often been erroneously asserted. Rather, Hume's *'Copernican Revolution* , according to which the world is to be constructed from experience, i.e., described by means of the elements of experience, is carried out precisely in terms of causality, as the relation between causality is reversed. It is no longer assumed that causality is to be attributed to the things-in-themselves, a causality that can be discovered by certain rules presented as instruments of inquiry. Instead, what causality means in each instance becomes clear only when the criteria that characterize the specific kind of induction are indicated. At first Hume replaced 'necessary' relation with 'constant relation' in his analysis of the concept of causality, thus removing the idea of an absolute principle 'lying behind the phenomena'. But in doing that, little was done for empirical research, for an event is not repeatable in its full concreteness. Therefore the concept of causality is formulated in such a way – and this is a decisive step forward – that it no longer refers to individual phenomena but rather to *classes of events.* Because of this, the problem that now comes to the fore is the determination of the principles, according to which the grouping into classes is to be carried out, in order that the laws of nature can be formulated in the most uniform and simple manner possible. This, in turn, makes two things possible: (1) the emergence of the idea of the *spontaneity of thought*, given its classical formulation in the *Critique of Pure Reason*; and (2) the first promising step toward a theory of induction. The basic task is to make explicit the implicit

[60] Newton, Optics, [Eng.tr.1931], pp. 400–402.

presuppositions hidden in scientific induction. Hume himself, like Leibniz, gives *contact action* as the most important principle implicit in the concept of causality. Every action at a distance must be mediated by a continuous transmission of the action. Furthermore, the need for constancy also applies, namely that sufficiently similar causes will have similar effects and that bodies and events remote from one another will exercise no noticeable effect upon one another.

We cannot treat here in detail the most recent modifications that these postulates of contact action and constancy have undergone in modern physics. However, we should like to clarify our present discussion by pointing to the stages in the formation of scientific concepts and, together with them, in the formulation of the laws of nature. These are described below by one of the most important contemporary mathematicians and theoretical physicists, Hermann Weyl[61]:

1. Dissection of the three-dimensional spatial reality into single partial systems (bodies or things), each forming an intuitive spatially isolated and relatively constant unit. In its behavior each is considered as *independent* of the others, unless progressive analysis calls for corrections. Closely connected with this is the dissection of the four-dimensional spatio-temporal reality into single isolated *events* that form natural intuitive units.
2. The conception-of an intuitively experienced event as having come about by spatio-temporal coincidence and *amalgamation* of several simple phenomena (each of which would produce other perceptions than the phenomenon as a whole if the others were 'erased' or replaced by 'normal conditions'; e.g. the sun setting behind a gold-edged cloud).
3. Apperception of the 'being-so', bringing out the characteristic features (self-insufficient parts) of the phenomena. Upon this procedure is based the grouping together of similar things, the subordination under concepts, in one word: *classification*. Such classification will correct itself as the wealth of our experience increases. It will thus learn to distinguish better and better the truly essential from the inessential and progress to the formation of more and more 'natural' classes. A concept is the more essential the more connotations it entails according to the evidence of experience, namely the more characteristics not contained in the concept itself are empirically found to be common to the objects falling under the concept.
4. We are not satisfied with intuitively isolable elements but interpret a series of properties, which always appear together as an indication of a concealed something. This leads to *hypothetical elements*, such as atoms, forces, electro-magnetic field, etc. Moreover, we learn to interpret not only the observable properties but also the reactions that occur if one system is brought together with another as manifestations of such hypothetical elements and of their intensive and quantitative values. (Reactions instigated at will are the essence of *experiments.*) Finally, we do not hesitate to dissect hypothetically even the intuitively simple, e.g. the white sunlight into the colors of the spectrum, or the acceleration of a planet into the partial accelerations brought about by the sun and the other planets. *It is* evident that along with the dissection the *synthetic principles* also have to be established according to which the elements unite into a whole (e.g., formation of the resultant of forces).

[61] Hermann Weyl, *Philosophie der Mathematik und Naturwissenschaft* [Eng.tr. *Philosophy of Mathematics and Natural Science,* Princeton 1949] in *Handbuch der Philosophie*, Munich 1927, p. 107, [pp. 145–146].

The first three steps still belong to the pre-scientific stage.

Of course we recognize that the primitive conception of causality as *substances* acting upon one another is inconsistent with such a demonstration of the principles governing pre-scientific and scientific thought, for the entities representing the acting substances, i.e., the solid bodies, when viewed in this way, are broken down into moments that are *not independent.* Since Galileo, the '*dissecare naturam*' has become the principle of inquiry in what are (for this very reason) the *abstract* natural sciences. Hence, faced with the concept of 'lawfulness', the concept of 'causality' is pushed increasingly into the background.

The contemporary conception is well characterized by Bavink's formulation[62]:

> To explain natural phenomena in causal terms means to bring them into lawful relationships with other phenomena, relationships whose validity is conceived as somehow logically justifiable, even when the laws are not yet actually justifiable. The direction of the causal relation then tends to coincide in some way with the meaning of the logical relation (ground – consequence). However, physicists as well as philosophers have always and repeatedly regarded order in time as determining this tendency because in by far the greater number of cases, and just in those of greatest practical importance, the general laws which enter into the question are of the kind which appear to connect the momentary values of certain magnitudes with their changes with respect to time, a point mostly clearly recognized with regard to the fundamental laws of mechanics. It is in this way that time actually comes into the content of almost all, or at least of the most important and frequent, causal judgments, and this has led to the error of including it in the form, the structure, of the causal relationship itself, instead of in its content.
>
> By far the most important for us are those causal judgments which allow us to predict in advance a certain course of events; much less frequently do we need those which allow us to conclude from a state of affairs at one point or another, what is happening simultaneously elsewhere. But in principle, these two cases of 'dynamic' and 'static' causality, as we might call them, have no priority one over the other. The essential feature, the conclusion from a determined A to a co-determined B, is in both cases exactly the same, only that in one case time occurs among the necessary variables, and not in the other case.

However, the postulate that causal law is to be replaced by functional law reflects not only a turning away from the primitive conception of causality as substances acting upon one another (where the mental image of the experienced effect of living force appears) but also the need for the *mathematical formulation* of the laws. This part of the postulate becomes especially decisive for its significance within the social sciences, above all in economics. The dominant conception here has been that one could speak only of exact laws, where a numerical relation between magnitudes is specified; otherwise it was just a question of rules. We can easily see that the idea of absolutely exact laws also originates in pre-scientific conceptions, since it is based upon the conviction that the laws represent the principles of a divine plan of creation which, being perfectly rational also has to be perfectly exact. A famous

[62] B. Bavink, *Ergebnisse und Probleme der Naturwissenschaften*, (*The Natural Sciences*, partial Eng.tr.) 5th ed., Leipzig 1933, pp. 74–75; [S. Drake ed. *Discoveries and Opinions of Galileo*, pp. 237–238 N.Y.1957].

example of this conception, which gave a powerful impetus to the seventeenth-century investigation of nature, is found in the words of Galileo[63]:

> Philosophy is written in this grand book, the universe, which stands continually open to our gaze. But the book cannot be understood unless one first learns to comprehend the language and read the letters in which it is composed. It is written in the language of mathematics, and its characters are triangles, circles, and other geometric figures …

However, the most recent developments of physics, as shown by *quantum theory*, have made it clear to natural scientists that the 'exactness' which is the goal of the methods of measurement can by no means be understood as a completely precise determination of the objectively existing magnitudes of nature. A thesis of the absolute exactness of the laws of nature interpreted in this way is just as untenable as the assertion of their absolute rigor (rational necessity). What remains is the discovery that natural phenomena can, for the most part, be described by means of relatively simple mathematical functions. In this regard we may point once more, and most emphatically, to the error of confounding the immanent 'exactness' that characterizes the deductive relationship – essential for the mathematical method – with empirical 'exactness'.

Thus, from the standpoint of logical exactness, the statement, 'The attraction of two masses is inversely proportional to the cube of their distance' is just as valid as the statement that, following the Newtonian principle of gravitation, asserts the inverse proportion of the attraction to be equal to the square of the distance. Hence, two deductive systems can be set up side by side, in complete conformity, except in the manner just specified, and no one can decide on the basis of internal analysis which of the two explanations of nature is to be preferred. The decision about this can only be made by means of experience based on observation. Today, only a handful of natural scientists will doubt that this is the case. But things are not so clear in the social sciences, where there are a number of complications, which will be discussed later. When dealing with laws, then, we must very carefully consider what kind of facts are to be regarded as criteria for their verification or refutation. In particular, we will have to determine whether the law whose validity is to be tested is to be confirmed by (relatively) isolated observations, or only in conjunctions with other laws, as, for example, the law of falling bodies can only inform us about the actual behavior of falling bodies in conjunction with the laws of friction and air resistance. Problems of this kind especially affect the analysis of the method of ideal types in the social sciences.

By clarifying the concept of law in general, and the law of causality in particular, the old conflict between a *deterministic* conception and an *indeterministic* one presents itself in a different light as well. As will become clearer in what follows, the debate about determinism should by no means be viewed as an essentially homogeneous set of problems. Various levels of problems are interwoven, and they must be *isolated* if we are to assess their significance for the methodology of the social

[63] *Opere* ed. Alberi, III, p. 355 (cited in Weyl, op. cit., p. 242).

sciences. The basic level of these problems is defined by the deterministic conception of the law of causality as a law of reason with *a priori* validity. Against this, as we have just explained, Hume had already shown that there could be no question of a rational necessity in the case of the law of causality. Further, the most recent developments in the natural sciences have led to the conclusion that there are certain finite limits even to the unequivocal physical characterization of facts, which forms the precondition for every investigation of cause. Thus, determinism, in its most exaggerated (transfinite) formulation, must surely be regarded as conclusively outdated. Modern formulations of the principle of causality by natural scientists show, by their modesty, the great change that has taken place here because of compelling results in the natural sciences. Philipp Frank states the law of causality as follows: "Every state can be determined by relatively few variables of that state[64]". This statement is by no means proclaimed as a directly evident principle of reason, but rather as a maxim of the investigation of nature, which has been confirmed up to now. However, corroboration lies not in a direct application of the principle, for that leaves the kind of phenomena to be linked as well as the more precise circumstances of the links themselves entirely undetermined, but rather in the discovery of particular causal laws. In keeping with this Frank[65] remarks: "In practical life we never trust in the general law of causality, but in our knowledge of the particular relation." Therefore the causal *principle* has appropriately been called the *general form of particular causal laws* (Wittgenstein),[66] which was intended above all to express the fact that it cannot be coordinated with them. The statement: 'A phenomenon of kind *U* is the cause of a phenomenon of kind T' thus requires supplemental specification concerning the character of the empirical relationship in order to have empirical content, i.e., in order to be testable.

Just as the statement that a certain phenomenon has causes can only make scientific sense when the nature of these causes and their relationship are characterized in a certain manner, so the concept of *chance* and thus the absence of causal lawfulness can only make sense in relation to a specific causal law or to a specific group of such laws. As the treatment of the problem of accountability in the second part of this book will clearly show, it is especially important to be very clear about the causal relationships involved in each case, where independence from these relationships is to be understood as 'chance'.[67]

This *need for a supplement* which is thus inherent in the concepts of cause and chance, also extends to that of *explanation*. If 'explanation' is defined as 'incorporation in a general context of experience' – and this is the definition which presents itself as soon as the prejudice of the existence of a single *causa efficiens* is overcome – then one will recognize that this concept only gains a clear-cut methodological meaning by

[64] Compare *Kausalgesetz*, p. 242.

[65] *Ibid.*, p. 287.

[66] *Tractatus*, p. 172.

[67] For this, compare the lectures of Reichenbach, Mises, Waismann, Feigl, at the first conference on epistemology in the exact sciences in Prague (1929), reprinted in *Erkenntnis*, vol. I, 1930.

specifying the system of reference, that is, by determining *out of what* and *by means of what* an explanation is to be made, in other words, what data and what general laws are to function as the basis of the explanation. As can be readily seen, in the *choice* of such a reference system, however, lies the crucial decision for the method of an explanatory empirical science, and consequently the investigation of the material motives, i.e., those oriented toward the goals of knowledge that influence this decision in one sense or another, forms an important task for methodology. In the second part of this book, we shall have to find a way to account for the epistemological fact that while in physics the epistemologically relevant frame of reference underlying an explanation appears to be determined in a relatively homogeneous way at every stage of research, in the social sciences, typically, a considerable variety of types of explanation compete with one another.

If I may close with a few words to clarify the relation between *'explanation'* and *'description'*: The confusion here stems primarily from the fact that the term 'description' is not used unambiguously. For on the one hand, we speak of the description of individual things, conditions, processes; on the other hand, we also speak of description (descriptive analyses) of *types*, kinds, species. We can easily see here again that fusion of empirical statements and conceptual analyses that we have dealt with repeatedly above. 'To describe a thing or a process' means to state how certain positions are occupied, and such specifications can be extended *ad indefinitum*. However, 'to *describe a type'* means *'to give a definition of a concept, which displays a specific number of firmly delineated features and besides this, an undetermined number of undetermined features, which taking additional experience into consideration can be determined at a later time.'*

According to this statement, from which it follows that the description of phenomena is no less empirical than their explanation, the relation between these two operations can be described as follows: the subject of descriptions is the *individual* qua *individual*; they indicate that specific places are occupied in a specific manner. An explanation, on the other hand, incorporates individual phenomena into the order of 'general relations', i.e., it pinpoints certain *phenomena in the environment* of the fact to be explained – mostly those prior in time – with the attribute that phenomena of this kind are 'generally' linked with the facts to be explained, in the same way.

To be sure, this *caesura*, which is analogous to that between fact and law, is made relative by the fact that qualities, strictly speaking, contain implicit reference to the general relations of experience[68]; nevertheless it is essential for scientific-theoretical analysis, which must always presuppose a certain relatively fixed sediment of pre-predicative experience. This will have to be taken into consideration in what follows, especially in our analyses of concept formation in the social sciences.

[68] Compare above, section "Basic Philosophical Considerations".

4. Life and Consciousness

The debates concerning the relation between *inanimate* and *animate* nature, as well as between the *physical* and *psychical*, can demonstrate especially forcefully the confusion that arises from the intermingling of factually different strata of problems, although from the viewpoint of the psychology of knowledge they may belong together. We wish to clarify this initially in terms of the first-named range of problems, which form the subject of the *debate on vitalism.* Within the framework of the present work this controversy deserves special attention, because the conceptual motif that has come to occupy an increasingly central place in it, namely, the psycho-vitalistic argument, also plays an important role in the controversy about method in the social sciences. The following presentation will therefore seek first to show how most of the vitalists' other arguments had to be discarded as untenable.

In a first approximation, the question under discussion in the *vitalist debate* can be made more precise as follows: are vital phenomena *sui generis* or are they merely highly complex physico-chemical phenomena, that is, phenomena explicable entirely by the methods of physics and chemistry? Vitalist doctrines affirm the former alternative, mechanist theories the latter.[69]

Now when we examine the *vitalist argumentation*, we will have to ascertain first that these by no means seek to deny that physical and chemical processes *also* play an important part in explaining the phenomena of life. No vitalist who is to be taken seriously will deny, say, that studies of the optical structure of the eye, the lever activity of the joints, diffusion and osmosis, and the chemical analysis of organic compounds, and the chemistry of colloids have made substantial contribution to the investigation of vital processes. What the vitalists deny was that this kind of research was *sufficient* for an understanding of vital processes. In order to substantiate their claim, the vitalists offered various arguments. We shall refer first to the older and now outdated ones.[70]

The first argument concerns the *complexity* of organic compounds. During the early decades of the nineteenth century it was asserted that chemists would never be able to produce organic matter artificially (synthetically). A special *vital force* was thought to be required, an agent the creation of which is beyond human capacity. This theory was refuted in 1828 when Wöhler produced synthetic urea. Soon other organic substances, like acetic acid and alcohol, and, recently, hemin, a principal component of hemoglobin, were produced synthetically.

Nevertheless, vitalists did not admit defeat even though they had to make suitable modifications of their thesis. They now argued that the *fact* of synthetic production of organic substances in the chemical laboratory was to be conceded, but that the *manner* in which these syntheses occurred was entirely different from their production in nature. The classic example adduced was the transformation of carbon

[69] The expression 'mechanistic' in this context has come down from a time when it was believed that 'mechanistic' could be used interchangeably with 'physical'.

[70] Here we follow closely the excellent exposition in Bavink, *Ergebnisse*, p. 293 ff.

dioxide into sugar, of particular importance in plant life. From the fact that yeast was indispensable in this process it was concluded that it contained the vital force required to start the process. But this assumption also proved untenable. Zymase, isolated from dead yeast, is able to bring about fermentation. It is now established that catalytic effects are involved, and, although the role of catalytic agents in chemistry is in need of further explanation, every chemist knows that catalytic processes are by no means confined to the vital sphere.

Besides this first *biochemical* group of vitalists, there is a second group that supports its thesis by reference to the *specific mode of functioning* of living bodies or their constituents, cells. They stress nutrition, growth, reproduction, and response to stimuli. In opposing them the *mechanists* tried to provide mechanical analogues of these processes. To this end, great ingenuity was exercised in the construction of 'artificial cells', hardly distinguishable from natural cells with respect to observable moments. The analogies, to be sure, cannot be carried very far, because the vital processes copied in this way certainly do exist in different interconnections, but they are not without significance, since they caution the vitalist against hastily claiming certain kinds of phenomena as peculiar to life processes.

We cannot discuss in detail here the vitalists' various other arguments, which are intended to call attention to the special features of life processes, for example, their *irreversibility* or the phenomenon of *death*.[71] By way of summary we can state that as far as the authoritative vitalists among biologists and philosophers are concerned, this kind of argument has receded more and more into the background, as almost every assertion of this kind was experimentally or theoretically refuted by the mechanist opponents. By contrast – and here we return to our point of departure – more recent vitalism does not see the characteristics of life in certain individual phenomena, but rather in *complexes of phenomena* to which, accordingly, only a *teleological approach* can do justice. As the most eminent representative of this *neovitalist* tendency (which for reasons to be indicated immediately below can also be called *psycho-vitalism*) we should mention Driesch, whose well-known arguments for the autonomy of life form the core of vitalist doctrine today.

Among these arguments we should mention first the principle of *'organic regulation'*. Using experiments on sea urchins' eggs, Driesch showed[72] that fragments of widely differing form and size were able to regenerate complete and typical organisms. This seemed to provide a significant example of *teleological* regularity, for the initial states might vary to a great extent and yet the developmental idea 'residing' in the sea urchin always led to the same result. On the other hand, Driesch argued, we cannot conceive of a machine that, after small parts had been removed, could restore its previous form and function by utilizing surrounding materials.

---

[71] *Ibid.*, p. 315.

[72] Especially in Driesch's Gifford Lectures delivered at Aberdeen University, 1907–1908 and published first in English as *The Science and Philosophy of the Organism* (Aberdeen 1908–1909 and later in German as *Philosophie des Organischen*. See Section A, Part I, B: "The experiments on the egg of the sea-urchin", pp. 59 ff. of the English edition. Kaufmann cites the 4th German edition (Leipzig, Engelmann, 1928).

Concerning this argument Bavink remarks quite aptly[73]:

> We know only so much from these experiments, namely that organizing effects are exerted by certain groups of cells, particularly in the vicinity of the blastophore upon other cells, and that this appears to take place according to quite strict laws, so that the result may be calculated ahead if the conditions are correctly controlled. This whole thing by no means gives the impression of purposiveness, but rather of purely mechanical causality. On the other hand, it seems just as certain that we cannot imagine machines, which, as Woltereck says, 'stimulate themselves, regulate themselves, construct themselves, gear themselves up, and multiply.' The very idea is absurd, as is clearly shown by increasingly close observation of the incredibly complex relations within an organism. However, when examined more closely, this only proves that we can by no means compare an organism with a *machine*; it does not prove that organisms cannot be understood in terms of physics and chemistry, for the physico-chemical 'sphere' includes things other than machines. There are systems of so-called 'mobile equilibria' in physical chemistry that constantly maintain a certain state of equilibrium in spite of continual changes in the relative positions of particles. They also reconstitute this state after the particles are removed. We may doubt that this comparison helps us understand the nature of organisms better than the comparison with a machine (there are many things both for and against this) but in any case this illustration does prove that the failure of the machine comparison by no means allows us to conclude that nothing at all can be achieved here with physics and chemistry. In order to maintain this, we would have to know far more about the possible physico-chemical relations within these complicated organic systems than we actually do know.

Driesch considered his '*proof from the genesis of equipotential systems*' (*heredity*) a second conclusive argument for the *autonomy* of life.[74] He thought it absurd to suppose that a machine could divide itself into several parts from each of which a whole could be regenerated. Here too there must be regulative forces guiding biological processes toward a predetermined end.

Bavink's objection to this thesis[75] seems altogether appropriate to us and therefore we want to quote it verbatim:

> Driesch again incorrectly ties this mechanism to the image of the machine in order to show then that this will not do. We can readily conceive of any number of physico-chemical systems which, when divided again and again, still retain their 'Gestalt', that is, they always retain the relative disposition of their parts toward one another, even though we cannot conceive of any such 'machines'. The physico-chemical systems mentioned above, a suspension, for instance, fulfill this condition too. To be sure, they do not divide themselves of their own accord. But that is not the point. Lehmann's fluid crystals, on the other hand, very clearly showed the capacity for fusion in the reconstitution of the original form (the counterpart to the division of cells) that was realized in Zur Strassen's experiment. It is quite evident that here and in the problem of morphogenesis we are basically faced with only one problem, for, as we have already remarked above, the 'gene' of heredity and Spemann's 'organizers' are in all probability siblings. We know just as little about how they carry out their morphogenetic activity as about how they divide quantitatively in chromosome division, while preserving themselves qualitatively, obviously in order to replenish themselves again quantitatively. However, as Goldschmidt in particular has shown in detail, we can readily invent hypothetical physico-chemical systems that display similar behavior. We need only

[73] *Ibid.*, p. 373f.

[74] *Philosophie des Organischen*, Abt. A, Teil I, ID.

[75] *Ibid.*, p. 373.

to assume that, on the one hand, substances are present in such systems which possess the capacity for assimilation in the sense mentioned above, and that, on the other hand, as a consequence of the changing equilibrium state, which varies with the relation of surface to volume, division necessarily takes place when a certain magnitude is attained. Such a system already possesses a whole series of properties, which, according to experience, are attributable to living cells. Of course it would be saying much too much if, as a result, we tried to maintain once again (as the old mechanists did) that everything had thereby been explained in physico-chemical terms. However, such a consideration is in any case sufficient for us to understand that Driesch's second proof is just as unsound as the first.

The general, fundamental conception, found not only in Driesch but in almost all neovitalist doctrines, maintains that life phenomena must be explained in a manner such that the principal insights of physics can also be exploited to explain life processes, but that a specific vitalist principle must be invoked *with* these, and that means a *teleological principle.* But now, just as classical physics attaches preeminent significance to the principle of the conservation of energy, so the main thrust of vitalistic attempts at explanation has been directed toward making the distinctiveness of life processes comprehensible in such a way that active, vital factors intervene in the physico-chemical processes underlying the life processes *without changing the balance of energy.*

Sound objections can also be raised against this conception, but we cannot deal with them in detail within the present framework.[76] Nor we can deal with the transposition of the vitalist controversy into the sphere of genetics and thus take up questions concerning the origin of life, where the following two alternatives can be stated: Is life as old as the world? Or did life arise from inanimate matter through *spontaneous generation* (*generatio aequivoca*)?

In the present work, however, we cannot pursue biological problems for their own sake. What especially interests us are the difficulties in the vitalist controversy that arise from a lack of precision in posing the question. For the question which forms the main theme of the vitalist controversy – "Can the processes of life be explained completely by means of the laws of the sciences of inanimate nature?" – lacks precision in two respects: First, the concept of explanation as such requires supplementary clarification as long as it is not specified as to what the explanation is to achieve or, in other words, under what conditions an explanation is to be regarded as adequate. In relation to our problem the following alternatives emerge: Should we demand of the explanation only that any given variations of the life phenomena to be explained be coordinated with certain variations of physico-chemical phenomena in such a way that for every vital change a corresponding 'cause' in inanimate nature be shown? Or should we also require of the explanation that it specify the path pursued by the *synthesis of life* from inanimate matter, in order to produce thereby the experimental proof that physico-chemical fasts are not only the necessary, but also the sufficient conditions for life? A similar question arises within physics itself as soon as one asks to what extent a physical explanation contains the possibility of experimental confirmation. As it well known, the latter is not possible

[76] Compare Bavink, op. cit., p. 374 ff.

at all for a large number of astronomical hypotheses, particularly in the astronomy of fixed stars or only very indirectly. The mechanists can point to the fact that a considerable number of explanations (in the first sense) exist for life processes, while against this the vitalists can argue that the generation of life *solely* from inanimate substances has not been achieved until now.

The second point for which the formulation of the fundamental vitalist problem offers no resolution is the question of whether the reduction to physico-chemical laws is supposed to refer to presently known (presently valid) physico-chemical laws, or to *possible* laws established at some future date. This distinction is of special importance for what follows as it produces a clarification which affects the form assumed by most discussions of the vitalist debate.

The problem corresponding to the first alternative can be rendered more precise as follows: Can physiologico-biological laws be derived from the general principles of physics in the same way that special physical and chemical laws are derived from them? Or to put it another way, does a hypothetico-deductive system, dominated by the principles of modern physics also embrace physiology and biology? Now one can deny this and still hold that there is just as little reason to draw a sharp dividing line between physiologico-biological research and research in the physico-chemical sphere, as between physics and chemistry, although for a long period during the development of those two sciences such a division seemed likely. Here one could argue, say, in favor of the assumption that the laws of physics as presently known are still not general enough to embrace the life processes, and yet entertain the hope that the progress of science will bridge the gap existing between these two domains today as happened in the case of physics and chemistry, or in the subdisciplines of physics, for example, in mechanics and optics.

But the second alternative does not in fact refer to the reduction of physiologico-biological *laws* to physico-chemical *laws*, but rather, as one will recognize upon closer examination, to the reduction of physiologico-biological *concepts* to physico-chemical *concepts*[77] In order to understand that, we must reflect on the following: When we describe a biological process, for example, metabolism, reproduction, fertilization, or growth, as is done in biology textbooks, we find that as a rule such a process is characterized as an event in the external world. (Thus fertilization is described as the spatial union of the spermatozoon with the ovum). So, looking at this from a purely morphological point of view, the life process characterized is not distinguished from any of the other events in space and time that are subject to explanations in terms of physics. However, on the other hand, there can be no doubt that the neovitalist approach – and that of older vitalism too, insofar as it operates with a specific vital force – alters our perspective considerably by introducing an element which, even if it is not inherent in every single elementary vital phenomenon, still inheres within the nexus of these phenomena. That Reinke's *dominants* and Driesch's *entelechy* are not physical forces is an assertion essential to the basic

[77] For this compare Carnap, 'Die physikalische Sprache als Universalsprache der Wissenschaft', (The Physical Language as an intersubjective Language) [Eng.tr. Max Black, *The Unity of Science*, London 1934, pp. 52–66], *Erkenntnis* II, pp. 432–465, p. 449.

position of neovitalism. Given this ambiguous perspective, it is indeed understandable why neovitalism tries principally to explain the role of vital factors in such a way that they become purely regulating factors, as it were, merely filling in the gaps in physical lawfulness. In this way it tries to eliminate the conception that extra-physical forces intervene in physical lawfulness.

Thus one sees that the 'philosophical' questions, strictly speaking, in vitalist debate do not raise the issue of whether, given the present state of research, the integration of biology into physics and chemistry can be carried out, or better whether important indices for such an integration already exist, but rather whether this is possible in principle, or whether such an integration would appear to be excluded from the outset because the categorical apparatus of physics does not contain the fundamental concepts of biology.

However, as meaning can only be linked to terms in general (and hence to 'dominants' or 'entelechies') by coordinating models of experience with them, we can readily understand why every consistent vitalism that maintains the inadequacy of the categorical apparatus of physics for biology must become a *psycho-vitalism*. For if the data of external experience are included within the system of physics and if no sources of knowledge exist other than external and internal experience, then any phenomenon that cannot be grasped by means of external experience, or only inadequately so, can only be characterized with the help of internal experience, unless, of course, it is seen to be transintelligible and thus beyond all scientific knowledge. Thus it is obvious that vitalism characterizes the relation between the physico-chemical and the specifically vital factors involved in life phenomena in terms of an analogy with human behavior, even though this analogy, insofar as it is explicitly formulated at all, appears generally to be linked with certain reservations.

Let us take human speech acts as an example. These acts also have their physical side and can be described and explained coherently in physical terms. But in spite of this, the description or explanation cannot be conceived as description (explanation) of speech acts as speech acts. Rather, such a description will have to make reference to the *meaning* of what is said, that is, to the *data of inner experience.*

At this point we are obliged to define clearly the relation between the *causal* and the *teleological* method,[78] which is of the greatest significance for the range of problems involved in the vitalism debate. To begin with, we must state here that the juxtaposition of causality and teleology as opposites is misleading insofar as the causal approach is *contained* in the teleological. In analyzing purposive action we can see two causal relations: (1) the causal relation between consciously posited facts (means) and the facts resulting thereby (ends); (2) the causal relation (motivation process) between the, possibly emotionally charged, conception of ends and the positing of means. It is especially important to keep these two causal sequences distinct; for in the mixing of 'end' as the content of our conceptions and 'end' as a real phenomenon that becomes still more obscure through the reification of cause and effect lies one of the main roots of dangerously mistaken metaphysical and

[78] In the *Methodenstreit* in the social sciences, this question plays an important role.

scientific (pseudo-scientific) doctrines in which it seems that the real end is the 'creator' of the means.

However, frequently juxtaposed to the '*subjective end*' just characterized is an '*objective end*' that does not arise solely out of the positing of ends, i.e., out of the purposes of certain persons, but rather is claimed for all human endeavor and action as something to strive for. This normative (ideal) quality of the end as given then appears as cause with respect to the human action directed towards its realization, so that here, *prima facie*, the causal sequence (or its recognition) seems to be reversed, since it is apparently by means of the as yet unrealized effect that its causes, namely the striving toward realization and the actions proceeding in this direction, are realized. This is the *Aristotelian* notion of *entelechy*, which again plays a big role in recent speculative philosophy, especially insofar as it is oriented toward biological facts (Driesch). In what follows we shall see that the idea of the objective ends (goals) as a given has various roots, of which the most important lie in the realm of social conventions and codifications. The objectivity attributed to them is conceived as a validation of values that exist independently of the will of the individual. While the problems connected with it will therefore have to be treated in the next section, where we shall analyze the concept of value, we can already state here that, on the basis of the insights just mentioned, one can by no means speak of an inversion of the causal sequence, as the conceptualized 'end' does not correspond to the end actually realized.[79]

Now we will best understand the sense of the controversies that have developed with respect to the application of the causal or teleological method, if we first try to determine what *attainment of knowledge* is *reserved* for the teleological method, according to its proponents, as opposed to the causal method. Here we want to distinguish two cases. The first case occurs when, in explaining facts by means of the teleological method, we do *not know* (for the time being) the *end* that is supposed to be served by the facts to be explained, rather we only assume that the objects in question are purposive creations or that the processes in question are goal-oriented. For example, that is the situation of the archeologist who in the course of excavation comes across an object he assumes is a human creation (*artifact*) without, however, knowing more specifically the purpose it was supposed to serve. On the other hand, in the second case, one is *also* supposed to know the purposes which the objects serve or according to which the processes occur.

To begin with there is no such thing as 'absolute' utility. Rather, utility is always related to more or less specific ends. The concept of utility with respect to an end is a *relational concept* and consequently one requiring completion. Therefore the heuristic function of any hypothesis concerning utility can only lie in making guesses or hypotheses as to what the basic notion of purpose might be, based initially on certain morphologically accessible traits of the object in question, and then we can verify whether the remaining traits fit this interpretation freely. A similar thought process is also in evidence when we draw conclusions about the frame of mind of

[79] For this, compare the detailed discussion in the next section.

the actor from certain traits of an event interpreted as human action and we then test it to see whether this frame of mind manifests itself in other traits of the action.[80]

From what has been said, one can already recognize that as far as the division between the causal and the teleological method is concerned, the role played by inner experience in the latter is very significant. Of course, such a division presupposes first that the concept of causality has been limited to that of causality in nature, i.e., the relations between spatial-temporal events; still, this is almost always the case in pertinent debates on method. If this is not done, then the teleological method is only a special case of the causal method, and as will be shown later, understanding is only a special case of explaining. Therefore the greater achievements which the teleological method can attain, as compared to the causal method, can consist only in the contributions that inner experience makes to it.

Of what do these contributions consist in the two cases we have differentiated, or rather, as we can say now, in the *two stages* of teleological explanation? The role played in the first case by inner experience is roughly the same role played in the empirical sciences by the assumptions that world events are lawful events from which, accordingly, inductions can be made. The fact that a thing is regarded as an artifact, or a process as goal-directed, without one first knowing what the goals are means nothing more than that one is *looking for* goals and that one will therefore seek to interpret the individual parts of a thing (of a machine, for example) or also the individual phases of an event as symptoms of such purposefulness. In a cosmology oriented to a theological statement of faith, all events can, in principle, be comprehended as symptoms of the idea of the plan of creation, and thus of the world's purpose; and then from the results of such considerations, further inference will be drawn concerning predictions about future world events. An especially striking example of this is astrology. But here too it must be carefully borne in mind that nothing can yet be stated about the world from the idea of a world plan as such. The contrary impression arises only because as a result of other considerations, the plan is thought to be already characterized in some specific manner.

However, once one has arrived at a complete specification of the ends and an ordering by rank of the means (for example, the development of the greatest possible simplicity), then no further reference to inner experience is required. The assertion that certain processes take their course in a purposive (goal-directed) manner means nothing more than that certain effects are attained under certain conditions. But this state of affairs must not be interpreted as if there were a lawfulness of ends as well as a lawfulness of means. As soon as one has freed oneself from the attendant anthropomorphic conceptions regarding the causal and final forces, one will see that in principle there is no difference between causality and finality. Thus the two statements: "the inclination of the two optic axes toward each other serves for binocular vision," and "without the inclination of the optic axes toward each other, binocular vision would, *ceteris paribus,* be impossible", mean the same. Let us clarify what

[80] Compare Kaufmann *Die philosophischen Grundprobleme der Lehre von der Strafrechtsschuld* (*The Fundamental Philosophical Problems of the Doctrine of Guilt in Penal Law*), (cited hereafter as *Strafrechtsschuld*, Vienna 1929, p. 89 ff.

has been said with the important example of the relation between *organ structure* and *organ function.* Certain functions of organs, i.e., modes of behavior, are known and from this inferences concerning their structure are drawn. It would be wrong to treat this cognitive process in terms of causality, as if cause were here explained by effect, while in causality, on the contrary, effect is explained by cause. The root of this erroneous interpretation lies in the fact that peculiarities of organ structure can often be *discovered* because certain determining grounds are still required to explain the function. However, this kind of interpolation of causes is by no means confined to the vital sphere but is, rather, a general principle of research.

Now we are able to take a position on the frequently raised question especially in Kant's investigation of teleological judgment in his *Critique of Judgment*[81] as to whether ends are *found* in phenomena or *placed* in them by us in our thought? Here the following is to be noted: What we 'find there' is a complex of especially close mutual interactions of the parts of an organism or its spatial substrate, and the fact that to a great extent the causes of various 'functions' can be spatially localized. For this reason, an approach that takes the functions as its starting point and from these functions draws inferences with respect to the morphology of organs has proven especially fruitful. In principle, we are faced here with the same state of affairs as in the general question, namely whether laws are contained in the phenomena of nature or whether they are placed there by the thoughts of the human mind. As we have seen, every law is a hypothesis (in the broadest sense), and thus an ensemble of assumptions; its objectivity, however, lies in its real verification, i.e., in the fulfillment of predictions directly or indirectly based upon it.

What we have just explained about the character of the teleological method, by using the example of the relation of organ structure to organ function, can also be applied to an analysis of the idea of the universal goal that is supposed to govern all vital events, namely the *preservation of the species*. Here we must ascertain that the following two statements mean the same: 'The specific behavior of an animal or plant organism serves the preservation of the species in question' and 'Without that behavior the species in question would, *ceteris paribus*, be doomed to an early death'. The accuracy of such a statement is verified by testing experimentally or at least in a thought experiment, the consequences that a variation of behavior would have for the preservation of the species.

We can now summarize our reflections in the following *conclusions*:

The idea of a teleological method implies that inner experience helps to explain phenomena, and the distinction between the teleological and causal methods is based on the '*use or non-use of inner experience*'. However, there is no such distinction between the causal and teleological approaches; the latter is rather a specific kind of causal aspect which can be compared with other such aspects in terms of its heuristic usefulness.

Finally, we must still ask under what circumstances we would draw upon inner experience to explain certain *phenomena of the external world*. As we know, primitive people tried to interpret phenomena by using an analogy to inner experience, as

[81] In his investigation of teleological judgment.

Lévy-Bruhl and others have shown, but to their increasing disappointment they soon became aware that no reliable *predictions* could be made in this way.[82] The apparently trivial answer to our question is that the inclusion of inner experience in an interpretation, i.e., the assumption of a psychical agent, will be done where experience has proven it valid, but it is precisely modern animal psychology that has shown that we must be very careful in setting up psychical analogies.

Because of the dominance of psycho-vitalist arguments on the vitalist side, the vitalist debate is closely connected with the second great debate, which we will have to treat in this section, namely the debate that has raged about the so-called *psycho-physical problem*. An analysis of this problem shows that because of a similar confusion in the various levels of questions such as we identified in the problem of vitalism, research has been complicated to a considerable degree. In this context, then, the two questions that have often been entangled need to be distinguished:

1. Can the laws that link psychical facts with one another be reduced to *laws* about processes in the external world and thereby ultimately to laws of abstract natural science (physics)?
2. Can the concepts of psychical facts be reduced to the concepts of physical facts and then further to the concepts of physics? (The genetic paraphrase of the question plays a special role here: how can the psychical arise from the physical?)

This manner of posing the problem and its pertinent treatment contains a peculiar mixture of legitimate scientific requirements for the greatest possible continuity in following the course of events as well as the most complete specification possible for laws of coordination (laws of covariance) between the physical and the psychical *and* the old speculative dogma that the effect must be contained in the cause. The analogy with the seed of plants and animals provides a familiar example. From the two theses – 'the physical and the psychical are essentially different' and 'the essentially different cannot affect each other', together with their corresponding antitheses – we can form four basic positions with respect to psycho-physical problems:

1. the psychical and the physical are essentially different and do not interact (doctrine of psycho-physical parallelism);
2. the psychical and the physical are essentially different and do interact (doctrine of psycho-physical causality);
3. the psychical and the physical are not essentially different, because the psychical can be reduced to the physical, as it is 'essentially' physical (materialism);
4. the physical and the psychical are not essentially different, because the physical can be reduced to the psychical as 'essentially' psychical (spiritualism, absolute idealism).

The leading exponent of the doctrine of *psycho-physical causality* in modern philosophy is Descartes, who, to be sure, maintained a sharp distinction in essence

[82] Compare especially Levy – Bruhl, *Das Denken der Naturvölker* (*Thought of Primitive Peoples*), translated from the French by Friedländer, 2nd ed., Vienna 1926.

between the *substantia extensa* and the *substantia cogitans*, but in spite of that assumed an *influxus physicus* through the pineal gland. However, precisely because this thesis of the *influxus physicus* could not be consistently accommodated within the great Cartesian system, it was soon regarded as a solution adopted only out of embarrassment. Thus the Occasionalists, Geulinx, Malebranche and others, who depended on Descartes to a great extent, founded *the doctrine of parallelism*, the leading proponent of which was Leibniz.[83]

Materialism flowered in France in the eighteenth century and in Germany around the middle of the nineteenth. The doctrine of behaviorism, closely related to materialism, has gained in significance in the last two centuries. It found its logically sharpest formulation most recently in *Physicalism* (Neurath, Carnap).[84]

Finally, *Spiritualism* had its most important representative in Berkeley. Mach's positivism which for the most part differs from Berkeley's 'absolute idealism', still shows an affinity with it with respect to the psycho-physical problems, insofar as it conceives of things of the external world as complexes of psychical elements (sensations) and consequently, in the absence of an essential difference between the physical and the psychical, considers the psycho-physical problem non-existent.[85]

If now, after this sketchy survey of the four main tendencies in the treatment of the psycho-physical problem, we take a position ourselves, we must first of all remove the *speculative prejudices* contained in the theses and antitheses we have presented. In line with this we must first establish that, as soon as the speculative idea of causality as an active force is abandoned, the conflict between the doctrines of psycho-physical causality and parallelism turns out to be just a pseudo-opposition, because 'interaction' is nothing but a 'lawful correlation between groups of phenomena.' For the modern conception of lawfulness, therefore, the contrast 'parallelism' and 'interaction' loses its meaning, because what always remains decisive is that the phenomena of one group can be determined (calculated) once we know the phenomena of the other group. It is obvious, that in the sense of the contemporary conception of causality, there exist relationships between physical and psychical phenomena: take the anxiety attendant upon cardiac arrhythmia as an example of the effect of the body on the psyche (from the physical to the psychical) and resultant action on the part of the will as an effect in the opposite direction [from the psychical to the physical].

Nonetheless, the conflict we have mentioned still plays a great role today in the *Methodenstreit* in psychology and the sciences of mind which is understandable in view of the following considerations. From the assertion that no causal relationship

[83] Well-know is the analogy of the clocks – which incidentally can already be found in Geulinx. The synchronized course of two clocks can be explained either through the mechanical dependence of the one on the other, or through continual regulation, or also through the circumstance that both were set at the same time to begin with and have the same motion. Leibniz considers the third possibility to be the one that is realized in a pre-established harmony). The same would hold for the relation of body and soul.

[84] Compare below. Part Two, section "The Social Sciences and the Natural Sciences".

[85] Compare his *Analyse der Empfindungen* [Eng.tr. *Analysis of Sensations*, (1914; new ed. 1959)], 6th ed., Jena 1911, especially pp. 1–30 and p. 289 ff.

exists between psychical and physical phenomena it follows, persuasively, that physical facts are only to be explained in terms of physical facts and psychical facts in terms of psychical facts. The first of these postulates has become so dominant for the natural science of the last centuries that it is regarded as 'self-evident' today among natural scientists, while the corresponding postulate for psychical facts stands at the center of a scientific controversy. After all that has been said in the preceding sections, there is no particular need to substantiate the fact that what is involved in both instances are proposals for procedures, hence a suggestion for *regulations* [*Festsetzungen*] (conventions), which are wrongly presented in the guise of *statements* [*Feststellungen*] (knowledge).[86]

Reflections concerning the problem of the *essential difference between psychical and physical phenomena*, hence the question of the relation of psychological concepts to those of natural science, go further. This much is clear at the outset: the assumption that there is a difference *toto coelo* between the two groups of phenomena, and especially between stimuli and sensation, requires revision. For as soon as we have freed ourselves from naive realism and have realized that the 'nature of the world' cannot be comprehended in any way other than through an analysis of experience, then it becomes clear that the things and phenomena of the external world are woven from the material of experience. To be sure, Mach's doctrine, according to which things are complexes of sensation, is not tenable, and this is because he does not take at all into account – or only very inadequately – the spontaneous element in knowledge. He therefore arrives at an inconsistent conception of the *I* (the thinking subject), which has since been repeatedly criticized quite aptly.[87] In this regard, a more thoroughgoing analysis will have to start from the fact that the *physical is regarded as a correlate of external experience* and the *psychical as a correlate of inner experience* and that therefore the relation between the physical and the psychical is to be characterized in conformity with the relation between external and inner experience. Here, however, pertinent analyses prove especially difficult, not only because it is not possible to ascertain the absolute primacy of one or the other of these two kinds of experience, but also because they cannot be isolated as being independent of each other. Not only does inner experience refer back to inner experience, but external experience refers back to inner experiences as well.

First, with regard to *external* experience as an *element involved* in inner experience, we can easily see that not only memory and reproduction but also free imagination refer back, with respect to their contents, to facts derived from experience of the external world in the narrow sense, thus to sense data or to facts drawn from the experience of one's own bodily states (feelings). This insight forms the central theme of the doctrine concerning the relation of sensation and reflection developed by British sensualism [sense-data empiricism], in which the primacy of sensation over reflection is declared. This is also taken into account in the above-mentioned designation of the data of perception as the data of *constitutive* experience, in the extraordinary refined analyses of Husserl.

[86] Compare above, section "Basic Philosophical Considerations".

[87] Compare, for example, B. Hönigswald, *Zur Kritik der Machschen Philosophie,* Berlin 1908.

On the other hand, though, it should not be overlooked that all external experience, insofar as it identifies and differentiates objects, already *presupposes* the possibility of retention and reproduction, thus, of *inner* experience. From this it follows again – in conformity with the results of the reflections presented in the first section– that the appropriate way of treating the psycho-physical problem consists in unearthing the various *strata of experience.* Here insight into the interrelations of the strata is gained by ascertaining their reciprocal implications. The most difficult part of this task, to which Husserl devoted himself especially during his last years, lies in differentiating the various strata of the concept of time, which have to be taken into consideration in this investigation.[88]

All this already points to the correct conception that the physical and psychical cannot be coordinated with each other, as if an independent physical object corresponded to an independent psychical one, but that the physical must be contrasted with the *psycho-physical*. Descartes' ideas, to the effect that thinking about the '*ego cogito*' involves a pure *substantia cogitans*, apparently – but only apparently – contradict this conception. For if we consider that the *ego cogitans* is at the same time the *ego agens*, then the appearance of this independence vanishes. Then we recognize that the disengagement of the psychical from the complex of the psycho-physical is, when considered formally, an abstraction not unlike the abstracting of a color from its embeddedness in physical objects.

To make this clear requires a profound and difficult analysis of *thinking*, which appears as the psychical in its purest form. We cannot deal with this in detail, but must confine ourselves to indicating the decisive point in a few words.

In the sense of the findings just presented, cognitive acts are never considered independent, but belong rather to a psycho-physical unity as acts 'carried out' by this unity; as already emerges from our reflections in section "Basic Philosophical Considerations", the nature of cognitive acts is such that they comprehend an 'object', something beyond consciousness, something transcendent. In the perception of a house, in thinking about one's mother, the 'perceived house', the 'mother thought about' are by no means components of the cognitive acts that refer to them. Rather, it is precisely in these acts in which they are comprehended that they are comprehended as independent of the acts. Franz Brentano, who was the first to make this essential state of affairs clear, invented the term '*intentionality*' for it, in order to express the 'directedness' of the act towards its object. Since then this designation has prevailed in philosophical terminology. The concept of intentionality, the elaboration of which may well be called one of the most important results of more recent philosophical thought, today forms one of the crucial points for the descriptive analysis of psychical phenomena.

[88] Until now [1936], of Husserl's profound and comprehensive analyses only the lectures of the year 1904 have appeared under the title *Edmund Husserls Vorlesungen zur Phänomenologie des inneren Zeitbewusstseins,* ed. by Martin Heidegger in *Jahrbuch für Philosophie*, vol. IX, Halle a. d. Saale 1928; however a large work devoted to these problems is in preparation with the collaboration of Eugen Fink [Cf. Part One, section "Basic Philosophical Considerations"].

The analysis of an act can either be directed toward the act as a whole, thus determining above all that it is an event taking place in the inner time of a certain human being; or, it can disregard these real aspects of the act and limit itself exclusively to its *content* (meaning), thus considering only the intentional (objective) aspects of the act. We have sufficiently clarified how this is to be understood in the investigations of the concept of the meaning of judgments in section "Basic Philosophical Considerations" and section "Logical-mathematical Thought". Accordingly, wherever one believed that the psychical was being comprehended independently, the isolation of the contents of thought had to be borne in mind as guiding conception. Also included in this *isolation*, as we have determined, is an *idealization*, namely, the *presupposition of completely clear thought* (whose significance for logic we established above), so that confused thought appears linked to the body, while completely clear, 'pure' thought is assumed to be free of all connection with the body. Here is one of the essential conceptual themes in the treatment of the mind-body problem in neo-Platonism, in scholastic philosophy and in Leibniz's doctrine of the monads.

These considerations will also lead to an understanding of the *emotional* acts of consciousness, i.e., those laden with *feeling,* for example, those of the *will*, where, under 'feeling' in the narrowest sense, the experience of one's own bodily states is to be understood. What is, strictly speaking, psychical in the will is the *projection of one's behavior.* Hence the question, what is being willed? is nothing other than the question, what sort of behavior is the individual thinking about realizing? The deep emotion of the will and the causal relation in which it is brought to 'external action' allow the psycho-*physical* character of this act to emerge more strongly than in the case of pure thought, where the bodily state does not appear to be brought out in experience.

This has a very important, direct bearing on the theory of the social sciences in that the opposition between the will as a psychical phenomenon and action as a physical phenomenon is incorrect. *Will as well as action* are *psycho-physical phenomena.* A sharp distinction between the two only occurs, first, because action – or more correctly, the bodily motions belonging to action – can be directly observed by an *alter ego*, while the will cannot; and secondly, only actions produce direct effects in the external world. For this reason actions are usually characterized in terms of their typical effects. It is especially important to recognize clearly that action is never solely an external process. Thus that it can never be interpreted simply as a physical process. The opposite conception has created a great deal of confusion, with considerable practical consequences. Thus, the really unclear distinction, which plays such a great role in criminal law, between the external and the inner facts of a case are closely related with this.[89]

The insight that the psycho-physical presents itself as a unity should also further our understanding of the phenomenon of the *unconscious*, which has recently been subjected to systematic exploration especially by *psychoanalysis*. For the more recent 'philosophy of life', which found its most profound expression in the works

[89] Compare Kaufmann, *Strafrechtsschuld*, p. 83 ff.

of Bergson[90] and was elaborated in the last decades particularly by Scheler[91] and the *philosophical anthropologists*[92] influenced by him as well as by Bergson and in part by Kierkegaard[93] this insight forms one of the most important conceptual themes. From such analyses the theory of science can hope to gain progressive clarification of the fundamental epistemological question of the relation of bodily experience to external experience on the one hand and to inner experience on the other. We have to forego even the sketchiest presentation of the central thought of Bergson, Scheler, and the younger philosophical anthropologists. We can do so because these conceptions, for the time being, have had only a limited effect on the treatment of methodological problems in the social sciences. In any case, recently a *soul-mind dualism* has emerged, where the soul is conceived as a sort of center of life *(vital soul)* and has been opposed to pure mind. Here the concept of mind is generally understood as a concept with *value content*, and thus all the prejudices enter into it that are linked to the value concept in general.

5. The Concept of Value

In treating the problem of the 'nature of value' we wish, to begin with our considerations concerning the transcendence of the world. Carrying out our critique of naïve realism, we designated this as *'immanent transcendence'* in order to emphasize the relatedness of all being to consciousness, and have recognized that even so-called ideal objects participate in this relatedness of consciousness, that therefore naïve Platonism (for which, to be sure, Plato himself cannot be made fully responsible) is subject to the same objections as naïve realism.

Now it will be seen that the pre-critical conception of absolute transcendence, and the conclusions drawn from it relating to the character of knowledge, find a precise analogy, in a great part of value theory, as transmitted in the history of doctrines, and even in its modern form. If –according to this conception – the human *sensibility* is 'affected' by transcending things, by which a more or less faithful copy of what has Being is produced, so here too, allegedly, human *feeling* is affected by transcendent values, the comprehension of which, according to this doctrinal view, is presented as passive acceptance. Just as the evidence of perception serves as assurance of the existential truth, so, correspondingly, the evidence of feeling serves as assurance of the truth of value. Therefore our critique of the principles – especially those of the

[90] Especially his *Évolution creative,* [Eng.tr. *Creative Evolution*; also appearing in German under the title Schöpferische Entwicklung] has had an enduring intellectual influence.

[91] A brief summary is contained in the work published shortly before his death, *Die Stellung des Menschen im Kosmos* (*Man's Place in the Cosmos*), Darmstadt 1928. There are said to be extensive investigations on philosophical anthropology in the papers Scheler left behind.

[92] The *Gesammelte Werke* were published by Diedrichs in Jena.

[93] Thus in Martin Heidegger's *Sein und Zeit (Being and Time)* (which first appeared in the *Jahrbuch für Philosophie*, vol. 8) we can see, besides the influence of Edmund Husserl, that of Kierkegaard.

concept of evidence – will find similar application here, and it will be recognized that the question of the 'nature of value' points to *an analysis of the meaning of value judgments.* The by no means inconsiderable additional difficulty of this rational reconstruction – as compared to that of typical empirical judgments – lies, however, chiefly in the traditional confusion over a number of different questions, independent of each other – even if not pair wise – which we will have to distinguish from each other terminologically, in the imprecise form in which they have played their role in the history of ideas, and then seek to render them more precise, and finally seek to answer each separately.

1. *The problem of the transcendence of values*: Is there a sphere of values which exists 'in itself' which, though it can be comprehended – to a certain degree – by human reason, still remains totally unaffected by the fact of this comprehension (value realism, Platonism); or is this sphere of values nothing else than the hypostasizing of certain emotional acts of human beings (value psychologism)?
2. *The problem of the independence of values*: Is there an independent sphere of values beside the sphere of real being; or – subjectively phrased – is there specific knowledge of values besides knowledge of that which has (real) Being?
3. *The problem of the absolute status of values (absence of presuppositions in knowledge of values)*: Are there value judgments whose validity has no presuppositions – supposedly, like the propositions of logic and mathematics – or are all laws of value based in a analogous manner on hypothetical assumptions of value, as the laws of nature are based on hypothetically assumed principles?
4. *The problem of the relational character of value concepts*: Are value judgments judgments of qualities, or judgments of relations; is a quality of value attributed in them to the (valued) objects, or is a certain relation between the (valuing) subject and the (valued) object asserted in them?
5. *The problem of the hierarchy of values*: Is there a highest value or a plurality of highest values, such that all other values derive their validity from them?
6. *The problem of the objectivity of values*: Can we speak of the truth of value judgments in the same sense as of the truth of judgments of factual Being which latter imply intertemporal and intersubjective agreement about the series of verification? Are value 'judgments' true judgments at all?

The clarification of these problems has to be carried out by means of a rational reconstruction of the meaning of value judgments, and it is to this that we now turn. To be sure, in actually carrying out such a rational reconstruction the *sequential order* of the problems, as they are listed above, cannot be preserved. It will only re-emerge in the formulation of the results.

Let us begin with the consideration, whether in a value judgment, something else, that is, *more*, is asserted, than a *specific effect* of the valued object on the valuing person, and on the majority of the human beings who come into contact with the object. Is, for instance, the aesthetic value judgment 'The picture P is beautiful' equivalent in meaning with the judgment: 'The picture P arouses general (disinterested) pleasure'? A little deeper consideration will show that this 'translation' of the value statement cannot be accepted as completely appropriate. For if the two sentences 'The picture P is beautiful' and 'The picture P arouses general pleasure' had the same meaning, then the sentence 'The picture P is certainly (or, probably) beautiful, although it does not please me, nor most other people' would be contradictory. Yet it is not commonly understood that way. Instead, we would interpret it to mean that though the picture does not please most people (including the person making the judgment), who only possess a slight understanding of art, it would please the few who are truly experts. This would point back to a definition of the following sort: 'A work of art is beautiful insofar as it gains the approval of those who truly understand art', or, 'an action is good insofar as it has the approval of those who think righteously.' But would not such a definition be circular? Aren't those who truly understand art characterized solely by the fact that they judge art correctly, and those who think righteously by the fact that they make the right moral judgments, and thus approve of what is deserving of approval and disapprove of what deserves disapproval?

The consideration above already permit us to see that the main problem – or at least one of the main problems – within the framework of the analysis of value judgments, lies in the characteristic *specific element of correctness* contained in them. We now shall carry out such an analysis.

Let us depart from the concept of *correct assertion*. An assertion is correct only when the state of affairs about which something was asserted, is indeed such as was asserted, i.e., when the assertion fits coherently – in the manner characterized above – within the complex of experience. After our critique of naïve realism, we can speak of the 'correspondence of the assertion with the facts', without fear of being misunderstood. The formal core of this concept of *theoretical correctness* thus is 'correspondence'. But even in this context an ambiguity still adheres to our concept. For, as the goal of intending to make an appropriate assertion, the verification of the assertion at the same time represents the realization of that goal. Thus here, there is a second kind of correspondence, that we will call appropriateness to '*goal fulfillment*' or appropriateness to 'fulfillment of purpose'. Accordingly, with a view to this goal of finding the truth, the *theoretical* correctness (validity) of the assertion – its correspondence with the facts – forms the criterion for the goal fulfillment (its *practical* correctness) of an act of assertion.

Now, how is the relationship of these two 'aspects' of 'correctness' shifted for a human activity which is oriented toward other goals than finding the truth? Here, theoretical correctness is not always the criterion for practical correctness; it is thus possible that action proves appropriate to fulfilling a goal although, or even because, it is based in false assumptions. For example, the following judgment relates to one such case: "I would never have undertaken this dangerous venture, which I have now accomplished entirely successfully, had I foreseen the difficulties and dangers involved." Still in the overwhelming number of cases, experience teaches us that fulfillment of the goal of action tends to increase with the degree of correctness of the assumptions contained in the design. For this reason, predictions of success in action, by an observer who knows the design, or can infer it from the initial stages of the action, most generally depend upon whether, in his opinion, the action in question is guided by correct assumptions.

Such an evaluation can, in turn, itself be subject to evaluation by a third party with respect to its theoretical correctness, and here objections of two sorts are possible: If we call the acting person A, the evaluator of A's action B, and the evaluator of B's evaluation C, then we can formulate the objections as follows: (1) B has comprehended the goal of A's action correctly, but has wrongly evaluated the appropriateness for goal fulfillment of the means applied; (2) B has not evaluated the action of A appropriately, because he has not understood A's goal, and therefore has examined the appropriateness for goal fulfillment of his action, with respect to goals which A did not set at all.

If B now doubts whether he has understood A's goal, then as a consequence of this doubt, he will perhaps arrive at the following judgment: 'If A pursues goal G, then his action is correct.' This means: According to B's conception, A's action is appropriate to its goal with respect to goal G, but B does not know whether A does in fact strive to attain goal G. Through this the false appearance might easily arise as though the relation between the setting of a goal, and appropriateness to goal fulfillment (practical correctness), were the relation of that which conditions a thing, to that which is conditioned by it; while in fact the concept of *appropriateness to a goal* is a *relational* concept (thus an *incomplete* concept), and the specification of the goal belongs to the definition of this concept. The false interpretation just indicated would correspond to the following [false] interpretation of the (directional) concept 'right': 'If P_1 sits right of P_2, then P_2 sits left of P_1', where the conjunction 'if – then' is conceived as empirical condition. For in both cases we have the confusion of *empirical judgment* with *concept analysis* discussed at length above. That the goal-related character of the concept of *correctness* is so often overlooked, however, has other special reasons too. One of these lies in the confusion of 'theoretical correctness' with 'practical correctness' already mentioned. Let us once more clearly state the meaning of these two concepts. We designate a judgment as 'theoretically correct' – independently of reference to any goals – when it 'corresponds with the facts'. However, once the goal is set of arriving at a correct judgment J of a certain kind (theoretical goal), then every correct judgment J_1, which represents a step in the cognitive process leading towards J_2, is also practically correct. Although theoretical correctness in this case forms a criterion for practical correctness, it does not coincide with it, because it does not refer to the same goal. Due to the confounding of these two, the declaration of truth as an 'absolute value' has come about.

In other cases, again, the goal-relatedness of practical correctness has not been recognized because the goals in question were 'self-understood' and therefore frequently did not come to our attention at all. Using a term applied by Max Scheler,[94] we can speak of *unquestioningly* given goals. Such goals are, for instance, the preservation of life, of freedom, of health, the attainment or preservation of a position of respect within a narrower or broader social circle, the avoidance of pain. The concept of goal is of course to be understood in such a broad sense that the prevention of certain events can also be set as goal.

---

[94] *Die Wissensformen und die Gesellschaft* (*The Forms of Knowledge and Society*) 1926, p. 59.

However – and with that we arrive at an important new point in our reflections – this ‘taken-for-grantedness’ of goals disappears whenever these come into *collisions* with other goals. We speak of a collision between two goals G_1 and G_2 of a person P if (or insofar as), the realization of G_1 and the realization of G_2 by P are incompatible. Here we have to distinguish between *direct* and *indirect* collisions. The first will be the case if the incompatibility presents itself without reference to the causal nexus of each of the two goals. Thus the two goals of being at a certain place P_1 at a certain time, and at another place P_2 at the same time, are directly incompatible. We would speak of an ‘indirect collision’ however if in general experience, the accompanying circumstances of one of the two goals frustrate the achievement of the other goal. In both kinds of collision, we must decide which goal we wish to pursue at the cost of relinquishing the other goal. However, the collision between two goals which can be (relatively) readily isolated, is only the simplest case of collision between goals; in the face of the circumstance that human action is oriented to a – more or less ordered – multiplicity of goals, very frequently we have to make a choice between the realization of one goal and the realization of a group of other goals; and here the situation is often complicated even more by the fact that for both eventualities of realizing the goals, there are different *degrees of probability*, which have to be taken into account. For this it follows that in many cases the question of appropriateness to goal fulfillment arises not for an isolated goal but for a *system of goals*. For systems of goals, what we have said concerning their ‘taken-for-grantedness’ also holds true; it will frequently happen that such goal systems are not apperceived at all, due to which the false interpretation of the ‘correctedness of human action with respect to such a system as absolute correctness’, becomes understandable.

The preceding considerations can be applied to the setting of goals itself. For rather rarely are goals set which are to be realized ‘at any price’. Therefore we can speak of the ‘correctness’ of a goal, and measure this by how the goal fits into the system of other goals. Here not only the collision of goals has to be taken into consideration, but also its counterpart, the (positive) *goal-interrelation* which exists between two goals, whenever the realization of one goal presupposes or entails the realization of the other.

In all our entire reflections up to this point the concept of practical correctness of an action, or the setting of a goal, referred to the goal system of the person acting or setting the goal at the time of the action. In what follows we shall speak of the actor and include the person setting the goal. The evaluation of an action, either by the acting person himself at a later point of time, or by a third person, as incorrect, in this sense thus indicates that it does not fit into *this* system of goals. Now a complication arises due to the fact that in both cases, an evaluation frequently results, consciously or unconsciously, on the basis of *another goal system*.

Let us first consider the case of this *shift of the basis of evaluation* – referring to our own past action. Here, on the one hand, not infrequently – and mostly inadvertently – the goal system held at the time of evaluation is taken as the basis of evaluation. On the other hand, even where reference to a goal system that faithfully reflects the conditions at the point of time [of the action] is explicitly intended, another goal system, to be characterized more precisely immediately below, is frequently substituted for the goal system that determined the design and carrying out of the action. The latter substitution takes place most frequently when the reproduction of the relevant design for action, on which the evaluation is based, shows that the motivation was evoked by a system of goals which only has validity for an extraordinary psychological situation – for example, violent passion or desire. In such cases the evaluation of the correctness of one's own action is usually made relative to the *normal* goal system of the acting person. The appropriateness of the action with respect to goal fulfillment is evaluated with reference to that goal system which would have been determining for the acting person during the period of time in which the action occurred – that is, according to his conception at the time at which he is making the judgment – had he then had presence of mind. Among these goals, those which concern conduct toward one's closer or broader [social] environment (family, professional community, religious community, political community) play a special role.

Appropriateness with respect to these goals very frequently forms the criterion for evaluating the correctness of one's own past actions. Here the person making the judgment is aware that his conduct, should it be evaluated, would also be evaluated in conformity with such criteria by other members of his circle. In this case he relates his behavior to goals which for the most part are *unquestionably* given to him, preselected by moral precepts, by tradition, and perhaps also by hereditary disposition; an in addition he knows himself to be in far-reaching agreement in this with members of his social environment, beyond which he will hardly transgress in his thinking.

With what we have already designated, to a first approximation, an '*intersubjectively valid*' system of goals that is to a large extent uniform for certain social spheres. This is then regarded as so self-evident by the overwhelming majority of the members of the community that the correctness or incorrectness of an action relative to his system of goals is considered to be simply correctness (or incorrectness) as such. Here the following must be added: the actor is aware that the result of the evaluation of this action by members of his social circle will frequently bring consequences with it which will closely touch on the other partial domains of his overall goal system. If, for example, his action could be judged as 'illegal' or 'immoral', then there is a chance that certain members of his circle would behave toward him in such a manner, that his (external) freedom or his capacity to earn a living would be restricted. Therefore, if he judges such conduct relative to *these* goals, he too will say that it was incorrect. If recognizing incorrectness of our own past conduct is accompanied by painful feeling, we call this '*remorse*'.

Concerning the evaluation of the actions of other human beings with respect to their practical correctness, the following must finally also be said: even where we endeavor to evaluate such action with reference to the goal system of the actor at the time of the action – for example, in certain historical analyses – it will be seen that the evaluation will be made with reference to more or less general goal systems, and indeed must be made in such a way, because the goals of the person in question are only accessible to a small degree to the person making the judgment. *Atypical (abnormal)* behavior is therefore not 'measured by its own standards' in most cases, even when there is an effort 'to do it justice'. Therefore a critique of the 'correctness' of such evaluations will frequently maintain that they have been based on an *inadequate goal system.*

The results of our reflections carried out so far sought to shed light on the various meanings of the concept of 'practical correctness' of human action and with that on the criteria for the 'correctness' of evaluation of such action; at the same time, they have shown the *relational character* of all these concepts of correctness, so that the idea of an absolute, i.e., relation-free, correctness has proven to be a *contradictio in adjecto.* We will soon recognize that this result has decisive significance for the solution of the question concerning the meaning of value judgments as such, but for this purpose, we will first have to analyze the *meaning of the goal concept*, which we could only provisionally assume to be well-defined in our previous reflections.

To begin with we can propose the following formulation: "I set a certain state of affairs as my *goal*, if I decide to *endeavor* to *realize* this state of affairs, i.e., to direct my activity in such a manner that the chance of its realization becomes as great as possible." To be sure, in keeping with our exposition above, we find here that the setting of most goals is only *conditional*, that the endeavor to realize the state of affairs will only go so far as it can be carried out, according to certain assumptions, with the investment of certain means and the avoidance of certain secondary effects.

Setting a goal means a *resolve about future conduct.* It says: through my conduct I want to effect the occurrence (non-occurrence) of a state of affairs of a certain kind at a certain place. We cannot explore in detail the difficult question whether, or to what extent the resolve (decision, will), is to be comprehended as an independent phenomenon, or falls into the sphere of judgments – judgments about one's own future action. Instead, in what follows we will operate with these concepts, regarded as synonymous, as basic concepts and seek to clarify for ourselves to what extent other concepts in the emotional sphere can be traced back to these.

Let us begin with the concept of *choice*. A resolve is called an act of choice if its content is: "Realization of the state of affairs S_1 and nonrealization of the states of affairs S_2, S_3, ... S_n, because of their – direct or indirect – collision with S_1." We can say then that one *renounces* S_2, S_3,...S_n *in favor* of S_1, (that one *sacrifices* S_2, S_3, ... S_n for the sake of S_1).

It is important to note that the object of choice is not a thing but a condition (state of affairs) – expressed more correctly, the chance of *bringing about a condition*, or a multiplicity of empirically correlated chances. So, say, the 'choice of a dish in a restaurant' means the resolve to eat that dish and not the available alternatives on the menu. Now we could say that basically every resolve is an act of choice, as every decision to realize a specific state of affairs entails renouncing the realization of other states of affairs: however only such resolves are customarily called 'acts of choice' for which the renunciation is *consciously* made. To be sure, in view of the various levels of consciousness, this is variable.

Most closely linked with the concepts of resolve and choice, are those of *wishing* and *preferring.* 'I wish the state of affairs S' means 'I would realize S if I were capable of doing so'. 'I prefer condition C_1 to condition C_2,' means 'In case I had to choose between the two conditions, I would decide in favor of bringing about, or else maintaining, C_1 and sacrificing C_2 for its sake'. The relation existing between wishing and willing, is most strikingly expressed in the confrontation of 'willing' with '*mere* wishing'.

Understanding the complex of relations just presented, has been obscured by intermingling the contents of the acts analyzed in an uncritical manner with the *lived experiences of the bodily states* that generally accompany these acts. Thus the experience of the nerve impulse, regarded as the starting point for the bodily movement belonging to the action, was considered to be a constitutive element of the act of willing. To characterize the wish, on the other hand, the 'feeling of pleasure' was adduced, to be sure without in general specifying whether the feeling of pleasure was *associated* with the mental image of the realization of the desired condition, or whether the feeling of pleasure accompanying the realization of this state, was part of this 'mental image'.[95]

But taking into consideration these lived experiences [*Erlebnisse*] in the definition of acts of will, and acts of wishing, leads us to *inadequate* definitions. This is, because the question whether an act of will or an act of wishing of a certain content is occurring at a certain personal-temporal location, never aims at ascertaining whether the mental image of these contents is accompanied by certain nerve impulses, or 'emotionally laden' in a certain manner. Thus the question, 'Did you wish for the condition C?' means nothing else than 'Would you have decided to realize C, if you had had a chance of realizing it?'

But with respect to the problem of *'feelings of pleasure'* (*'feelings of displeasure'*), the following is to be noted: it is a general, but nonetheless erroneous conception that the pleasure – or else, displeasure – content of a feeling is an internal quality of this feeling. In fact, the judgment that a certain feeling, or a feeling of a certain kind, was a feeling of pleasure, states nothing else than the assertion, that the occurrence or continuation of a feeling of this kind is desired.

In the same way, the designation of a feeling as a 'feeling of displeasure' corresponds to the judgment that the disappearance of this feeling, or of a feeling of this kind is desired.[96] Thus it does indeed depend upon what kind of feeling it is, whether its continuation or disappearance is desired, or whether neither of these wishes is aroused, but just because of that, the judgment concerning the occurrence of one of these two wishes, is still not a judgment concerning an *isolated quality of feeling*. However, that judgment forms the criterion for assigning a feeling to one of the two classes 'feelings of pleasure' and 'feelings of displeasure'. Now whether there is a wish to have, or not to have, a feeling of a certain kind, whether it is a feeling of pleasure or displeasure, does not depend solely on the internal quality of the feeling, but on the overall psycho-physical situation, within which it appears to occur. The fundamental insight that the *quality* of feeling and the *tone* of feeling (pleasure-displeasure), can vary independently of each other – this insight cannot be affected (changed) by the fact that biological-psychological automata exist,

---

[95] A painstaking analysis of the concept of wishing is contained in Sander, *Allgemeine Gesellschaftslehre*, Jena 1930.

[96] Compare Kaufmann, *Strafrechtsschuld*, p. 18.

which in many cases produce an extremely close empirical connection between feeling and the arousal of desire (wishes). The opposite conception is a special case of the erroneous commingling of *experiential* simplicity (or complexity) with *structural* simplicity (complexity); the latter can only be judged after the completion of a rational reconstruction. It must also be taken carefully into consideration that the division of feelings into feelings of pleasure and displeasure does not represent a dichotomy, as by no means every feeling is either pleasurable or displeasurable. Rather there appear to be a large number of feelings, which we would want to call *neutral* feelings, feelings that are neither associated with a wish in one direction, nor in the other.

From these considerations flow the following important consequences:

1. It is not proper to define the concept of the wish (endeavor) with the help of the allegedly elementary concept of feeling of pleasure; rather, on the contrary, the concept of wish (endeavor) enters into the definition of 'feeling of pleasure'.
2. The 'apodictic character' of propositions such as 'The occurrence (continuation) of pleasant (pleasurable) feelings, and the non-occurrence (disappearance) of uncomfortable (unpleasant) feelings are desired (wished)', 'The pleasant is preferred to the unpleasant' results from the fact that we are not dealing here with propositions about reality, but with the analysis of concepts.
3. The eudaemonistic doctrine, that all endeavors of human beings are directed toward procuring pleasure and avoiding pain, has its main source in the failure to recognize the relation between endeavor (or also wishing) on the one hand, and 'pleasure' or also 'displeasure' (pain), on the other. For this doctrine seeks to comprehend in terms of 'rational grounds', i.e., by pure reflection, why the bringing about (or preservation) of conditions of a certain kind, and the avoidance (disappearance) of conditions of another kind, is typically or 'correctly' the goal of endeavor.

   Here the following aspects are confounded:

   (a) The fact that in the act of striving, the *mental representation* of attaining the desired condition is often linked with the mental representation of feelings of pleasure.
   (b) The fact that very frequently the attainment of the condition for which one strives, is linked with feelings of pleasure, where it may either be this condition as such, or else the consciousness of having attained it by one's own powers which is a source of pleasure.
   (c) The connection between the concepts 'endeavor' and 'pleasure', according to which 'pleasure' is necessarily, i.e., by definition, *something one strives for.*

From this confounding arise the two basic errors of the *eudaemonistic doctrine.* The first consists in misconstruing empirical findings (see (a) and (b)) as apodictic statements; the second error, closely related to this, consists in inverting the relation between the concepts of 'endeavor' and 'pleasure', from which it is then concluded, that 'in its essence' all striving has pleasure as its (ultimate) goal. But *de facto* the goal of striving is by no means always, or even in the majority of cases, the occurrence or continuation of a feeling with a specific coloration. The striving for knowledge, for power, for wealth is not a striving for pleasure, as would have to be the case according to the eudaemonistic doctrine.[97]

4. Feelings of pleasure and displeasure can no more serve as criteria for 'practical correctness', than the perceptions (sensations) can for theoretical correctness. To be sure – similar to perceptual judgments – certain judgments about emotionally charged positions will form nodes in the system of evaluation of practical correctness, but these must never be seen as 'absolute', or 'ultimate' in principle, criteria for these evaluations.

With these assertions, the preconditions for an understanding of the *meaning of value judgments* have been created; for the judgment 'D is valuable' states nothing but that it is correct to endeavor (wish for) D, and the judgment 'D is more valuable than E' states that it is correct to prefer D to E. From this, the root of the 'apodictic validity' of the following statements emerges directly: "It is correct to endeavor to achieve the valuable", and "It is correct to prefer the valuable to the less valuable". These are tautologies. The real problem, however, hidden in this, is the problem we have discussed in our analysis of the concept of practical correctness, namely: what are the criteria for judging whether D is correctly subject to endeavor, or D is correctly preferred to E?

It emerges directly from our reflections that these criteria refer to statements of the relative positions of D and E in pre-established systems of goals (systems of desires). In making these statements, we must take account of the ambiguities of the goal systems to be chosen as the basis of evaluation discussed above. However if a *certain goal system* is presupposed as the constant system of reference, the value of D, and the preferability of D to E, depend *solely* on the qualities of D and of E (in the broadest sense, including their typical effects). This easily leads to neglect of the relational character of the concept of value and to the false assumption that objects possess 'value qualities' which justify their respective value 'objectively', i.e., without reference to a subjective system of reference.

To remove the roots of this error was made still more difficult by the following circumstance: – In the value judgments of pre-scientific, and also of scientific language, it is not only conditions which appear as 'possessing value', but also *persons and things*. Thus persons are also designated as 'of moral value' ('virtuous', 'noble', etc.), and therefore the question arises whether value judgments of this kind can be incorporated (fitted) into the schema of meaning for value judgments, which we have just characterized. This is in fact the case.

The criteria for whether a thing can be called 'beautiful' result from the observation of the thing, and for whether a human being can be called 'good', from the direct or indirect observation of his external and inner conduct; and from here the bridges to the system of goals (or wishes), which in this case forms the reference system for evaluation, can be built.

---

[97] For this compare the critique of eudaemonism in Scheler's work *Der Formalismus in der Ethik und die materiale Wertethik*, 2nd ed. Halle a.d.Saale 1921, p. 245 ff.

Finally, in this connection we must also indicate the following point: the analysis of value judgments found itself confronting a difficulty which is by no means slight: that the meaning associated in *language usage* with value concepts (good, beautiful, etc.) *vacillates* greatly with respect to the *emphasis* placed on the aspect of '*correctness*'. We wish to illuminate, by means of examples, what is meant here: when during the enjoyment of a dish we declare it to be 'good' (tasty), then in most cases we do not wish to assert anything more than that at just this moment it tastes good to us; the assertions that the same dish will again taste good to the person on subsequent occasions and that it will also taste good to other human beings with normal sensations of taste, only appear to be linked to the statement as background meaning. Therefore, any challenge to the evaluating person to offer a reason for his value judgment will be regarded as misplaced, and will be rejected with the reply 'because it just tastes good to me'. No infrequently there will be a similar response where certain phenomena of nature, say, clouds illuminated by the setting sun, are described as 'beautiful'. Here the companion who reacts to this value judgment with the question: 'Why is that beautiful?' must be prepared for the reply: 'If you do not feel (perceive) it, I cannot explain it to you.' But already the matter will be typically different in the case of those aesthetic value judgments which refer to the human body – and especially the human face. Still more far-reaching justification will frequently be felt to be required for the aesthetic judgment of works of art and the moral judgment of human actions, or of persons acting or having acted in a certain way. Here rational reconstruction can reveal in what manner the various points of view, on the basis of which judgments are made, interpenetrate. Thus for instance in the judgment of a painting we may ask what *task* the artist set for himself in the creation of the picture, and then evaluate, on the one hand, the idea, and, on the other, the actual achievement in solving the task. That is why the question of *originality* also frequently plays a great role in the aesthetic judgment of works of art. Often in aesthetic judgments, the evaluation of the achievement overwhelms the evaluation reached on the basis of unreflecting sense impression to such an extent, that the representation of the ugly is adjudged as 'beautiful'. In such cases we say – quite imprecisely – that the ugly has been ennobled by art.

In the *ethical* evaluation of human actions – or also, of actors – the evaluation of the achievement (attainment) on the basis of either individual or collective goal systems, and evaluation of the [moral] disposition and qualities of character which are manifested in the action typically interpenetrate. However, when looked at more closely, the disposition and qualities of character in turn appear to be sources of behavior of a certain kind. The various ethical systems differ, above all by the goal systems they use as the basis of evaluation, and the aspects forming the criteria of evaluation (for example, personalistic ethics, eudaemonistic ethics). Clarity on this point is important for many social-scientific analyses.

According to these considerations, we can say: Among the sentences which appear linguistically as 'value judgments', we will discover sentences that either only assert a momentary emotional relation on the part of the evaluating person toward the object evaluated, or on the other hand, assume a *typical* relation of this kind to exist for the person making the judgment, and possibly also for a larger or smaller circle of other persons. A deeper analysis of the pertinent problems, which seeks to comprehend the 'nature of values' need not refer to such statements at all for they are obviously either individual judgments or general *judgments about psychical facts.* Instead, the essential problem in the philosophy of value arises – as we have already emphasized at the beginning of this section – only where the 'objective correctness' of taking such positions is asserted. Then it must be shown that this 'correctness' is a relational concept, which requires speaking of the goal system as one of the *termini relationis.*

Now we have created the conditions for answering the questions posed at the beginning of the section, concerning the 'nature of value', and thus we want to formulate our answers as concisely as possible:

1. The Problem of the Transcendence of Values
The idea of an absolute transcendence of values can hold up under reflection just as little as a similar idea with respect to real and ideal objects in the 'sphere of Being'. Our reflection, carried out with respect to this in section "Basic Philosophical Considerations", find an analogous application here. It cannot be said, however, that 'value' is simply a hypostatization of wishes, endeavors and sets of preference so that what is valued simply corresponds to what is in fact wished or striven for, while that accorded higher value corresponds to what is in fact preferred. For only those among such expressed attitudes, or opinion-telling acts which display a certain degree of intertemporal and intersubjective consistency, will be considered as correct, as positions that actually represent values. Here an extensive analogy with the incorporation of the data of perception into the context of experience readily suggests itself.
2. The Problem of the Independence of the Sphere of Values
In seeking to maintain that there is a sphere of values independent of the sphere of Being, the idea of absolute transcendence is usually also involved. But even if the latter has been relinquished, and therefore independence is merely conceived in such a manner that a special source of knowledge is claimed for the recognition of value, this still is not tenable. Against the assumption of a specific 'evidence of feeling' which accords to an ancient doctrine – in the recent decades elaborated especially by Franz Brentano,[98] Max Scheler,[99] Dietrich von Hildebrand,[100] Nicolai Hartmann,[101] – seeks to justify the intuitive comprehension of values, the same objections arise, to begin with, as against evidence as an isolated truth-criterion in general. It does not go to the basic issue of the question: what actually is supposed to be revealed by this evidence? Thus it closes itself off to understanding axiological (practical) correctness, which is defined only relative to certain goal systems (systems of wishes) and changes its meaning as these vary. Accordingly, we can speak of the independence of the sphere of value, only

[98] Brentano, 'Vom Ursprung sittlicher Erkenntnis', Scheler, *Phil. Bibl*, vol. 5.

[99] See Scheler, note 97.

[100] von Hildebrand, 'Die Idee der sittlichen Handlung', *Jahrbuch für Philosophie,* III 1916, p.16 ff.

[101] N. Hartmann, *Ethik*, Berlin 1926.

insofar as by this we understand its relatedness with specific acts of endeavoring, wishing, preferring (the so-called *attitude-taking acts*).[102]

Thus all concepts of value are reducible to 'concepts of Being', i.e., defined in terms of them, and the *'value judgments' are subject* in principle to the *same procedures of verification as 'judgments of Being'* – insofar as they do not turn out to be definitions. To be sure, due to the usually very slight degree of clarity of the attitude-taking acts, the sort of confirmation required for value-judgments will, as a rule, only emerge after rational reconstruction.

3. The Problem of the Absolute Character of Values (Absence of Presuppositions in Knowledge of Value)

With the removal of the basis for assuming a specific sort of evidence that grasps values, and with the insight into the reducibility of value judgments to judgments of Being, the thesis of the absence of presuppositions in knowledge of values, and the existence of an intuitively comprehensible sphere of values is also disposed of. But in the confused conception of absolute values, the *idea of absolute goals* is most frequently retained, and for this a clear meaning can be secured. For under absolute goals we usually understand such goals as satisfy both of the following conditions, or, – as language usage vacillates – at least one of these:

(a) Such goals are – to a large degree – goals which can be isolated, i.e., they are not bound to one fixed place within a goal system, where they would be inseparably linked to other goals; and especially, they are not merely mediate goals for certain final goals, but are themselves final goals.
(b) They are unconditional goals, i.e., goals whose realization will be pursued 'at any price'.

Although conditions (a) and (b) are frequently both fulfilled together – especially where the satisfaction of a violent passion is presented as a goal – still they are logically independent of each other. Sometimes talk of 'absolute goals' relates exclusively or predominantly to criterion (a), sometimes, on the other hand, to criterion (b). However to the setting of an absolute goal – *qua* absolute – in the sense of criterion (a), the question about its correctness is not applicable, for this question, as we have shown, is directed toward ascertaining the relationship of coherence of the goal in question with other goals. This circumstance is usually not recognized when such goals are 'pre-determined' in such a way (especially by tradition), that it is difficult to become clearly conscious of the freedom to set or change goals. This freedom only emerges into the foreground of awareness when the goal is questioned. 'To question a goal', however, means 'to weigh it against other goals', and thereby it loses its absolute character in the sense of criterion (a). This must be kept in mind when we are tempted to identify 'absolute goals' too hastily with 'correct goals'.

4. The Problem of the Relational Character of Value Concepts

We have already established that value concepts are relational concepts, and that the appearance of the opposite arises only because the reference systems of goals

[102] The dualism of the acts of taking a position and acquiring knowledge is developed by Diedrich von Hildebrand, following Husserl.

very frequently remain outside the sphere of clear awareness. As the attribution of a (positive or negative) value-predicate to an object on the basis of a constant reference system, only depends on the properties of this object, (apparent) value judgments frequently appear in the following form: An object is valuable when it possesses the features F_1, F_2, etc. Thus one can declare: 'The body of an adult human being is beautiful if it approximates certain measurements.' It will readily be seen on the basis of our previous reflections about the meaning of empirical judgments, on the one hand, and definitions, on the other, that such a statement if *not a judgment* but a *definition*, and that it is a question of interpretation, whether it is understood as a nominal definition or a real definition. On most occasions the latter will be the case. But even then one could not say correctly, that a concept of *value* is defined by that definition, but only that in the definition the 'objective' criteria – those attributed to the object – for the recognition of a certain synonymous value predicate (for example, 'beautiful'), are given. For if the definition is to be consistent with what we 'essentially' mean by a value concept, then it is inadequate if it does not define the concept as a *relational concept* of the kind just characterized.

In this connection, the vacillating position of the real definition, characterized in our second section, forms a source of dangerous confusion. For on the one hand such a definition of value, *qua* definition, is regarded as irrefutable; on the other hand, *qua* assertion, that with the term in question generally a certain meaning is associated, it is regarded as containing knowledge. In addition, increasing the confusion still more, there is the circumstance that concepts of value are 'really' comprehended as relational concepts, and that in them reference to a system of goals (wishes) is implicitly included in the intended meaning. However, this relational character is misinterpreted as a conditional relation in the manner described above.

The most important example of this amphibole, from the history of ideas, is offered by Kant's categorical imperative: "So act that the maxims of your will could always hold at the same time as the principles of a universal legislation."[103] The clarification of the character of validity attributed to this can be found in the "second remark to the deduction from the categorical imperative," which states: "Pure reason is practical of itself alone and it gives (to man) a universal law which we *call the moral law*." We add the decisive passage to this:

> The fact just mentioned is undeniable. One need only analyze the sentence which men pass upon the lawfulness of their actions to see in every case that their reason, incorruptible and self-constrained, in every action confronts the maxim of the will with the pure will, i.e., with itself regarded as a priori practical; and this it does regardless of what inclination may say to the contrary. Now this principle of morality, on account of the universality of its legislation which makes it the formal supreme determining ground of the will regardless of any subjective difference among men, is declared by reason to be a law for all rational beings in so far as they have a will, i.e., faculty of determining their causality through the conception of a rule, and consequently in so far as they are competent to determine their actions according to principles and thus act according to practical *a priori* principles, which alone have the

[103] *Kritik der praktischen Vernunft* [*Critique of Practical Reason*], Book I, Part 1.

> necessity which reason demands in a principle. It is thus not limited to human beings but extends to all finite beings having reason and will; indeed, it includes the Infinite Being as the supreme intelligence.[104]

For exemplifying what we are expounding, it is crucial in the above passage that Kant wants to gain confirmation of the moral law by means of *analysis* of the judgment which human beings make about the lawfulness of their actions. His theory that all human beings, indeed all beings endowed with reason, will regard an action as corresponding to morals, i.e., as moral, when it corresponds to the criteria contained in the categorical imperative, is nothing else than recourse to the *consensus* which exists in each being with respect to the characterization of an action as 'morally good', insofar as they overcome *perturbationes animae*, the disturbing passions and inclinations; for the idea that also resonates here, that to the consensual agreement of conceptions of rational beings, is linked an impulse toward moral action, which could take full effect after overcoming the opposing inclinations, obviously does not play any role with respect to the question of validity. The empirical-psychological fact of impulses of a certain intensity – vacillating personally and interpersonally – can have no effect on the *a priori* validity in Kant's sense of the practical principle relating to the pure will.

Therefore we come to the result: the 'absolute' validity of the categorical imperative is *validity* by virtue of a *definition* which is based upon a – supposed – *consensus* with respect to the concept of the morally good. If we investigate further what Kant means by the capacity for action according to practical principles *a priori,* then we will recognize that by this he means a capacity to guide one's action by general ('formal') considerations, concerning the 'practical correctness' of action. The reference system for that practical correctness, however, is, as emerges clearly from, especially, Kant's *political* writings, the existence of a society in which peaceful coexistence is guaranteed and each person regards a certain sphere of freedom of the others as unavoidable. Practical correctness *a priori* in Kant's sense, thus is nothing else than appropriateness to goal fulfillment with reference to the goal of the preservation of human society, and its further development in conformity with the goal cited.

J.J. Fries,[105] and following him, Leonard Nelson,[106] in their further development of Kant's ethics and theory of law, have carried out the 'deduction' of their ethical and natural right principles with an orientation provided by the fundamental question: 'How is human society possible?' They thus asked themselves,

---

[104] The question of the relation of moral laws as laws of reason and the divine will, already forms a major problem of scholastic philosophy. Here the rationalism of Thomas Aquinas confronts the voluntarism of Duns Scotus.

[105] Compare *Philosophische Rechtslehre und Kritik aller positiven Gesetzgebung,* Jena 1803.

[106] Compare *Kritik der praktischen Vernunft*, with *System der philosophischen Rechtslehre und Politik* (*System of Philosophic Doctrine of Law and Politics*), Leipzig 1924. Also compare Erich Vögelin, 'Das Sollen im System Kants' (The 'Ought' in Kant's System) in *Gesellschaft, Staat und Recht. Untersuchungen zur reinen Rechtslehre*, Vienna 1931, pp. 136–173.

what conditions with respect to the inner and external conduct of human beings must be realized (by and large), in order that social coexistence of human beings can be maintained, and then set these conditions as practical postulates; from this followed the confounding of the relational and conditional, which has already been discussed.

The preceding explanation should have made the relational character of value concepts sufficiently clear. From this clarification emerges the heuristic postulate, to seek for the conceptually presupposed reference system of goals, wherever linguistic expression arouses the false appearances of value qualities inhering in the valued objects. This postulate will prove a true thread of Ariadne to guide us through the labyrinth of axiological conceptual confusion.

5. The Problem of the Hierarchy of Values

Here the considerations we have undertaken concerning the incompatibility of goals, on the one hand, and the character of acts of preferring, on the other, find analogous application. The 'correctness' of acts of preferring is always relative to a given goal system, structured in a certain manner. Even the (definitional) specification of a hierarchy of values reveals the amphibolies just discussed, which arise due to the confounding of 'validity by virtue of *definition*', 'validity by virtue of *consensus*', and 'validity by virtue of *justification* by reference to a generally recognized system of goals, or one assumed to be generally recognized'.

The futility of the attempts to find generally persuasive formal criteria for the hierarchy of values, was recently made evident in an especially impressive way by the fact that two thinkers as important as Max Scheler and Nicolai Hartmann, whose value theories have a very close affinity otherwise, arrived at *opposite* formulations here: for Scheler the fundamental value is the higher one, while for Hartmann it is the one for which a foundation has been provided.[107]

It must still be mentioned that even value judgments, in which a comparison of values does not find explicit expression, frequently are implicitly *comparative*, as the attribution of a positive or negative value predicate is made dependent upon whether a certain level of value has been exceeded, or has failed to be achieved. Thus a teacher usually bases the 'classification' of individual achievement – more or less consciously – on its level with respect to the 'class average'.

6. The Problem of the Objectivity of Values (the Domain of Validity of Value Judgments)

This problem is relieved of its difficulty as soon as we keep rigorously separate the question of the appropriateness of goals with respect to a given goal system, from the question of the choice of goal systems. Once the goal system is established as unequivocally structured, the judgment of the appropriateness of the goals can be objectively decided by means of a cognitive procedure (process of subsumption); but there is no logical path which leads from this 'objectivity' to a thesis in which generally valid goals are asserted. What are designated as such goals as a rule are 'self-evident', 'taken-for-granted' goals

---

[107] Compare Scheler, *Formalismus*, p. 92 ff. with N. Hartmann, *Ethik*, p. 230 ff.

which become comprehensive in terms either of the vital – in the narrower sense – or the psychic-mental constitution of human beings, and find their reflection in the conventions and customs of communal life. To determine the interrelation between the setting of goals by the same human being at the various stages of his life and under diverse given conditions, and further between goals of different human beings, who, as contemporaries, are associated more or less closely, and finally the goals of human beings in different historical epochs – to determine these is one of the most important tasks of research in social science. Such a research must presuppose these goal systems as determined to a large extent, in order to be able to comprehend and to predict social behavior.

The results of our reflections in this section will permit us to take positions on the cardinal question of the *Methodenstreit*: in particular, they will prove to be an adequate foundation for the analysis of the relation of 'is' to 'ought' (descriptive and normative methods), the postulate of value-free social science, and the principle of value relations in the cultural sciences.

6. Metaphysics and the Theory of Science

After having discussed specific methodological issues in the preceding sections, we shall now discuss the significance of methodological thought for science in more general terms. In the first place, this significance lies in reflections on the goals of scientific thought; secondly, on the road that leads to these goals. After further reflection, and especially when taking into consideration the results of our analysis of *a priori* thought, we come to the conclusion, in opposition to the prevailing view, that a sharp division between science on the one hand, and methodology as theory of science on the other, does not exist; for reflection on the goals and pathways of scientific procedure is an indispensable part of scientific thought itself. Therefore the debate about the significance for science of methodological discussion, can only involve the question, how far and how deeply such reflection should be carried out.

This insight makes it possible to take a clear position with respect to the typical objections, which have been raised against methodology. In essence, they amount to this: that because of the disproportionately great effort he invests in preparatory work, the methodologist never manages to arrive at the actual topics of science. Epigrams like 'Because he is continuously sharpening his knife, the methodologist never manages to cut anything' or 'Because he just keeps on saddling his horse all the time, he never gets to ride it,' express this view. It would be vain to reply to this, that the labor of sharpening knives is an efficacious preparation for cutting, as saddling a horse is for riding; for what is being alleged here is not the superfluity of any and all reflection within scientific thought, but the disproportionate intellectual effort in this reflection as compared to its material results. However, the unsound character of these objections against methodology can be shown by the following considerations:

The nature of theoretical reflection is not adequately comprehended if we conceive of it as a more or less significant secondary *process that accompanies* the attainment of scientific knowledge; rather, the higher clarity and distinctness, which is the result at which reflection aims, is also set as a postulate with every purely scientific posing of a question. Every science *qua* science is never merely concerned with the formulation of propositions which will prove themselves in their practical application; rather it wants to reach its proposition by way of *insightful grounds*, and this means it wants to achieve a clear insight into the interconnections of thought and being.

Because, however, these interconnections of thought and interconnections of being, have a multiplicity of levels, persistent reflection will not be content to stop at the point where, by binding a problem to a specific material, its structure is obscured, but it will seek instead to determine its degree of generality, and with that its full scope. It is only in this that the *quasi-independence* of methodological reflection lies, as contrasted with specific research in the special sciences. But we say *quasi*-independence, because we have recognized that universals are not *ante rem* but *in re*.

This leads us directly to the second decisive argument against the opponents of methodological investigation. This lies in that these opponents, in spite of their endeavor to remain as close to the facts as possible, indeed because of it, misunderstand the essential nature of facts; for they see in these facts, unstructured givens grasped by pure receptivity, which are supposed to form the altogether ultimate elements of knowledge, and thus they fail to notice how much theory is already contained in the facts. *The quarrel between theorists and the opponents of theory thus in no way can be characterized by the confrontation 'here theory – there the facts'; in fact this opposition should be formulated to read: 'here explicit presuppositions – there implicit presuppositions'.*

However, it must be emphasized as strongly as possible, that this argument is by no means intended to issue a proclamation of freedom for *pseudomethodology*, which unfortunately occupies a position of broad scope in the social sciences, and has contributed much to discrediting the theory of science and philosophical investigation in general. Its mode of procedure consists in invoking as ultimate arbiter, reason or feeling or existential condition, and thus dogmatically decreeing certain goals of research to be the sole essential ones, and certain paths of research as the only correct ones. Now, to be sure, speculative dogmatism can be called methodology insofar as its results, which consist in establishing goals and paths of research, are of the same kind as those of 'true' methodology; but in its entire tendency, this dogmatism is unmethodological, for methodological reflection means the posing of a *question of confirmation*; while every question of this kind is cut off by dogmatism.

Of course, it must not be overlooked, that the two concepts, confronted with each other here, of true methodology and dogmatic speculation, must be understood as limiting concepts. For the experience of the history of intellectual doctrines, shows that in the development of science and philosophy, both ways of thought are almost always closely linked. This can even be shown quite readily in Kant's philosophy, perhaps the most magnificent breakthrough of critical reflection in the history of human thought. The epistemological situation becomes still more difficult to survey due to the fact that dogmatic speculation hardly ever operates in a vacuum, but has a broader or narrower *fundamentum in re*, which moves it close to scientific hypothesis. What distinguishes it from the latter is not the content, but the claim to validity of the results, the claim to absolute, irrefutable validity. For if in all other respects the term 'metaphysics' can hardly be employed unequivocally, there is unanimity among almost all philosophical schools, and thus both among metaphysicians and antimetaphysicians, that the validity of metaphysical propositions is conceived as absolute, not requiring verification by experience and never to be disproved. (Thus the difference between metaphysicians and anti-metaphysicians is that the question of the existence of metaphysical propositions is answered affirmatively by the former, negatively by the latter.) Differences in opinion, however, arise as soon as the sources of this 'absolute validity' are to be determined. We have touched on this point already at the beginning of our first section, and here only wish to point to an especially important distinction, which leads to the confrontation between '*irrational* metaphysics' and '*rational* metaphysics'.

For the 'absolute validity' of propositions can either be conceived in such a manner that these propositions are not only exempted from confirmation by experience, but by any sort of confirmation at all; while, conversely, any experience that conflicts with them is invalidated. Such propositions, we call 'dogmas' and emphasize that they are comprehensible not by means of reason but by means of faith. Consequently, they are distinguished from the basic principles (axioms) of a science constructed as a hypothetical deductive system by the fact that their falsification does not only appear to be suspended, but is excluded in principle.

This claim to validity we want to designate as that of 'irrational metaphysics' and confront it to 'rational metaphysics', which seeks to secure its claims to supremacy over all empirical knowledge – a supremacy conceived in the same way as in irrational metaphysics – by pointing to specific sources of knowledge, as for example, the intuition of essence in Plato's sense. In pursuit of these efforts on the part of rational metaphysics to arrive at knowledge of true being, we then arrive at the formation of 'meaningless statements'(illusory statements).

We now have to investigate the important question of the *application* of the different varieties of metaphysics to scientific thought. In this we have to clarify, especially concerning the manner in which the intelligibility of 'true being' in thought is conceived in each case, whether it is assumed to be unknowable in principle (transintelligible) or only knowable under special circumstances – only by certain human beings or only under unusual conditions (for example, ecstasy or mystical meditation) – or finally, whether it is thought to be accessible to every human being in full command of his senses, if he only were to familiarize himself with certain presuppositions and (possibly) assume a certain basic approach.

A completely consistent agnosticism is only to be viewed as a limiting case, for it does not permit any possibility of making statements. Therefore, in practice, the assumption of any *transintelligible* always appears in such a manner that though human knowledge, or certain parts of human knowledge, like logico-mathematical knowledge, are indeed conceived as points of departure for thought about 'true being', and that, *per analogiam,* this can be described by means of such knowledge; but at the same time it is declared that the human intellect, in its finiteness, cannot progress beyond one stretch of the road which leads to the absolute. In this case the typical content of metaphysical statements is, on the one hand, a declaration of the limitation of scientific thought with respect to its kind (thought which is down to earth, this-worldly) and its extension, and on the other hand, an ordering with respect to rank within such thought (for example, according higher rank to *a priori* knowledge than to *a posteriori* knowledge, and to inner experience than to external experience).

However, more precise analysis will soon show that the relation between these metaphysical propositions and consequences drawn from them for scientific thought are by no means of the kind that the former were unambiguously *prior* to the latter. Rather the state of affairs is as follows: The task set for metaphysical thought is not different from the tasks of everyday thought or scientific thought – it consists in explaining certain phenomena which come into view – and the postulates of the greatest possible unity and simplicity of the explanation also are the same in both cases.[108] Furthermore, as we have emphasized several times already, prescientific and scientific experience also transcend the given, as every experiential rule projects an open horizon of results to be expected. This is true even for every ascertainment of facts, as the determination of their being-thus-not-otherwise [*So-und-nicht-anders-Sein*] contains within it a wealth of anticipations.

Accordingly, with respect to metaphysical explanation, we can ask quite analogous questions to those asked with respect to scientific explanation, namely: 'What is to be explained?' 'On what presuppositions does the explanation rest?' 'What requirements must the explanation satisfy?'

Now as far as the first question is concerned, we will begin with a brief and necessarily quite incomplete survey of the multiplicity of facts which already impress themselves upon the *primitives* requiring explanation. From observation of the inanimate external world, which, to be sure, only gradually is distinguished from the animate world, and the world of mental/spiritual things, emerge the general problems of the way things change in the external world; and correlate to this, that of constancy within change. Particular phenomena awaiting explanation include the regularities in the course of terrestrial and sidereal events, which already are open to simple observation – for example the rising and setting of the sun, the moon and the stars, the changes of the tides and the seasons. Among the facts of the animate external world that impress themselves on our awareness and demand explanation are the growth and decay of plants, as well as procreation, the birth and death of animals and human beings. All these phenomena also appear to be marked with value quality concerning their utility and harmfulness and the question arises from what do these 'value qualities' arise?

[108] This was impressively described by Ernst Cassirer for that part of metaphysical thinking which can be designated as 'mythical' thinking. See his '*Philosophie der symbolischen Formen*'. Part Two, *Mythical Thinking*, (Berlin 1925).

Of the lived phenomena of inner experience, those of action and of what we passively suffer, in the narrower sense, thus our experience of our feelings, and in the broader sense embracing our sensibilities as well; further the lived experiences of sleep and dreams, of joy and fear, of love and hate, all these are of special interest to us, and therefore especially in need of explanation.

Of social experiences, those appear above all to be of significance which influence the behavior of our fellow human beings, i.e., those concerning friends and enemies.

Added to these primitive experiences are those which already presuppose a *higher level* of the capacity for concentration, and abstraction, as well as critical reflection. Here the following need to be listed; first of all, the known facts of fallibility, of distortions of perspective, of the 'subjectivity' of sense perception, the distinctive character of conceptual thought and the phenomenon of language, mathematical thought and its application to knowledge of nature, the discovery of self-consciousness, the dualism of body and mind, the distinction between fantasy and reality, the characterological differences between prudence and a passionate nature, the aesthetic opposition of beauty and ugliness, artistic skill and artlessness, harmony and disharmony, and finally the teleological distinctions among expediency, purposelessness and unsuitability, where the human body appears as the model of efficacy.

Now whenever explanations of these are analyzed, it must be taken into consideration that not only do the materials to be interpreted consist of experiences, but the principles with the help of which the interpretation is to be carried out – the *schemata of interpretation* – are also derived from experience; for that which transcends experience cannot be characterized positively in any way. Therefore the content of metaphysical propositions must be composed of the elements of experiential knowledge, not otherwise than the propositions of everyday thought or of science. Thus the distinguishing feature must be sought in the kind of linkage that exists between these elements, and in this also lies the *root of the meaninglessness of metaphysical propositions.*

In order to recognize this, let us consider the third of the questions we have raised, namely, *what requirements are set for metaphysical interpretation*? We have just established, that the cognitive endeavor which is supposed to find its fulfillment through the metaphysical interpretation is in principle no different from that which is to be fulfilled by scientific interpretation. But what distinguishes the former from the latter are the exaggerated requirements – i.e., requirements incompatible with each other – imposed upon it.

The most important of these requirements are the following:

1. The propositions of metaphysics are to have universal validity.
2. In spite of the circumstance that the principles of the world are contained in them, they are not supposed to require confirmation through facts.
3. They are to possess a special inner evidence, that represents an autonomous criterion of truth, or else to be capable of being derived from evident propositions by means of rational (discursive) thought.

4. In contrast to the laws of natural science, which only specify conditional relationships between events, they are supposed to make the totality of all events comprehensible in terms of a *single* origin.
5. In contrast to scientific thought, they are not only supposed to offer the correct means for the achievement of given goals, but also point to the ultimately correct goals.

However, the point of departure of these requirements lies in turn in *specific knowledge*. Thus the demand for universality appears to be realized in logic and mathematics, which are valid for all objects generally. The knowledge conveyed by those sciences is regarded as a model for the alleged *a prioristic* character of metaphysics, and for its inner evidence. Ever since Plato, geometry, especially, has been adduced again and again as evidence for the possibility of making *a priori* statements about the world; that is why Kant, in his refutation of metaphysical dogmatism, had to show above all in his *Critique of Pure Reason* that the synthetic judgments *a priori* of mathematics could find no analogy in metaphysics. To be sure, as already noted, post-Kantian developments in theory of knowledge have shown that Kant's theory of mathematics was untenable in the point relevant here.

The model for the fourth of the requirements named, according to which metaphysical propositions were to shed light on the origin of all beings, and thus on the *prima causa* of the world, lies in the inner experience of the genesis of action in the will. Here too, as in inner experience generally, specific evidence is apparently functioning as an unmistakable sign of truth – but as we have seen, only apparently.

In the previous section we have already presented what needs to be said about the fifth postulate.

If we now test these postulates for their compatibility then we will recognize without any difficulty that it does not exist, that the postulates cannot be combined with each other without contradiction. The epistemological fact, however, that this incompatibility has been overlooked time and again can only be traced back to the failure to recognize that the *validity* of that knowledge which is always presented as the model for metaphysical propositions is linked inseparably with the *limits* of such knowledge. Thus the universality of logic and pure mathematics is a consequence of their formal character, which excludes gaining every and any material consequences from mathematical propositions. Their independence from experience is based on the fact that logical and mathematical propositions contain no reference whatsoever to the factual, and therefore cannot come into contradiction with any ascertainment of fact. Finally, the evidence which is attributed to these propositions lies in the complete clarity of the presuppositions, on the one hand, and of their links, on the other; but in this, again, an important limitation is contained, namely the renunciation of gaining 'new' knowledge during the various stages of the procedure due to hidden (implicit) presuppositions – i.e., new knowledge not contained in the presuppositions.

Metaphysics, however, wants to link universality with material content, and a priori condition with reference to facts, and, despite its requirement for the highest evidence of its propositions, relinquishes clarity where it establishes the claim that its objects have at least partial transintelligibility for human thought. It is readily understandable that these exaggerated requirements must lead to the formation of still more excessive – which however also means contradictory – concepts.

Requirements (4) and (5) are already inconsistent when taken by themselves. Let us begin by looking at the first of these, and then we will recognize that the concept of origin as exemplified in the experience of the relation between will and action must be conceived as an event in time, and therefore necessarily presupposes preceding events. The idea of a *prima causa* as an absolute beginning cannot be conceived with *clarity*.

It is quite similar with the concept of *absolute* correctness, as we have already argued in the last section. Because the concept of correctness is defined as a relational concept, every attempt to extract it from the nexus of relations leads to nonsense; the conception of absolute values, or also of absolutely correct goals, cannot withstand reflection. The status of the rest of the metaphysical concepts which have played such an important role in the history of philosophy is quite analogous, as, for example that of the 'Ding and sich' [thing-in-itself] or of 'objective spirit'.

The considerations just presented permit us now to give a clear answer to the question of the *application of metaphysical propositions* to scientific thought. At first, this question seems to be settled quite readily as a pseudo-problem by pointing out that meaningless propositions are not propositions at all. And this in fact is the position taken by *positivism* toward this problem. But what occupies us here is, after all, not only the epistemological question – in the narrower sense – of the content of metaphysical propositions – that, to be sure, is settled by the demonstration of their senselessness – but rather the question of the psychology of knowledge, or also, of intellectual history: what significance do metaphysical propositions have for the state of knowledge in the social sciences?

Therefore we have to consider the following: even if by means of the combination of mutually exclusive elements, no authentic concepts or propositions can be formed, still, propositions can be formed from the *partial conceptions* contained in them or from the *associated conceptions* linked to them (in view of the mode of obscurity which is characteristic of metaphysical propositions, these two spheres of conceptions cannot be clearly distinguished from each other). To be sure, what we will have here in most cases will not be true derivations, not deductions in the strict sense, but merely considerations of plausibility. The theses involved are frequently of great vagueness, often they even lend themselves to opposing interpretations.

One of the most important of these theses in the history of philosophy is that of the *perfection of the world*. Related to nature at first, this can be conceived in quite different ways, even when, as has usually happened, consistent lawfulness and the greatest possible simplicity were considered to be the 'attributes' of this perfection – today we would say, the elements of the concept of perfection. This will be recognized immediately once the concept of 'simplicity' is clarified. For then it will be seen that not even in mathematics is this concept so precisely defined 'from the outset', i.e., without more or less arbitrary conventions, that unambiguous decisions about the relative degree of simplicity of any two arbitrary given functions would result from it. We have here the characteristic case for the transition from pre-scientific to scientific thought that we believe we have already clearly defined the extension of a concept if we succeed in stating (concerning certain objects of knowledge) that without doubt they are subsumed under the concept, and of others that they are not. But this thin appearance of a well-defined character vanishes, as soon as we attempt to state explicitly the necessary and sufficient conditions for that subsumption.[09]

But it was precisely the *vagueness* of these principles, which based themselves on, or else appealed to, metaphysical propositions, which made it possible that at certain stages in the development of the sciences, they could be regarded as significant *impulses* for research. This was especially true for the creation of classical natural science, for thereby it became possible for the powerful thinkers of that time, especially Kepler and Galileo – half consciously, half unconsciously – initially to conceive the results of their syntheses of empirical observations, and fundamental reflections about the structure of experience, in terms of metaphysical propositions which were accepted by them as traditional beliefs, hardly to be doubted, in order then again to 'deduce' the results from the principles. This concordance of laws, derived from experience with the truths of fundamental beliefs, strengthened their faith in the validity of their work and their eagerness to carry on further research to an extraordinary degree.

Perhaps the most magnificent example of this linking of metaphysics and exact science is offered by Leibniz. Whoever studies his *Discourse on Metaphysics* will find the astonishing results Leibniz arrived at by metaphysical-theological speculation to be amazing – only so long as he does not understand the complex facts of the psychology of knowledge we have just briefly presented. This is true especially for the sources of Leibniz's conviction concerning the constancy of world events (which incidentally, from the time of High Scholasticism on, was considered to be an attribute of the perfection of nature), and the far-reaching inferences which he drew from this. As is well-known, these also stand in close relationship with those of his mathematical investigations, which led him to the discovery of the infinitesimal calculus.

But just one of Leibniz's most important insights, in which he was 200 years ahead of his time, namely, that space is nothing other than the ordering of bodies with respect to each other, allows us to see how independent, actually, the result he secured was from its alleged metaphysical presuppositions. In his third letter to Clarke, Leibniz argues as follows:

> Now from hence it follows, (supposing space to be something in itself, besides the order of bodies among themselves,) that it is impossible there should be a reason why God, preserving the same situations of bodies among themselves, should have placed them in space after one certain particular manner, and not otherwise; why every thing was not placed the quite contrary way, for instance, by changing East into West. But if space is nothing else, but that order of relation; and is nothing at all without bodies, but the possibility of placing them; then those two states, the one such as it now is, the other supposed to be the quite contrary way, would not at all differ from one another. Their difference therefore is only to be found in our chimerical supposition of the reality of space in itself.

[109] The lack of clarity just described is also responsible for most of the paradoxes of logic and mathematics. Compare Kaufmann, *The Infinite in Mathematics,* ch. 7 p. 190 ff.

> But in truth the one would exactly be the same thing as the other, they being absolutely indiscernible; and consequently there is no room to enquire after a reason of the preference of the one to the other.[110]

Leibnitz's great rival Newton, on the other hand, though as pious as Leibnitz and taking his departure ultimately from the same theological premises, arrives from these at precisely the opposite result, one that has momentous significance for this system, in which 'true' and 'relative' motion are distinguished on the basis of the facts of dynamics (for instance, centrifugal forces). For his theses:

> Absolute space, due to its nature and without relation to external objects, remains always the same and immovable;
>
> Absolute movement is the transition of a body from one absolute place (the part of space which the body occupies) to another absolute place,[111]

is based on the doctrine, influenced by the theology of Henry Moore, of space as God's sensorium, which therefore must be conceived as independent of things. As is well-known, this basic dogma of classical mechanics was invalidated only by Einstein's theory of relativity.

The above should have made clear that the 'consequences of metaphysical propositions', which appear as guiding ideas of scientific investigation, can be separated completely from their alleged presuppositions. But *de facto* this isolation is rarely carried out in such a radical manner that the claim of absolute validity of these guiding ideas would thereby also be shaken. This is shown especially strikingly in, for instance, the history of the principle of causality.[112]

Therefore, wherever propositions with a claim to being irrefutable, or in any case of possessing universal validity, appear *within the framework of scientific procedure*, whether these appeal to their metaphysical origin or declare that they are laws of reason, it is the first task of the critique of method to subject these claims as such to its critique, and to reject them. But this critique by no means decides anything with respect to the *empirical validity* of the theses under discussion, and therefore anyone who, after carrying out this critique, considers his task completed is doing incomplete methodological work.

With this in mind, in the following analyses of the theory of the social sciences we will rigorously maintain the separation of these phases of methodological critique and thus avoid the danger of losing sight of the true problems in the resolution of pseudo-problems [*Scheinprobleme*]. This will be especially important when we discuss the doctrine that certain research methods are the sole correct ones, and where the question of absolute values forms the subject of controversy. When the latter is the case, then – in the sense of the analyses of our previous section – the chief methodological task will consist most frequently in clearly grasping the underlying goals, whereby the character of value judgments, as acts *in which a position* [*or attitude*] *is taken*, will emerge sharply.

---

[110] *Hauptschriften* (Main Works) edited by Cassirer, I., p. 135.

[111] *Naturalis philosophiae principia mathematica*, pp. 6–12.

[112] This can be shown in particular through the fiction of Laplace's demon, by which determinism received its classical formulation.

7. Proposal for a Universal Methodological Schema

As conclusion of our general methodological analyses, we wish to exemplify the significance of methodology further in terms of one of the most important of its tasks, namely the *preliminary design of more or less general schemata of problem structures.*

For if we wish to comprehend the substantial issues of the methodological dispute [*Methodenstreit*], we must not operate with sloganistic characterizations of the contending research tendencies, such as 'research in natural science versus research in the 'human' science', 'individualistic versus universalistic method', 'value-free versus normative science', without investigating whether certain modes of procedure are unambiguously characterized by these terms; for an examination of the history of doctrines will show that every single one of these terms urgently requires to be much more precise. Such, however, can only be regarded as having been successfully accomplished, when the specific properties of the methods being characterized appear firmly defined. But what is required, not only for the treatment of questions concerning methodological principles of the kind just mentioned, but also for the evaluation of heuristic divergences of a more particular kind, is an overview of the *degrees of freedom of the methods.*

In the construction of such *schemata of pure possibilities,* the significance of methodology for scientific research may perhaps be seen most clearly. But precisely here, methodological thought is in the greatest danger of turning into metaphysical speculation, because this domain of pure possibilities may be conceived as a sphere lying beyond the experiential world.

The following schema is conceived as *universal schema of the theory of science*; the schemata which are to represent the structure of specific problem domains would have to be inserted at their proper place within it.

The first point of departure of a systematic ordering of scientific research is gained by confronting the *posing of problem* with *working on problems*. The former corresponds to the question: 'What does one want to know?' and the latter to the question 'How is this knowledge to be gained?'; thus posing the questions can be designated as setting the *goals* of knowledge, working on problems as indicating the *paths* of knowledge.

If we further recognize that the attainment of knowledge is not an original genesis of knowledge, but takes its departure from what is already known,[113] we will understand without difficulty, that both in characterizing the posing of certain problems as well as in characterizing certain treatments of problems, the specification of what is already known, and is to be utilized in the process of gaining knowledge, is required. Thus the following schema of questions is obtained:

1. What knowledge is presupposed by the posing of the problem?
2. (Additional) knowledge of what kind will count as solution of the problem?
3. In the treatment of the problem, what kind of knowledge is (a) presupposed, (b) acquired?

[113] See above, section "Basic Philosophical Considerations".

If this division is related to the *stages of research*, in which a function, to be defined more precisely, is assigned to knowledge, then we arrive at a further subdivision – uniform for the individual stages – when we undertake to classify knowledge itself according to its chief varieties. Here we have to distinguish:

(a) knowledge of facts;
(b) knowledge of essence (knowledge *a priori*);
(c) knowledge of laws (assumptions).

Let us briefly summarize the principal *results* of our reflection about the relations between these three kinds of knowledge: Knowledge of facts instructs us that at a certain – spatio-temporal or personal-temporal – place something specific has occurred or is occurring. 'Just now a stone is falling from the roof of the house next door', 'Yesterday I reflected for an hour about Goldbach's conjecture', 'Last Friday the Foreign Minister gave a report about the League of Nations' – these are sentences expressing knowledge of events which have taken place or are taking place. The indication that at a certain place just such an event and not something else is occurring or has occurred, thus the indication concerning a now – here (there) – thus or at that time – here (there) – thus presupposes on the one hand knowledge of the positional order, (of where and when), and on the other hand, knowledge of the possible occupation of the various points of this order (the how), as knowledge that has been required and is reproducible; and it is this knowledge that we want to designate as knowledge of essence. It is a prerequisite for all knowledge of facts. That in spite of this, knowledge of essence too 'originates in experience', and therefore can only be characterized by pointing to an example – that, as we have already recognized, only seems paradoxical as long as the multiple stratification in the building-up of experience has not yet been comprehended, and knowledge has been totally identified with explicit predicative knowledge. Common elements of all knowledge of essence are the capacity for identifying and distinguishing, as well as for generalizing and particularizing; they form, as we have established, the core of the formal *a priori* of logic and pure mathematics.

Knowledge of facts and knowledge of essence, however, do not exhaust knowledge, for they do not contain statements about *future events*; but it is part of the meaning of at least all those sciences which are not historical,[114] that predictions are possible. What is still lacking is 'knowledge of laws', which is supposed to make prophesy of future events possible, as well as statements about past events which have not been observed. To be sure, knowledge of laws is not autonomous knowledge; instead, since the laws are to be confirmed by the facts, it is basically assumptions about facts. As in scientific thought, however, these assumptions have a function coordinated with the knowledge of facts, to distinguish these is justified in a proposal for a methodological schema, even when we have grasped that all knowledge of fact contains implicit general assumptions.

[114] Concerning the relation of prediction and historical reconstruction, see below Part Two, section "The Way to Overcome the *Methodenstreit*".

By confronting *predicative* with *pre-predicative* experience, we have gained sight of a further principle of the classification of knowledge that is of fundamental importance for the understanding of scientific thought, namely the distinction of various *levels of clarity and distinctness of thought.* In accordance with the terminological tradition within modern philosophical schools,[115] an object of knowledge will be designated as *clearly* comprehended when it emerges as identifiable within the totality of experience, and therefore is distinguishable from any other objects of knowledge; however, an object is to be called *distinctly* comprehended when its elements and the kind and manner of their connection (synthesis within the object of knowledge), can be *clearly* comprehended within that object of knowledge. From insight into the existence of a multiplicity of degrees of clarity and distinctness of knowledge, emerges the task of clarifying, or making more distinct, knowledge which is as yet unclear or indistinct; and this is to be carried out by reflection on such knowledge.[116]

A basis for further classification is formed by the various *possibilities of ordering* the knowledge which enters into the procedures of inquiry. Important, to begin with, are differences in relative position; for research procedure is not an aggregate of cognitive elements without order, but a process with a determinate direction. Consequently the kind of ordering (position) an element of knowledge receives within the procedure is codetermining for its function within it. For the inductions which it grounds can depend on the manner of its connections with other elements of knowledge, connections which are determined by this ordering.

Another class of ordering principles are principles of *order of derogatio* (order of rank) of the elements of knowledge. We define this concept as follows:

"We wish to say that in a research procedure, an order of derogation of certain elements of knowledge K_1, K_2,... K_n is prescribed if for any three of these, K_i, K_k, K_l, the following is true: (1) In case K_l and K_k are incompatible with each other, then there is a prescription which decides whether K_i derogates K_k or K_k derogates K_i. (2) If K_i derogates K_k and K_k derogates K_i, then K_i also derogates K_l." That a known fact is derogated means here that it is regarded as not decisive for the designated complex of knowledge; this does not necessarily mean however that this particular knowledge is negated, or even subjected to doubt, but merely that it may be reinterpreted; thus, for example by drawing on either real or merely supposed facts (disturbances, changes of data), until then not considered, it can be brought into conformity with the known fact that derogated it. Recall, say, the changes in the quantitative theory of money, by introducing the rate of circulation. Known facts whose capacity for derogating within the framework of a procedure is excluded altogether, we can designate as absolutely valid facts of knowledge *for this procedure*. In case such an order of derogation of the known facts is fixed within the framework of a scientific procedure, we wish to call the position which a certain known fact occupies within this procedure its *status*. Therefore this concept is

[115] It was initiated by Descartes and carried forth in the work of Leibniz as well as Spinoza.

[116] A series of works of Husserl, as yet unpublished, especially his *Logische Studien*, which he was kind enough to let me see, treat this theme in analyses of fundamental importance.

initially defined only for known facts within the process of working on problem; for only here can we speak of a capacity for derogation in the narrower sense. But it can correspondingly also be applied to those known facts which form the solution of problems, if we consider that the process of gaining knowledge – insofar as it is directed towards acquiring knowledge of facts – can never in principle be considered as concluded. Therefore every solution of a problem achieved can be relativized with respect to the ongoing process of gaining knowledge. Thus insofar as this relativization is carried out, we no longer see in certain known facts simply the solution of a problem but only a step along the path to the acquisition of knowledge immeasurable in its full extension. It can then very well be specialized by a rule of derogation, which experiences shall lead to relinquishing, or at least modifying, the 'solution'.

Now, to be distinguished from the order of derogation of known facts within a scientific procedure is another, 'ordering by rank', which we wish to call *'selective ordering'*. We arrive at this concept through the following considerations: The typical pre-scientific situation of knowledge, from which the posing of a scientific problem emerges, is that we want to know the 'essential' [facts] about a more or less vaguely designated domain of knowledge. The treatment of this 'essential' is the task formulated more or less sharply in posing the problem; the 'essential' becomes the theme. Then the decisive question is, according to which viewpoints is the 'essential' selected in each case? This is treated – above all with a view to historical investigation – in Rickert's theory of the value-relation [*Wertbeziehung*]. We will have to deal more closely with it in the second part of this work.

From the principles of choice constitutive for the theme, the heuristic *principles of selection* – in the narrower sense – are to be distinguished; they are frequently indicated by those constitutive principles. They determine the treatment of the problem, i.e., they single out the 'essential facts' from the store of available facts of knowledge, to be acquired according to rigorous rules. With these, in a narrower sense, methodological principles of selection, we will also have to deal in more detail later. Here we only wish to mention that the points of view from which the selection between the various kinds of problem treatments is made, are not difficult to typify; we usually speak here of postulates, and accordingly distinguish postulates of *unity*, *simplicity* and *purity of method*. The procedural sense of these postulates, however, is only rarely apprehended with perfect clarity.

Let us look back at our reflections up to this point. We have characterized scientific inquiry (in a first approximation), as the acquisition of additional knowledge, and accordingly distinguished three stages in it: (a) the designation of the knowledge to be acquired, in which both this designation as such, as well as the possibility of acquisition, presuppose already existing knowledge (posing the problem); (b) the setting up of an ordered chain of judgments, which contain knowledge not already (explicitly) present in the posing of the problem (working on the problem); (c) as final link in this chain, a judgment which contains the knowledge intended in (a) (solution of the problem).

Secondly, we have distinguished from each other, three characters of knowledge contained in every science, indeed in every authentic judgment; namely, knowledge of facts, knowledge of essences and knowledge of laws; thirdly, we have brought into view the varying degrees of clarity and distinctness of knowledge; and, finally, fourth, have fixed the three kinds of ordering of knowledge: – order of position, order of derogation, order of selection.

The schema of variation that can be constituted on the basis of these distinctions, which contains those elements of a theory of science which are independent of the content of knowledge, and therefore also of the sources of knowledge in each case, do not as yet, however, permit the incorporation of those questions – and they occupy an important place within the methodology of the social sciences – which we can comprehend under the title 'methods of the natural sciences or the human sciences'. In order to do this, we must complete our schema by including the distinction between the two sources of knowledge, *external* experience and *inner* experience. Once this is done, however, it is possible to survey possible procedural differences within the social sciences, by means of the relevant substitutions. Therefore we gain an important guide toward the discovery of the material core of the doctrinal views discussed in the *Methodenstreit.*

This will become even clearer if we proceed now to develop a *schema of the objections* which can be raised against a method: (a) objections against the *way the problem* is posed: (1) it is contrary to sense (contradictory); (2) it is ambiguous (for instance, it does not comprehend enough); (3) it is based on false presuppositions; (4) it is uninteresting; (5) it is unfruitful.

1. This objection denies that a problem which appears to exist actually is a problem. It maintains that as soon as one tries to clarify for oneself, what the meaning of the question is, that is contained in the way the problem is posed, i.e., what criteria will determine the answer – then one will discover that such criteria have not been given at all, that therefore no question has in reality been posed. Here the reproach of unclear thinking is raised, which falsely believes it has a grasp on an object of knowledge. As there is also no reply to a (contradictory) question, we do not wish to distinguish it further from a simply meaningless pseudo-question, from pure nonsense.
2. Here the objection is directed against the inadequate characterization of the new knowledge, the acquisition of which is the task set by the posing of the problem. Because of this inadequacy, the undetermined character of this knowledge, resulting from the indication contained in the posing of the problem, remains hidden. Subject to such objection, for example, is setting the task of explaining a phenomenon of a certain kind in a case where it has to a great extent been left undetermined on what basis and by what means the phenomenon is to be explained.
3. Here it is maintained that a part of the knowledge contained in the way the problem is posed is false knowledge (pseudo-knowledge).
4. This objection alleges that a different selection of topics should have been made, as due to the present selection, questions which are not really essential have been posed.
5. Finally, this objection is directed against the choice of topics, because in it, questions are posed which are unsolvable in practice, or at least too complicated.

It is similar with respect to the objections to the *treatment of the problems*. Here too it can be maintained that among the relevant propositions, pseudo-propositions can be found, or that not all the propositions are logically compatible with each other, that though the knowledge contained in this treatment is authentic, the desired solution will not result from it, that false knowledge (pseudo-knowledge) is contained in the treatment, or that it is too clumsy (inelegant). However, we will not speak of an uninteresting treatment of the problem, as analogous to an interesting posing of the question; as the (more or less interesting) goal of the inquiry has already been set, therefore the treatment of the problem can only be judged according to its adequacy to that goal. On the other hand, objections to the treatment will be directed primarily against the ordering of the knowledge available for the solution of the problem, and especially against the manner in which, and the extent to which, the various kinds of knowledge are utilized. Here, for example, we will find the objections against the degree of abstraction of an investigation, and especially objections against 'excessive formalism'.

Finally, objections against the solution of a problem can consist in charging that the supposed solution is called a 'pseudo-solution', and this either for logical reasons (inconsistency, circularity), or because 'confronting it with the facts' shows, that it does not fulfill what has been demanded by the posing of the problem. Finally, objections can be directed against the claimed status of the solution; such an objection obtains when, say, it is denied that a solution possesses the apodictic and perfect validity claimed for it. Such claims, as we have seen, can generally be traced to false interpretations of knowledge of essences, or knowledge of laws; we will encounter this repeatedly in the second part of our work.

The universal scheme offered above will serve us as a guideline in our examination of social theory, even on those occasions when we do not explicitly refer to it.

Part Two. The Dispute over Method in the Social Sciences (*Methodenstreit*)

Preparatory Remarks

One question that has become inflamed again and again is the debate over method [*Methodenstreit*] in the most varied domains of research, is whether a group of more or less sharply delimited domains and results of inquiry have any scientific character at all; and if that be the case, whether they constitute an independent science, or group of such sciences. Let us leave out of consideration the quarrels about the hierarchy of faculties of former times, which largely took place without any theoretical basis, and turn our attention exclusively to those conflicts over method, which were guided by a viewpoint of a systematic classification of the sciences: then, the most important issues in theoretical characterization of the social sciences [as sciences] are the following:

1. Are there, beside the *natural sciences*, independent *human sciences* [*Geisteswissenschaften*], or is such a division solely a symptom of the backward stage of development at which these so-called human sciences still find themselves today, a stage wherein speculative constructions replace the exact knowledge still lacking? This is the thesis of naturalism, most recently in the clarified form of 'physicalism', which takes into account the results of recent research in the natural sciences.
2. Is it not the case that all human sciences (and therefore all social sciences), since they are sciences of human beings – whose specific character lies in that they are beings endowed with feelings and intelligence are only subordinate disciplines of *psychology*? As we have already noted [cross reference to page 40 of the original manuscript] this question grew especially acute for logic, which was also conceived of as a human science.
3. Are the human sciences in their entirety *sciences of Being* [*Seinswissenschaften*], or *normative sciences*, or should a division be made within the sphere of the human sciences into sciences of Being and normative sciences?
4. Are there independent *generalizing* human sciences, or are all human sciences *historical* sciences and therefore subject to the general principles of inquiry that govern the science of history? Supporting this is the entire experience of the mental life of human beings as historical experience, and thus the development of the human mind [*Geist*] can be studied only within the framework of its history: Therefore there is no possibility of transcending the domain of the historical in research in the human sciences.
5. Have the sciences which are usually subsumed under the name '*social sciences*' such far-reaching methodological properties in common that a comprehensive terminology will appear to be justified from the viewpoint of the theory of science?

The fury with which these battles over method have been waged is comprehensible only in terms of *extra-theoretical* motives, but we shall not concern ourselves with these in this work. We are interested in the rational grounds which are brought forward for the superiority of any one method, *not* in the *emotional background* reasons for making such claims. In what follows, we shall deal more with the former; here we wish only briefly to characterize the typical course of such struggles over method.

The postulate of the *unity of science* [*Einheitswissenschaft*], as it can be established either for the entire domain of the sciences, or at least for the entire domain of the empirical sciences, or even only for a partial domain of the empirical sciences – say, the human sciences – is confronted by the postulate of the *autonomy* of the individual sciences in question, which demands *purity of method (avoidance of syncretism in method)*. What is regarded in the [unity of science] thesis as ultimately belonging together, is considered to be *toto coelo* different in the antithetical view; and in so doing, both parties usually call on *logic* to serve as arbitrator between them.

And indeed this appeal to logic is quite appropriate for both parties and promises success, in a limited sense. For the proponents of the idea of a total or partial unity of science, logic as theory of identification and differentiation, is called upon to bring to light, in a rational reconstruction of the intended meanings, what is common to the various modes of inquiry; thereby logic shows the absurdity of radical differences, which would exclude comparison between different modes of inquiry. But, with equal justice, logic is deployed upon the field of battle against the exaggeration of what is held in common. Such exaggeration leads to neglect of important differences and only too frequently the consequences are ambiguities, which appear in language as *equivocations* and turn up in the conclusions as *quaterniones terminorum*, thus as fatal errors in thought. To hunt out such errors in scientific concept formation, and to remove them, forms one of the most important tasks of methodology, and in this sense the postulate of purity of method, and the prohibition of syncretism of method, have often proved to be a trail-blazing achievement for the progress in scientific thought, especially in the last half-century, But, as we have yet to see, not infrequently methodological requirements have been set up in the name of this postulate, which in no way can claim necessary (logical) validity.[117]

We have characterized this confounding of problems of clarification with empirical problems in the preceding general section, and have pointed to the fact that from the viewpoint of the psychology of knowledge they form one of the main roots of exaggerated claims of validity made for certain assertions. In the analysis of the *Methodenstreit*, we will soon find numerous examples of this; at the same time we will find that only after the requisite distinctions have been made here, will the authentic methodological problems be fully revealed. Among these problems, we will recognize that of a clear comprehension of the goals of knowledge, as well as of the principles for selecting the paths of knowledge appropriate for the given goals; here, to be sure, we will find that the division between the goals of knowledge and the paths of knowledge is by no means rigid. In setting the order for the main lines of the following investigation, we wanted to be guided by the cardinal questions in the *Methodenstreit*, that we have just emphasized.

1. The Social Sciences and the Natural Sciences

The assertion that a range of problems, work on these problems, and their solution, are all brought together under the title of 'social sciences' – or also 'human sciences' [*Geisteswissenschaften*] – only deserve the name of science insofar as the method of the natural sciences is used in them, is based on two groups of principal arguments, which, to be sure, are not always kept sharply separated in the actual *Methodenstreit*.

The first group of arguments, which we can call the *behavioristic* arguments, is based on the fundamental thesis that objective knowledge is possible only of objects of the external world. For while events in the external world could be observed in the same manner by all normal human beings, no one has access to the soul or the mind of an *alter ego*. Therefore the knowledge of experience gained by one human

[117] See below Part Two, section "Social Laws and Ideal Types"

being through introspection could not be verified by means of experience of the same sort on the part of fellow human beings. With that, the decision about the scientific character of propositions concerning the life of the soul of any *alter ego* would already have been decided negatively; for the concept of science already contains that of objective knowledge. And the criterion of objectivity is intersubjectivity, especially the intersubjective communicability which allegedly is lacking here. For only such judgments about our fellow human beings which refer to their bodies would be verifiable intersubjectively. By this the paths and the goals of the social sciences are supposed to be prescribed. Their point of departure would have to be observation of 'external' human behavior and its biological and physiological determinations. Its goal, however, would have to be formed by the discovery of *the natural laws of human action*, by means of consistent observation of typical human behavior, observation organized in accordance with the principles of statistics. Accordingly, social research would have to take its departure from the results and the methods of those of the natural sciences which deal with the human body, and above all the human brain.

In most of these arguments the scientific method is thought to be represented by physics, the objectivity of which appears to be guaranteed by its 'formal character'. The great achievement of physics lies in its having organized prescientific data of sensation into a scientific system by means of an appropriate system of coordination, and this would justify the hope that a similar ordering could be carried out for those aggregates of prescientific knowledge, which are called the 'human sciences' or also the 'social sciences'.

From this point of view Fechner's research in psychophysics[118] was greeted with great expectations half a century ago [i.e., 1880], especially his famous *basic law of psychophysics*, which specified a logarithmic function between stimulus and sensation; to be sure, since then these expectation have been fulfilled only in very small measure. A similar [limited] state of affairs holds with respect to the physiology of the brain; a solution of the problem of the psychophysical relationship had been expected.

If the arguments of naturalism concern the criterion of the *objectivity* of knowledge, which is claimed for thought in the natural sciences alone, then the second group of naturalistic theses centers on the claim that it is exclusively the methods of the abstract natural sciences that are capable of furnishing *exact* knowledge, knowledge in the fullest sense. All other disciplines of knowledge, and therefore also specifically those of the social sciences, would only be able to set up rules or to establish tendencies, insofar as they do not 'purify' themselves by adopting these methods; thus they remain stalled in the antechamber of scientific knowledge.

Above all this claim is supported by pointing to the mathematical method, which alone would make it possible to achieve exactness, and guarantee it. Therefore quantification – the replacement of intensive magnitudes by extensive magnitudes – appears to be the true criterion of scientific character. In our First Part we already

[118] It hardly needs to be pointed out that the significance with which the word "psychophysical" is used in the present work does not coincidence with the use Fechner makes of it.

pointed briefly to the speculative roots which gave this idea such tremendous power that it was expounded by the most significant modern philosophical thinkers with all their powers of persuasion. Hermann Cohen furnishes a striking example of this. For Cohen, intensive magnitudes provoked "the most forceful challenge to pure thought."[119]

Besides this naturalism that took its orientation from physics, *biologism*, which works with the analogy of the living organism has played a major role in the various disciplines of the social sciences. However, we do not have to deal more closely with organismic arguments, because the idea of an end or purpose which form the core of the concept of the organism, and is also the basis of the analogies to the organism, is itself derived from inner experience; therefore, on closer examination, the biological-physiological arguments in the *Methodenstreit* turn out not to be naturalistic ones.

To the radical arguments on the part of the naturalists correspond the no less radical arguments of the *anti-naturalists*. But while the naturalistic theses are essentially uniform, those of their opponents split into diverse varieties, the most important of which we will now have to analyze briefly.

First those investigators have to be mentioned who have set themselves the goal of finding laws within the social sciences, which are strictly characteristic of them, and of establishing the social sciences as *nomothetic sciences*, independent of the natural sciences (e.g., Carl Menger[120] and his school). The idea of absolutely strict laws is decisive here, just as in the natural sciences. The fundamental line of argumentation which is typical of these scholars goes as follows: the attempt to subordinate the laws of the soul and the mind to the categories and laws of the natural sciences is condemned to failure from the outset. The investigator in the human sciences does not envy the natural scientist the exactness of his laws and his mathematical method, because this is based on the measurement of phenomena, and is only applicable to such phenomena which, due to their spatial nature, can be measured. Laws of this kind, however, are essentially different from those which are intended to grasp the essential nature of human beings. The natural scientist can *register* the validity of his laws, but he cannot in the end *understand* them. In contrast to this, the inner experience of one's own thought and feelings, and empathy for one's fellows, leads to laws, whose validity can be attained by direct insight, as they flow from sources that lie within ourselves.[121] From this it is concluded that the methods of the human sciences should be distinguished most sharply from those of the natural sciences. Above all a sharp separation should be maintained between *explaining* [*Erklären*] in the natural sciences and *understanding* [*Verstehen*] in the human sciences.

---

[119] Ernst Cassirer, the most eminent among Hermann Cohen's students (the Marburg school) has freed himself from this one-sidedness.

[120] Carl Menger's two chief works are *Grundsätze der Volkswirtschaftslehre*, Vienna 1871 and *Untersuchungen über die Methode der Sozialwissenschaften und der politischen Oekonomie insbesondere*, 1885.

[121] See for instance Wieser "Theorie der gesellschaftlichen Wirtschaft" in *Grundriss der Sozialökonomik*, I Abt., II. Teil, Tübingen 1924, p. 12. Our next section contains a more detailed treatment of the concept of understanding. [verstehen].

Besides the scientists who see the task of the social sciences as the establishment of *laws*, there is another group of thinkers, such as Schmoller[122] and his school, consisting in large part of historians, who deny in principle the possibility of establishing laws in the social sciences. To support this thesis, they not infrequently maintain that in contrast to the external world, as a sphere of rigid causality, the psycho-physical world, and consequently also the social world, would have to be the domain of *free will*, and therefore of causal indeterminism. Though it might be possible, by careful study of historical events, to find rules or tendencies within the development of social groups, nevertheless to regard such rules as strict universal laws would mean to misunderstand the essential nature of human behavior. Thus we see that this group agrees with the naturalists, since they too recognize the distinction between strict laws in the sphere of nature and mere rules in the sphere of the mind; but from this the opposite conclusion to that of the naturalists is drawn. For while the latter demand that the method of the natural sciences also be applied to the sphere of social knowledge, the *historicists* reject this method in principle as *a priori* inadequate for the social sciences.

Finally, among the historical schools which oppose naturalism, the Southwest German school of neo-Kantians (Windelband,[123] Rickert[124]) must also be named; they defend the following thesis: Though the greater part of the natural sciences tends towards *generalization*, while the greater part of the human sciences tends toward *individuation*, nevertheless, there are also *generalizing* (*nomothetic*) human sciences, as well as *individualizing or idiographic* natural sciences (e.g., geology). The goal of the nomothetic sciences is to arrive at statements which are as general as possible. This tendency is expressed in the structure of their concepts and the formulation of their laws. On the other hand, the goal of the idiographic sciences lies in the selection and bringing forth of especially significant facts, where the nature of the significance appears determined by the concept of the *value relation*, which we will have to analyze more closely.

After this brief presentation of the principal theses of naturalism on the one side, and anti-naturalistic doctrines on the other, we want to take our stand with respect to the arguments raised by the two sides. In so doing we shall always have to distinguish sharply – in the sense of the reflections of the First Part of this work – between assertions as such and their claims to apodictic or exclusive validity; accordingly, when we have invalidated the latter, we will not maintain that the theses as such – which method is the correct one, which means, is to be applied with advantage – are to be rejected *a limine*.

Let us begin with the thesis that the natural sciences alone have *objectivity*. The variant in which this doctrine appears today, is behaviorism, which is the relevant form for the methodology of the social sciences and also much more sophisticated

122 See for instance Schmoller's *Grundriss der allgemeinen Volkswirtschaftslehre,* 1st. ed., Leipzig 1900.

123 See for instance his Strassburg rectoral address "Präludien", 3rd. ed., Tübingen 1907, pp. 355–379.

124 See for instance Rickert's *Die Grenzen der naturwissenschaftlichen Begriffsbildung,* 2nd. Ed., Tübingen 1913.

in comparison with the eighteenth and nineteenth century naturalism. Within behaviorism itself we have to distinguish between an older and a more recent doctrine, which diverge broadly in their argumentation.

The *older behaviorist* doctrine, whose best-known representative is J. B.Watson,[125] originated at the beginning of the century in America, in the field of psychological research and especially in animal psychology. According to this behaviorism, psychology must limit itself to external facts – thus above all the bodily behavior of the research subject – both in its descriptions as well as its explanations, for only such facts can be verified intersubjectively. Here we have a radicalized position with respect to the controversies that go far back into the nineteenth century about the value of *self-observation* [introspection] for psychological inquiry. As a reaction against hasty psychic-mentalistic interpretation of animal behavior, obtained by 'empathy', as well uncritical operation with 'psychic forces' as explanatory principles, the achievements of this approach are most praiseworthy; but, as critical analyses have already and repeatedly established, it has never succeeded in carrying out its basic program with any kind of consistency. Just at the most important points behaviorism no longer remains purely a description of physical (bodily) phenomena, but instead the body is treated as *living body* [*Leib*], as the expressive field for interpreting psycho-physical events. But be that as it may, the decision about the significance of the behaviorist method, as propagated by Watson and his school, is a question of fact; new discoveries for instance in the physiology of the brain, linked to new possibilities of observing brain processes in living human beings and animals, could shift the research situation greatly toward behavioristic doctrine; for, in principle, the possibility of explaining the psycho-physical by means of the physical exists no less than that of explaining colors and sounds by means of transverse or longitudinal oscillations.

From the viewpoint of the theory of science, a more recent behaviorism is much more interesting: the doctrine called '*physicalism*' by its spokesmen. Its two most important representatives are Rudolf Carnap and Otto Neurath, who developed the fundamentals of their doctrine in a series of articles in *Erkenntnis* and other philosophical journals, and defended them against objections.[126]

---

[125] Compare Watson *Der Behaviorismus*,(German translation) Berlin 1930.

[126] The following works must be mentioned: Carnap: 1, 'Die physikalische Sprache als Universalsprache der Wissenschaft', *Erkenntnis,* 2, 432; 2. 'Psychologie in physikalischer Sprache', *Erkenntnis* 3, p. 107; (Psychology in the Physicalistic Language of Physics): 3. 'Ueber Protokollsätze', (On Protocol Sentences), *Erkenntnis* 3, p. 215; 4. *Logische Syntax der Sprache* (Logical Syntax of Language), *Schriften zur wissenschaftlichen Weltauffassung,* vol 8, Vienna 1934; 'Les concepts psychologiques et les concepts physiques sont il foncierement different?' (Are concepts of psychology and concepts of physics fundamentally different?) *Revue de Synthese*, vol. X n. 1 pp. 43–53. Otto Neurath: 1. 'Physicalisme' in *Scientia* , November 1931, p. 117; 2. 'Protokollsätze', *Erkenntnis* 3, p. 204; 3. 'Einheitswissenschaft und Psychologie', Heft (Number) 1 of the collection *Einheitwissenschaft* (Unified Science), Vienna 1933; 4. *Empirische Soziologie, Schriften zur wissenschaftlichen Weltauffassung.* Vol. 5, Vienna 1931. For what follows also compare the notable critique of behaviourism by Roman Ingarden, 'L'essai logistique d'une refonte de la philosophie', *Revue Philosophique* 1935, pp. 137–159.

The basic thesis of this doctrine, as it relates to our main theme, goes as follows: every verifiable statement of psychology can be translated into a statement of physics – without any change in its content. This thesis can be dissected into the two assertions: that every statement of psychology can be translated into a statement about space-time events, and every statement about space-time events can be translated into a statement of physics.

The second thesis is not directly connected with the problem of behaviorism; therefore we will disregard it for the present, and following the language of physicalism, call every statement about space-time events a physicalist statement.

The crux of the argument of physicalism lies in the analysis of the *meaning of sentences*, which leads to the result that the meaning of a sentence is determined by its *truth conditions*, which are determined by the pertinent *control statements*.[127] Accordingly, an (illusory) sentence is without meaning if it is in principle not verifiable through observations, and two sentences have the same meaning only when they are true, or false, under the same conditions. In our clarification, we will counterpoise two examples to each other, which are taken from an excellent exposition of physicalist doctrine, recently appearing in French, by one of its adherents, C.G. Hempel[128]:

1. *Physicalist sentence*:

   The temperature in the physics laboratory in such a location today at 1 o'clock is 23.4 °C.

   Examples of control sentences for this statement:

   A mercury thermometer with a Celsius scale shows a coincidence between the top of the mercury column at the 23.4° mark.

   An 'alcohol thermometer' shows another precisely established coincidence.

   The pointer of a galvanometer connected to a thermocouple makes a certain swing if the element (couple) is placed at a certain location at the time indicated.

2. *Psychological statement*:

   Paul has a toothache.

   (a) Paul cries and makes gestures of such and such a kind.
   (b) Asked, 'What ails you?' Paul utters the words 'I have a toothache'.
   (c) Further investigation reveals a decaying molar with exposed pulp.
   (d) Paul's blood pressure, his digestive processes, his reaction show such and such changes.
   (e) In Paul's central nervous system, processes of such and such a kind are taking place.

The argument then states, briefly summarized: the control sentences for any given sentence of psychology are all sentences with space-time determinations: the [given] sentence can thus be traced back to (translated into) these: it is an abbreviated

[127] Compare the analysis carried out in our Part One, section "Basic Philosophical Considerations" on the 'meaning of sentences'

[128] C. G. Hempel, 'Analyse logique de la psychologie', *Revue de Synthese*, vol. 10, no. 1, pp. 27–42, p. 33.

formulation of these sentences. Pointing to introspection as counter-evidence is not valid, as statements based on introspection are only verifiable (intersubjectively controllable) when they permit a behavioristic interpretation, and thus only then possess the character of scientific knowledge. This is then a presentation of the fundamental thesis of physicalism; let us now turn to critical analysis of this doctrine.

The main objection directed against this argument – aside from others to be mentioned later – is that the control sentences of psychological statements are not exclusively physical statements – even in the broadest sense understood by physicalism. For the purpose of giving the reasons for this objection, a fundamental remark about the concept of the control sentence as it is understood by Carnap and the scholars close to him, must be made. Carnap defines this concept as follows:

> Given a synthetic sentence (i.e., a sentence about facts), called E, two cases can be distinguished, to be sure, without seeking thereby to draw excessively sharp boundaries between them. In the first case, E can be controlled by direct experience. In the second case we have to control E indirectly, in that we derive from E and other – scientifically accepted – statements certain directly controllable sentences (control sentences for E).[129]

Basing ourselves on this definition of control sentences which includes a completely appropriate characterization of the verification process, let us now investigate which control sentences can typically be employed for assertions about social facts. Let us take the simple state of affairs of a conversation between A and B: A has directed a question to B, and the assertion is made – let us say by A himself – that B has understood A's question. In what manner now will the assertion be controlled in the great majority of such cases? Obviously, by the *answer* B gives. If he answers appropriately it will be regarded as confirmation that he has understood that he has been asked, and also that he has understood the *meaning* of the question. A control sentence for A's assertion will thus state: "B will give an appropriate answer to A," and the "scientifically accepted sentences" needed for the derivation of the control sentence, will in the first place be insights about the kind of relationship holding between thinking and speaking; or also – if one conceives the external facts contained in the speech act (body movements) in isolation – the relationship between psychological facts and physicalistic facts. Now for the theory of science it is a fundamental difference – indeed, that between method in the natural sciences and method in the sciences of the mind – whether in the interpretation of the observations made, i.e., in their incorporation within a complex of experience, general statements which concern the *coordination between physicalistic (external) and psychological facts*, are *also utilized* or not. As is well-known, in the period of *mythical* thinking, natural science was also pursued by the psychological method; it was thought that standing behind the individual phenomena of nature, there were 'nature gods' who determined the course of these phenomena willfully; therefore people sought to discover this will (by means of oracles) and to influence it (by means of

[129] Carnap (5) p. 44.

prescribed magical acts). But the predictions arrived at in this manner did not prove correct, or successful often enough, and so it came to "*depriving*" nature of its soul, in one sense, and to the separation of nature from psychic-mental phenomena, whose description and adequate explanation (i.e., leading to useful predictions) could only be accomplished by comprehending them as events of the same kinds as one's own psychic acts given in inner experience (introspection).

That this fundamental methodological divergence was overlooked by the physicalists, or neglected as inessential – whereby an overextension of the idea of *unified science* resulted – probably has its deepest roots in the mistaken conception of the structure of experience, which originated as the *sensualist* tradition and has only been overcome progressively, but still not radically enough, in recent times. What is involved here is the failure to recognize the epistemological fact – a failure already criticized in the first part of this book – that all that is given in thought, thus all about which statements can be made, already is incorporated within a complex of experience and thus within general propositions, i.e., laden with interpretation, that thus the 'pure given' as the correlate of pure receptivity is a fiction. But according to this conception – as we have shown above – all of prescientific and scientific knowledge is an epitome of the logical linking of atomic sentences (statements) in which, according to sensualistic prejudice, the elementary propositions are characterized as the simple reproductions of external observation, as 'perceptual protocols'. This then leads to the following interpretation of statements about other human beings: All knowledge about one's fellow men arises by means of establishing links between observations of their body movements, and thus all control statements must be directed toward such observations, and accordingly the thesis, that sentences about one's fellow men have an added meaning that goes beyond that, is uncontrollable, unverifiable in principle, and thus unscientific (metaphysical).

Not many words are required to refute this argument. Though a human being cannot directly experience the thinking of another human being, he can assume hypothetically that a process *analogical* to his own thought goes on at another 'place', and this hypothesis can prove right in the same way as a physical hypothesis about objects which are not accessible to direct observation, e.g., atoms. To be sure, there is the important difference here that the contents of other psyches [*das Fremdpsychische*] are *in principle* inaccessible to direct comprehension (observation), while in the case of space-time objects, under certain circumstances such access appears to be only technically impossible; yet in spite of that the content of other psyches is not a *qualitas occulta*. For we know very well what the content of other psyches is, namely something psychic, i.e., *something like one's own psyche*; and there is no difference in principle between making one's fellow man understand what 'red' is and making him understand what 'thinking' is. Both have to be 'shown', where this 'shown' is an invitation to assume a certain psychic attunement.

Now to be sure, the doctrine of atomic sentences was not accepted by Neurath, and also abandoned by Carnap in recent years; but the assertion has been retained that the content of knowledge of all propositional statements about an alter ego is confined to findings or assumptions about his body; although now the main obstacle in the way of a correct analysis of these statements had been removed. Not until the most recent paper of Carnap[130] was a decisive turn presented here, although he does not appear to be fully aware of its import, as he recognizes expressly the paper of Hempel cited above (which appeared at the same time) as a clear presentation of the basic ideas of physicalism, as developed by Carnap and Neurath; and he describes his own recent presentation merely as an attempt to consider the physicalist thesis from a new point of view and to render it more precise.[131]

Here the identity of meaning of psychological and physicalist sentences (containing solely space-time specifications) is no longer asserted; all that is claimed is that physical and psychological sentences are *equipollent* (of the same content), i.e., are mutually derivable from each other, and that therefore one can be replaced by the other. This divergence (or change) is of the greatest significance, for according to Carnap's definition, one can speak of mutual derivability not only when each of two statements can be deduced from the other by the application of logical-mathematical rules alone, but also when beside these rules *extralogical* rules are applied, namely rules which are based on generally recognized laws of nature. In the first case, Carnap speaks of *logical* equipollence, in the second case of *natural-law* (physical) equipollence, and he claims that every psychological statement is equipollent to a physicalist statement by natural law equipollence.

Before we test this claim, it is necessary to point emphatically to the radical difference of logical equipollence from natural-law equipollence, a distinction which tends to be obscured by making the terminology similar. As follows directly from Carnap's definition, two statements have natural-law equipollence when according to the contemporary state of science it can be assumed that the facts asserted in the two statements always occur together. Here then, we are dealing with a 'matter of fact' [Kaufmann's English – Ed.] the assertion of which may prove true today and false tomorrow or vice versa. 'Logical equipollence' on the other hand is preserved against any such change, as it does not refer to facts as such at all, but to propositions and concepts. We have characterized this difference more precisely above in the confrontation of the concepts of implication and logical inference.

From this follows the critique of Carnap's application of the concept of natural-law equipollence in setting forth his grounds for physicalism. Carnap gives an example of his train of thought with that model clarity that is a characteristic of all his writings, with the statement: "Mr. A is angry at present." To simplify the presentation, a terminological specification is made

> Let us designate briefly as anger$_{ps}$ the psychic character of anger, the feeling of anger, the state of consciousness that is called anger. As experience shows, a certain bodily state (more precisely one would have to say, a certain class of bodily states exists, which appears, as one says, parallel to the state anger$_{ps}$. This bodily state we wish to designate as anger$_{ph}$. Then the two statements: (1) 'Mr. A is at present in a state of anger$_{ps}$.' And (2) 'Mr. A is at present is a state of anger$_{ph}$', are empirically equipollent, as according to the definition of

[130] Carnap (5).

[131] *Ibid.,* note p. 43.

'anger$_{ps}$' and 'anger$_{ph}$' the statement (3) 'When a person at any given moment is in a state of anger$_{ps}$, then he is also in a state of anger$_{ph.}$' is empirically true.

And now Carnap argues further: Statement (1) can be verified directly as well as indirectly, and indeed directly (only by means of introspection by A), and indirectly by any person whatsoever, including A. His method of indirect verification coincides with that of the verification (2) if one takes into consideration the parallelism maintained in (3) of the facts asserted in (1) and (2). But that for A a different sort of verification of (1) is available than for other persons; that – so says Carnap – is not of decisive significance: what is decisive is that a unified, intersubjectively controllable method of verification for (1) exists, and as this is the same as the method of verification for (2), then the distinction between the concepts 'anger$_{ps}$' and 'anger$_{ph}$' and the distinction between the corresponding judgments represents a reduplication that is unnecessary and for that reason was not made by ordinary language, not distorted by philosophy. Then according to Carnap, the situation is quite analogous for all psychological concepts and psychological propositions.

The obvious objection to this line of argument is that the formation of statement (3), which enunciates the natural-law equipollence of (1) and (2) includes within it reference to the *direct* verifiability of (1), so that it is incorrect to say the indirect verification of (1) coincides with the verification of (2). Rather, the indirect verification of (1) is composed of the combination of the verification of (2) with the application of (3), which latter statement is in principle withheld from physicalist transformation. Thus what may justifiably be said is merely the following: insofar as we are in possession of an empirical procedure which permits us to assume an unequivocal coordination between psychic events and space-time events – those accessible to external observation – we can first carry out these observations and put them in order – and thus for an extensive phase of the procedure operating only with space-time facts – and only at the end interpret the result psychologically. As a methodological postulate of physicalism – as it already was for the older psychophysics (of Fechner) – the aim could then be stated as follows: to seek to discover as many such parallelisms as possible, and seek to exploit them in the manner just indicated.

In many cases an orientation in conformity with this postulate may lead to advances in science, but it cannot play a controlling (decisive) role for psychology and for the social sciences as long as we cannot succeed in making even the simplest predictions about the behavior of other human beings *without* hypotheses that refer to their *thought contents* – predictions which can be made without the slightest difficulty *with the aid* of such hypotheses, and indeed are made all the time in social life. However that may be, with the breakthrough to the insight that psychological sentences and physicalist sentences are not logically equipollent, the illusion that physicalist doctrine is logically justified or as we would say could be grounded by rational reconstruction of the meaning of psychological statements – is in any case destroyed. The question, to what extent the methods of natural science prove suitable in psychology and the social sciences, can only be decided with the progress of empirical research in greater or smaller partial domains.

This critique, which is directed against the explicit theses of physicalism, must still be supplemented by an examination of the presuppositions implicitly contained in these theses. Above all, the requirement of *communicability* and *intersubjective verifiability* must be looked at more closely here. These two terms *cannot* be defined in accord with their application without using the concept of an *alter ego*. For, operating with the 'intersubjective unanimity of experience' must include operating with a plurality of subjects, because it is nothing else than an accord of a certain kind in thought-processes. But that in the language employed by physicalism, it is not 'thoughts' but 'sentences' that are compared with each other cannot disguise this requirement. For the content or meaning of a sentence is nothing else than the thought that is conveyed with the aid of the acoustical or optical symbol system. *Whatever is sensuously perceivable in a sentence thus has no content: and what has content, and therefore is the only thing of concern to investigations in the theory of science, is not perceivable by the senses.* To be sure, the characteristic thesis of physicalism, i.e., that all knowledge consists in a relation between sentences, represents essential progress vis-à-vis naïve realism which sees in knowledge a correspondence between thought and things, or events independent of thought. However, this physicalist thesis brings with it the serious danger of a confounding of two different concepts of objectivity, namely that of the intersubjective perceptibility of sign material, i.e., of a combination of sounds or figures, with that of the intersubjective confirmability of the state of affairs meant and communicated by the person employing the signs. Thereby it appears that the meaning (content) was attributed to the sign material as a *qualitas occulta*. How false this semblance is, we will clarify more completely in the next section. The anonymity of conventional signs must not lead us to a failure to recognize the epistemological fact, that the meaning of signs is a relation between psycho-physical subjects – the sender of the sign and the one who receives it – and that therefore substituting semantic terms for psychological terms does not eradicate this 'subjectivity'.

A further critical objection to behaviorism is formed by the findings presented in our Part One about the interconnection between external and inner experience, from which it follows that while the *topic* of the inquiry can be confined to the object of external experience (as in the natural sciences), yet every external experience contains inner experience within it. This remark leads us now to critique of the thesis, according to which even for statements about events in the external world, only physicalist statements can make a claim to objectivity and therewith to a truly scientific character.

After careful examination of the arguments by which this thesis is supported – by the way, this is hardly discussed anymore in the writings of the physicalists, but simply presupposed as *res iudicata* – I cannot resist the impression that here intellectual motifs are still very influential which were characteristic of older naturalistic conceptions, while they have already been superseded by Carnap himself in another context (in his study *Physikalische Begriffsbildung*). But whether this may indeed be the case or not, the second group of naturalistic arguments, which are now to be characterized, play an important role in any case in the *Methodenstreit* in the social sciences; therefore we shall have to consider them more closely.

At the center stands the more or less clearly formulated conception of *measurement* consisting in coordination of numbers which are independent of the subjective factors coordinating the sense data, and therefore intersubjectively valid from the outset, without requiring a consensus. Modern technology, with its self-registering instruments, apparently gives support to this conception, but only apparently, for in and by itself the instrument shows nothing at all. That it 'shows' something instead signifies that something can be read off from it; but among other things this reading presupposes optical sensation and communication of the result of this reading contains communication about optical data. For, indeed, it is not only sensations of color, of sounds, of smell, of taste and of warmth that are 'subjective', but also *sensations of length*. The opposite opinion which involves a distinction in principle between *primary* and *secondary* qualities, should today be defended least of all by physicists or by other scientists influenced predominantly by physics, insofar as it is just they who have succeeded to an ever higher degree during the last centuries in clarifying the implicit presuppositions of inquiry within the domain they are working on. But even if measurement were a process that is independent of subjective data, still it is not the case that all of physics would be absolved from such subjective data. For the physicist's procedure includes the scientist's approach and interpretation. When a physicist seeks, say, to explain acoustic data, he does not begin his procedure by measuring certain longitudinal waves, and he does not conclude it with the results of the calculations he has made on the basis of these measurements. Instead, his procedure is to be described as follows: Acoustics seeks to explain certain sense data, called 'sounds', by incorporating them within the general system of physics. In order to achieve this, acoustics does not stop at the phenomenal level of the data themselves, but coordinates them with oscillations which are wave motions. It operates with the latter, and finally applies the result to the original acoustic phenomena. Thus it subdivides the process of inquiry into three stages: (1) a link between the original sense data, and the physical phenomena, in a narrower sense, is created; (2) these physical phenomena are analyzed; (3) the result is applied to the acoustic data.

Now as a rule the second stage plays the major role, so that there is a tendency to trivialize the other two. Due to this, not infrequently the mistaken view arises that – remaining with our example – the subject matter of acoustics consists of *measurable* objects, to which, consequently, mathematics can be applied directly. The first and third stages of the research process, however, obviously contain all those 'subjective' moments which physics allegedly eliminates. Thus in fact the tremendous systematic achievement of theoretical physics consists in coordinating the phenomenal facts within various groups of phenomena and thereby a unified network of relationships is created. But this achievement is misunderstood if it is interpreted as though thereby the other levels were proven to be 'merely subjective'. That, with the help of that coordination, and only with its help, it has been possible to attain the envisaged goals of research – which, to be sure, are already conditioned in part by that epistemological fact – is itself a fact; to seek to justify it as logically or epistemologically necessary, will lead on to a false path from the outset.

The fact that physical inquiry cannot be separated from those 'prescientific' phenomena can be seen most readily when it is clearly recognized that it is precisely these phenomena that determine the main task of inquiry, and that inquiry must justify itself again and again with respect to how it solves this task. Thus optics has to take its departure from a series of prescientific facts concerning light (straight linear propagation, reflection, refraction, absorption, etc.) and the achievement demanded of it lies in incorporating all these phenomena within a unified complex foundation. The epistemological situation is complicated here, to be sure, in that frequently some phenomena to be interpreted only emerge in the course of the more refined observations, taken in connection with the scientific procedure itself; in the case we have mentioned, for example, the phenomena of double refraction, polarization and interference.[132]

These observations have been made above all in order to remove exaggerated conceptions about the '*exactness*' of the natural sciences, which have played an important role in the arguments of the naturalists as well as of the antinaturalists in the *Methodenstreit*. But with these remarks, we by no means intend to create the appearance of basically no significant differences between a general law of nature in theoretical physics and a more or less vague rule in the social sciences. Quite the contrary; the removal of that false ideal image is intended to sharpen our perception of the distinctiveness of physical laws and to make their specific features recognizable. The actual goal of these observations lies therefore in confronting laws of physics with the various types of laws in the social sciences, and thereby to obtain points of orientation for research procedures in the social sciences.

Let us now, as briefly as possible, point out those aspects of physical knowledge which – with the exaggeration just characterized – have led to the idealized image of natural law.

To begin with, we have to point here to the *unity* of the laws of physics (I) – and here the following moments can be distinguished: (a) *the unified basis of reference of physical laws*; (b) *the systematic* (*deductive*) *unity of the physical laws*.

On Ia. The unified basis of reference is characterized by this; that the phenomena of the various senses (e.g., optical, acoustic or thermal phenomena) are coordinated with spatial magnitudes, so that the empirical findings which the physicist has to make in his experiments are in principle confined to the observation of coincidences between a pointer and certain lines on a scale. Though as we have pointed out with respect to erroneous interpretations of this state of affairs, the subjective element of sense data is not eliminated by this, nevertheless, a unification takes place; the *reference system* is relatively *invariant* compared with the character of the *initiating phenomena*, although naturally their distinctive character finds expression within the reference system; thus, for example, optical phenomena correspond to other combinations of measured numbers on the scale than to electrical phenomena in the narrower sense.

On Ib. The systematic unity of laws of physics for which the unity of the base of reference furnishes a necessary but not a sufficient condition, lies in this,

[132] On this, see Neurath's instructive essay 'Principielles zur Geschichte der Optik', [Eng.tr. 'On the Foundations of the History of Optics', ch. 3 of Neurath's. *Empiricism and Sociology*, Vienna Circle Collection, vol. 2], *Archiv für Geschichte der Naturwissenschaften und der Technik*, vol. 5, pp. 371–389.

that altogether they are derivable from a small number of principles by *substitutions* for the variables. This can perhaps be demonstrated most impressively in terms of the 'principle of least action' of Hamilton.[133]

The second aspect to be emphasized in this connection is that of the *simplicity of physical equations*. (II) This is characterized by the following features: (a) a small number of parameters; (b) relations of a simple order between these parameters (in classical physics, including Einstein's theory of relativity, second order partial differential equations prevail; (c) a small number of constants; (d) functional continuity.

In relation to these points the following is to be noted:

On II-a. The small number of parameters in the laws indicates that in them only a few factors (classes of events) have to be taken into consideration. These prove to be *dominant*, so that other events can be disregarded to a large extent, without endangering the fulfillment of predictions. In what follows we shall have to recognize that one of the essential difficulties in establishing laws in the social sciences lies in this: that in them, such isolation cannot be carried out to the same extent.

On II-b. As far as the *simplicity* of the relations between the parameters is concerned, we have to be clear that every attempt to arrive at a precise definition of degree of simplicity, that would be in complete agreement with scientific practice, fails. Thus we can hardly decide whether an elementary trigonometric function e.g., y = sin x is simpler than an algebraic function of a higher order, e.g., the 7th degree; still every mathematician will probably declare a second degree algebraic function, or a function y = sin x, to be simpler that the function y = x sin x. That it is possible to express the relations between the variables which we have just declared to be the dominant variables in terms of functions which display such a slight degree of complication that it is possible to reach mathematical mastery of them in the time available, is an epistemological fact which does not permit further explanation, just as the small number of these dominant factors does not. To be sure, from the outset, scientific inquiry, as is understandable, searches for laws of such a simple nature, and in the choice between two possible formulations of a law it will, as a rule, give preference to the simpler one; but saying, that simple formulations are at all available to be chosen, cannot be derived from the research goals and research paths set by natural scientific thought. Here is one of the points where every conventionalist attempt at explanation and indeed every scientific attempt finds its limitation.

On II-c. The small number of *constants*, that is, of numerical factors, which appear in physical laws (e.g., gravitational constant, elementary quantum of action) is of great significance for physics, and it was one of the remarkable

[133] Hamilton's Principle states that among conceivable possibilities for a mechanical process that one will be realized in which a certain magnitude – namely the integral of time over the difference of kinetic and potential energy – is reduced to a minimum.

successes of research in physics, that it succeeded in reducing the number of independent constants so greatly. But for our reflections, this point is of little significance, and therefore we cannot discuss it further in the present context.

On II-d. Finally, in relation to continuity the following is to be noted: Here modern physics has removed the basis for one of the fundamental conceptions about the essential nature of natural law; for the statements '*natura non facit saltus*' or '*in natura non saltus, non hiatus, non casus*' had been guiding principles of scientific thought in the scholastic period, and in the classical physics of the seventeenth, eighteenth, and nineteenth centuries; this conviction appeared to have been definitely secured. The application of Newton and Leibniz's infinitesimal calculus to the natural sciences, which led to unforeseen results, appeared to furnish the most magnificent confirmation conceivable for these theses. But it has now been shown that this principle, formerly conceived to have unlimited validity, is only valid within certain limits of natural phenomena, and even within this domain its validity must be interpreted as merely the result of a great number of discontinuous processes distributed according to statistical laws. Yet in spite of this, in view of its close approximation to reality in so many cases, the idea of the continuity of natural events retains its great significance for inquiry in the natural sciences.

Besides the unity and simplicity of natural laws, which appear as characteristics of their inner structure, we now have to examine those aspects, which refer to the relation of law to the *domain of application*. Thus the question at issue is the suitability of those laws to the description of the reality of nature. Here we can distinguish the following points: the *domain of validity* (domain of application) of the laws (III) where (a) *temporal* domain of validity; (b) *spatial* domain of validity; (c) *density* of validity; (d) *material* domain of validity, should be considered separately; the *precision* of the laws (IV), and here again (a) the *degree* of precision; and (b) the *percentage of exceptions*, are to be distinguished. Point III refers to the question of the spatial, temporal and material limits of the domain within which the course of the events is a lawful one – in accord with certain specified laws. Point IV, on the other hand, concerns the considerations as to what preconditions must be fulfilled, in order for a domain to be regarded as governed by certain laws. These considerations become acute as soon as we have freed ourselves from the illusion of strict law.

Let us look at question (IIIa), the *temporal* duration of natural laws, and in close connection with this, the magnitude of the time span, for which predictions are possible. Here, in the sense of our reflection about natural law in general, we have to observe that every prediction can be justified only *relatively*, namely relative to the assumption of a certain uniformity in world events. This presupposed, we can say that those natural laws which contain a greater number of particular data, generally show a greater variability. This is assumed, for example, for the magnetic field of the earth, which changes quite rapidly. On the other hand, until now little reason has been found to assume that the most general laws of nature vary.

What has just been said about the temporal domain of validity of the laws can also be applied to their *spatial* domain of validity (IIIb). And here too, this is dependent upon the number of particular data contained in a law, in the same sense as in the temporal domain.

In order to understand what is meant by the concept of the *density of validity* (IIIc), the following has to be considered: that within a certain space-time domain a law is valid, according to which the occurrence of phenomena of type P is a sufficient condition for the appearance of phenomena of the type Q, allows no conclusion to be drawn as to the frequency of the applicability of that law within the domain. For then we know nothing as yet about the *frequency* of the phenomena. By 'density of validity' we will understand now 'the average number of cases of application of the law within the domain within a specified unit of time.' Thereby this is defined as a *statistical magnitude*.

The *material* domain of a law (IIId), finally, corresponds to the general delimitation of the definition determining the kind of events to which it is applicable. Comparison of the magnitude of two material domains of application thus is nothing else than comparison of the logical range of the concepts of those classes of facts to which the law refers. Accordingly, the 'relative magnitude' holding between two material domains of validity is only defined when the one includes the other logically.[134] (Coincidence of two domains is to be understood as mutual inclusion). In other words: the law with the larger material domain of validity is the more general law. From this it follows directly that a law with a greater (smaller) material domain of validity cannot have a smaller (greater) spatial or temporal domain of validity or a smaller (greater) density of validity than a law with a smaller (greater) material domain of validity. The task of determining the material domain of validity confronts the investigator who must specify to which classes of phenomena a certain formula is to apply; thus when, for example, it is a matter of establishing that not only mechanical masses (in accord with the law of gravitation) but also electrostatic entities (Coulomb's law) represent *models* of the formula $\frac{mm_1}{r^2}$

If the concepts of the temporal domain of validity, spatial domain of validity, density of validity and material domain of validity are related to the present totality of hierarchically ordered natural laws, then it will be seen that only relatively rarely are we forced to declare facts still inexplicable today, and thereby outside of known natural laws; even for those cases it is assumed that their incorporation will sooner or later be achieved. In this sense we can speak of a *universal* or *pervasive validity* of a system of relatively few 'highest' natural laws.

As far as the *precision* of natural laws (IV) is concerned, it can be noted, to begin with, with respect to the *degree of precision* (IVa) that though this is not, as we have already remarked, limitless, still with respect to a large number of phenomena it exceeds that measure required by technical practice; while in the case of a number of

[134] [Missing in the original edition – Ed.].

further phenomena, for example those of stellar astronomy, it is assumed that a refinement of measuring instruments could lead to the establishment of laws which would also permit a high degree of precision in computation. Finally, there are also groups of isolated phenomena which resist their inclusion within the laws of nature to a large extent: namely, the motions of single molecules, atoms and electrons; for as modern physics has shown, in contrast to older conceptions, the laws of nature of the 'macro-world' are not composed from independent laws of the 'micro-world', but are *statistical laws*, referring to a great number of individually incalculable elementary processes. Incidentally, in considerations of this kind, it must always remain clearly recognized, that after having relinquished naïve realism, the concept of precision of laws can only have meaning as correlative either to already existing criteria of confirming observation, or at least to such criteria that still remain to be characterized more closely.

While the testing of the precision of laws takes place in terms of those cases which are assumed to fall under the law, testing the *percentage of exceptions* (IVb) aims at estimating the magnitude of the ratio of the exceptions as against the cases conforming to the laws within a domain conceived as the domain of application of a certain law. What ratio is still assumed to be tolerable, i.e., not affecting the character of the '*domain of application*', is a question of conventional stipulation. We have already pointed to the fact that the conception of the validity of natural laws admitting of no exceptions had to be revised, and that therefore also a sharp distinction between laws on the one side, and rules and tendencies on the other, cannot be maintained; but even these gradual differences have to be given great significance for the theory of science.

Now it will be of considerable importance for the methodology of the social sciences to ascertain to what extent the characteristics of natural laws just briefly sketched also apply to these sciences, or at least to certain groups of them, and what consequences follow for scientific procedures from the divergences that will appear. However, the treatment of these questions still requires a good deal of further preparation; therefore, it will not be presented until our section "Social Laws and Ideal Types". First, just a few of the principal errors have to be rectified, which have obscured the understanding of the extent of the common ground between natural science and social science or else, the nature of their differences.

The first of these errors, with which we have already dealt in the first part of this work, is the assumption that the *exactness* of the natural sciences is characterized and guaranteed by the *deductive method*. For here logical coherence within the deduction, the 'internal exactness', is confused with empirical validity of the entire deductive system, the 'empirical exactness'. This error arises, as we have already recognized, in the following way: deductive inference is independent of experience insofar as its validity does not depend on empirical facts, but results from the logical relation between the premises and the conclusion. However, within the framework of scientific procedure, these premises themselves are already selected with a view to their applicability, and consequently this selection appears to be determined by experience, and indeed in many cases, by prescientific experience. Now this empirical content is accepted as established fact, while at the same time the assertion that the premises are subject to empirical verification is rejected by pointing to the *a priori* validity of the formal theory, i.e., of the deductive structure; in doing so, it is overlooked that this validity by no means coincides with the empirical validity of the premises. On the other hand, it is indeed characteristic of the method of abstract natural science, that within it deductive systems of an especially *simple* kind can be established, where to the aspect of simplicity just mentioned there is added another aspect: namely, that from a relatively small number of principles, the particular laws can be derived entirely by the substitutions of constants. No case is known in which any particular law that contains a larger number of constants diverges from the general system of natural laws, there, as it were, representing an isolated point within the sphere of lawfulness of nature. But this is by no means 'self-evident'.

The second error to be mentioned here, is the assumption that a fundamental difference between research in the natural sciences and in the sciences of the mind would consist in the fact that in the former, *experiments* and their results play a decisive role, while in the latter, experiments would hardly be possible, to any noteworthy extent. This conception requires correction in a number of respects. First it must be stated, and this has been emphasized repeatedly by natural scientists and philosophers of nature, that a great part of natural science research has to do without experimental confirmation. For example, this holds for almost all of the astronomy of the fixed stars. On the other hand, very frequently experiments can be conducted with human individuals and groups.

It is of fundamental importance to keep in mind that whether they were carried out with special technical equipment, say, in a laboratory, is not essential for the fruitfulness of a series of observations, made with the intention of gaining inductive insight. Rather, what is decisive is solely whether a constellation of phenomena is observed, which permits us to consider those groups of phenomena whose relationship is to be tested in relative isolation; and whether this isolation is produced by the planned employment of apparatus, or whether it is found to exist independently, plays no role for the conclusion drawn. With this we by no means wish to deny that the admirable technology of experimentation of the natural sciences has played a distinguished role in their development. What these remarks are intended to make clear is again nothing else than the impermissibility of interpreting comparative differences as distinctions of principle.

The third error, finally, to which we have to point, lies in the thesis that the rigor, or exactitude of the natural sciences from the outset cannot be postulates for the sciences of the mind, because causality is valid only for the inanimate world, the world without psyche or reason; while the sciences of the mind deal with men endowed with reason, and their associations are not governed by the principle of causality, but by the principle of freedom. Freedom, however, excludes predetermination, and with that, predictive calculations of the kind found in natural law.

In order to remove the basis of these arguments, we require insight into the ambiguity of the concept of freedom in which the ideas of prime cause, of psychic causality (motivation, i.e., causality due to freedom) and of responsibility interact. For this we can recognize that freedom of human action by no means presents itself as absence of cause. But as the various levels of meaning of the problem of freedom are significant for the *Methodenstreit* in a number of ways, we want to treat them comprehensively (in the section after the next), as soon as we have created the preconditions for their complete understanding.

If now – looking back upon the reflections carried out in this section – we want to ask ourselves what consequences flow from them for the evaluation of the doctrines, introduced at the outset, of either naturalism or anti-naturalism, then the answer will have to be: A series of prejudices were refuted by our reflections, prejudices rooted in an erroneous conception of the knowledge contained in natural science, and which form the basis of one or another thesis, or furnish the basis for their claim to absolute validity. The next three sections are devoted to other prejudices, influential in the debate on method, which concern the fields of psychology, axiology, and the historical sciences.

2. The Social Sciences and Psychology

(Meaning and Meaning-Interpretation)

It is the main goal of the analyses in this section to show that a major part of the controversies which concern the relation between the social sciences and psychology – and especially those which, according to their formulation, appear to contain unbridgeable opposed conceptions – can be resolved with little difficulty as soon as certain obscurities have been removed, which are linked to the conception of '*meaning*' '*meaning-context*' and '*understanding*'.

At the same time these reflections will furnish us with a striking example of the *typical affinities of meaning of equivocal terms* and show clearly the difficulties of resolving equivocations within such concept groups, which we want to call '*concept-families*'. The connection between the concepts of such a group is frequently such that the concept most impoverished in content is the element of meaning which all the concepts of the 'family' have in common; therefore we shall call it the 'core concept'.[135]

Now this is also the case with the concept-family designated by the term '*meaning*' [*Sinn*]; here rational reconstruction reveals that the core significance is that which is attributed to the term 'meaning' in connection with 'meaning of an act of thought'. In our first part, in the brief description of the concept 'act of thought' we have pointed out how we arrive at the meaning=content of an act of thought by abstracting from the occasional moments of this act; and prior to that, in our analysis of the meaning of a specific kind of acts of thought, namely acts of judgment, we devoted special attention to the clarification of this concept. Therefore, here we can

[135] I developed the account of this state of affairs about nine years ago, [1927 – Ed.] beginning with the concept of culture (within the framework of a series of discussions held by the *Kulturwissenschaftliche Gesellschaft* of Vienna. A similar analysis concerning the meaning of 'to know' and 'knowledge' is contained in the interesting lecture of Isaacs, 'The Logic of Language', published in the *Proceedings of the Aristotelean Society,* 1933, pp. 259–294.

confine ourselves to developing the relations between those meanings, equivocal with this core meaning, which have the greatest importance for us. We will also illuminate the degree of affinity between particular cases among these meanings, which may in some cases extend beyond this common property.

The principle of these relations is the following: If we attribute to any object a – more or less – definite meaning then with this is meant the meaning of cognitive acts which are empirically linked with this object in a manner that will be characterized more precisely immediately. These links we want to call '*symptom relations*' and to describe them briefly.

A state of affairs S_1 is called 'symptom for state of affairs S_2' if from the presence of S_1 the – past, present, future – existence of S_2 can be inferred. But that such inferences can be made indicates nothing else than that between S_1 and S_2 a real relation – an empirical connection – exists. As follows from the above definition, this real relation must not necessarily be of the kind that the symptom (epistemic ground) of a state of affairs coincides with one of its causes ('real grounds'). Instead it could also be the effect of this state of affairs, or as a consequence of shared causes, a regularly accompanying phenomenon. But even where we would have reasons for including a symptom of a certain phenomenon among its causes, an especially useful symptom must by no means be a particularly essential cause. From this results the meaning of the demand raised again and again in the various spheres of practical action, that in order to remove an evil, one must remove its causes and not its symptoms. In medical therapy innumerable striking examples can be furnished in justification of this demand.

Now the *significance [Der Sinn (die Bedeutung)] of the symptom* S_1 is nothing else than *the meaning of the judgment about* S_2 *which is based on knowledge of* S_1. We call such a judgment *an 'interpretation of* S_1'; the concepts 'symptom meaning' and 'interpretation' are thus correlative. We can understand without further difficulty that symptom relations are neither one-to-many relations nor many-to-one relations.[136] For on the one hand a state of affairs S_1 can function as a symptom for various states of affairs S_2, S_3, … S_n, and on the other hand each of these states of affairs S_2, S_3, … S_n can, if need be, also be inferred from other states than S_1. Furthermore, the following is to be noted: When we say that from the state of affairs S_1, the state S_2 is inferred, then – as rational reconstruction shows – this is not to be understood as though isolated knowledge of S_1 were the epistemic ground for knowledge concerning S_2. For as we have shown, every inference from one fact to another fact presupposes *general assumptions* about the connections between facts. The real relation, from which the symptom-relation arises, is not a relation between two specified individual facts as such, but between any two *arbitrary* selected facts of a given kind. The general assumptions on which the interpretation is based are called *schemata of interpretation*. We shall soon see that the explicit comprehension of the schemata of interpretation relevant for knowledge and grouping of social facts according to general points of view represents one of the most important problems in the theory of the social sciences. Here, for the time being, we want to make clear that the assertion, S_1 is a symptom for S_2, is incomplete as long as a schema of

[136] Compare for instance Carnap, *Logistik*, p. 46.

interpretation establishing the symptomatic connection has not been furnished. However, it is to be noted that S_1 can function as 'symptom' for S_2 on the basis of *various* schemata of interpretation.

If, given a schema of interpretation, S_2 cannot be inferred from S_1 alone but only from the combination (co-existence or succession) of various facts (S_1, T_1, U_1), then we want to say that the facts S_1, T_1, U_1 stand in a *meaning-context*. This is to be understood in such a manner that the state of affairs S_2 can only be inferred from relations of the named facts. that are to be specified. Thus, the meaning-context is a symptom for a meaning composed of various different facts, in this case for the meaning of a judgment which asserts the existence of the state of affairs S_2. Therefore it makes sense to say that the assertion that a totality of facts stands in a meaning-context, requires supplementary information as to *which* meaning is referred to, in order for this context to exist, and that furthermore the schema of interpretation which constitutes this context has to be characterized. We want to call facts, insofar as they stand within a meaning-context, *dependent symptoms*; facts, from which something can be inferred without the aid of other facts, we will call *independent symptoms*. From our general considerations concerning the context of experience of facts, however, it has emerged that the dividing line between independent and dependent symptoms cannot be sharply drawn, as the concept of fact (of state of affairs) itself by no means has precise contours.

From what has been said already, it emerges clearly that nothing else to be understood by 'the meaning of a fact', within the framework of an epistemic process, than the position (function) of that fact within this process. Finally it must be established that there is also a significance 'meaning-context' corresponding to the core-significance of 'meaning'. It is said that a *judgment* stands in a meaning-context with other judgments, if its meaning is supplemented in certain directions by the meanings of other judgments.

What is required for the analyses to follow concerning the general concept of symptom-meaning has now been stated. That, instead of 'symptom meaning' we frequently speak of 'symptom-significance' in order to specify the content of the first-mentioned term, must not lead to our failure to acknowledge the known fact, that the general concept of symptom is sufficient to discuss an essential part of the problems of the meaning of actions (especially the use of signs) as well as of products of action, i.e., of objects comprehended as bearers of meaning κατ' εζοχηυ [from the beginning].

Now in order to grasp the specific *differentia* which lead us to the bearers of meaning just mentioned, we first want to recall to consciousness the definition we gave of the concept 'symptom-meaning', according to which by the meaning of a symptom is to be understood the meaning of the judgment inferred from this symptom. The fact asserted in the judgments can either be a physical or a psychical (psycho-physical) fact; and according to the narrower significance of the term 'meaning' which we now have in view, a meaning is attributable to a fact insofar, and only insofar, as inferences with respect to psychical facts of psycho-physical beings – and thus to thinking, feeling, willing, to character traits, attitudes, etc. – are or can be drawn from it. Here we must take especially great care to avoid confusing the meaning of a judgment, which asserts the existence of psychical facts, with the meaning of the psychical facts asserted, as happens only too often. Thus, for example, when the person P_1, on the basis of the fact F_1, makes the judgment J_1, that at a certain point of time the person P_2 has made the judgment J_2, or is going to make it, then the meaning of J_1 and J_2 have to be kept sharply separate.

For our investigation of methodology in the social sciences it is important above all, to analyze those variants of the concept of 'meaning' and 'interpretation', which are connected with the comprehension of the meaning of *human action*. How is such meaning-interpretation carried out? Certain phenomena are comprehended as movements of a human body or as the effects of such movements – in the latter case the phenomena in question are symptoms of the preceding body movements – and these body movements are regarded as symptoms of a preceding project, appearing as a certain phase of the way toward the goal set by the project. As pure thought, the design is the specifically psychical element in the psycho-physical process which is called 'action', and the meaning of the project is the meaning of the action, where the word 'meaning' is used in its core significance according to which we speak of the 'meaning of a cognitive act'. The meaning of the project is designated as the *subjective meaning of the action*. It is 'the meaning which the actor links to his action' (Max Weber), or, expressed more precisely, the meaning to which the actor orients his action.

As the task of comprehending the subjective meaning of the action of other persons is included in all social science research, and as the numerous obscurities with respect to this comprehension form a main point of departure for methodological controversies, we must analyze this process in greater detail. Let us mention first that the comprehension of the meaning of our own past action is a task that frequently confronts us, especially when we can no longer recall the project of that action. It is, however, not always possible to make a sharp distinction between inferred and remembered elements of the meaning of one's own past action.

In grasping the meaning of another person's action, however, this 'access from within' is eliminated; it is a mediated comprehension in all cases. Therefore the investigation of the question, how the truth of assertions about the meaning of another person's action can be examined requires particular attention. Alfred Schütz has presented this range of problem for the first time in its full profundity, in his work *Der sinnhafte Aufbau der sozialen Welt* [*The Phenomenology of the Social World*][137]; our exposition of this topic which follows immediately below, is based on his analyses. One of the most important conceptual motives is his consistently developed insight, that the manner of verifying these judgments, and therefore also their meaning, depends on the *relative spatial-temporal situation of those acting and those making the judgment*. Schütz distinguishes here a relationship among consociates, a face-to-face relationship where the actor is bodily present to the person judging the action [*Umwelt*], a relationship among contemporaries where actor and the person judging the action are at a common time but not space [*Mitwelt*], and

[137] Schutz, Vienna 1932, cited hereafter as *Aufbau,* [to be distinguished from Carnap *Aufbau,* 1927 – Ed.].

the relationship with predecessors where the person judging the action cannot, on grounds of principle, enter either of these social relationships with the actor [*Vorwelt*]. In contrast to this, relationships among 'consociates' and contemporaries can merge into each other and also into relationships with predecessors. Instead of 'A belongs to B's world of predecessors', we can way 'B belongs to A's world of successors'.

To reveal the meaning of another person's action given in a *face-to-face* relationship, and to verify the pertinent judgments, an inexhaustible multitude of symptoms is available. At each phase of the observed action of others, the judging person can anticipate, according to his assumptions about the project involved, the subsequent phases of the action, and can find his judgment confirmed by the further course of the action, or alternately, requiring vision. Furthermore, from other voluntary or involuntary behavior – not directly concerning the presumptive goal of the action to be analyzed – especially from facial expression, inferences can be drawn; and finally, the acting person can be questioned about the meaning of his action.

However, this possibility of adapting the schema of interpretation to a continual succession of supplementary or corrective experience is lacking in the case of the relationship *among contemporaries*, to a lesser or greater measure, depending on the degree of contact. Therefore the interpretative schema is usually considerably more rigid; the interpretation is made according to a (more or less well-founded) pre-judgment with respect to the actor, which is supported either by direct knowledge of his person or else by assumptions based on his social positions (nationality, religion, class, family, function). In the latter case, it has no effect on the procedure of grasping the meaning of the action, whether the action is performed by the person A or by a person B who has a more or less similar social position. As has been noted, to interpret actions of contemporaries, closer contact with the acting person and, possibly, questioning this person about his action, remain in principle possible for the person making the interpretation. As a result, far-reaching unanimity between the interpretation of an action by contemporaries and consociates, and interpretation by those in the immediate environment can be reached in many cases (although of course the fact that the action to be interpreted was not directly experienced by the person interpreting it cannot be changed after the fact). In the interpretation of the world of predecessors by its successors, however, there is no possibility of establishing the relationship among consociates and yet, *tradition* can build bridges to this form of social relationship.

The gradations of mediacy of symptoms here sketched is by no means peculiar to symptoms from which the meaning of human action can be inferred. For, from the insight that no empirical judgment can be conceived as the pure registering of an immediate given, and that instead each such judgment contains a chain of assumptions within it, there follows as a direct consequence that the grounds on which the judgment is based are to be considered as symptoms of the validity of those assumptions. There symptoms can be ordered quite uniformly according to levels of mediacy. With respect to external experience, immediacy is represented by sense perception in this order of mediacy. But for the understanding of the interpretation of action of other persons, taking into consideration the scale of mediations is especially important because it renders understandable that typical shift of the interpretation problem, through which the task of grasping the *subjective meaning* of such action is replaced by the task of grasping its *objective meaning*. The question: 'What is – or else, was – the meaning the actor P connected with his action A?' is replaced by the question: 'What meaning is attributed to this action by the world of consociates, and possibly, by the closer or wider circle of contemporaries of P on the basis of the locally dominant rules of interpretation (e.g., in ordinary language)?' Then it is no longer a question of exploring the meaning of the action A, intended by P himself, but of ascertaining how actions of the kind A are typically interpreted by the environment of P – or else, correctly interpreted on the basis of the pertinent rules in force there. Thus we interpret the objective meaning of a sentence uttered by A, by specifying how it 'had to be understood' within the closer or wider environment of P.

This statement illuminates the concept of *objective meaning* and shows that it requires completion by the *specification* of the *relevant schema of interpretation*, and consequently its meaning varies with the variation of the schema of interpretation used. Before we turn to these problems, however, we still have to state some preparatory considerations about the ambiguity of the concept of subjective meaning. We have shown that the subjective meaning of an action is the meaning of the *project* of this action, which includes setting the goal and stipulating the way to be pursued, in reaching the goal. Now action very frequently serves not merely *one* goal, but a series of goals, which can be arranged in part side by side, in part one after another in succession. Of special interest in this connection are the goals *arranged in succession*. Consider, say, a *speech* act. The goal of the speaker is, in the first place, to communicate something to another human being, i.e., to let him know something. But with that, the goals he has set very frequently are not exhausted; rather by means of the communication, he seeks to induce the other to engage in certain actions, and this, because he desires that a state of affairs of a certain kind be realized by the action. What now is the goal of the action to be called? The transmission of the knowledge to another, the inducement of the other to engage in an action of a certain kind or the realization of the state of affairs what will be brought about by the action?

On the answer to this question depends the answer concerning the subjective meaning of the speech act as well; for the goal belongs to the meaning of this action. At first glance, the answer to this question does not seem in principle to involve any difficulties. For as, so it will be argued, the subjective meaning of the action is the content of the project, so the final goal of the action is determined by the 'span or breadth' of the project (Schütz).[138] The last of the goals indicated in the project is to be considered the final goal, those preceding it only the intermediary goals.

This argument is entirely cogent; however, it presupposes implicitly that in every project of action one goal is clearly delineated as the final goal, or that, by means of rational reconstruction, one of the goals set up in succession can unequivocally be shown to be the 'actual' final goal. But this will not prove the

[138] *Aufbau*, p. 62.

case for a large number of projects; the *goal horizon* of these will be an *open horizon* of ever less sharp contours that gradually become diffuse and in the course of further steps towards them, they constantly change. That is why not infrequently even determination of the meaning of *one's own* past *actions* must contain constructive element. In particular an effect of action that appears relevant to a retrospective view is frequently declared to be the 'originally set goal' and this with considerable arbitrariness.

Now it does not require many words to show that the degree of this arbitrariness rises immensely, as soon as the *objective* meaning of another person's action is to be comprehended by an *alter ego*. Earlier, in the analysis of the concept of 'practical correctness', we have pointed to the shift of meaning which results in the elevation of the action of others, because the person making the judgment refers the action to his own goal system instead of to the goal system of the present or past actor. Thus a major part of the judgments in the human sciences, as well as a major part of the statements in the natural sciences, are valid only *relative to a reference system that must be specified*, and the problem of the discovery of universal invariants, i.e., propositions the truth of which is not affected by the transition from one reference system to another, has become just as acute in the one domain as in the other. In the social sciences, however, this range of problems is rendered especially complicated by two further aspects. One of these lies in the dividing line between the immediate givenness of the meaning of an action for the actor himself, even in remembrance, and its merely mediated givenness for an interpreting *alter ego*. The second complicating aspect is characterized by the fact that in the human sciences (sciences of human behavior) the goals of knowledge usually are isolated to a much lesser degree from the system of practical goals, in the narrower sense, than is the case in the natural sciences, nor would such isolation be possible. This can be explained in part, though by no means exclusively, by a circumstance which frequently is the only one taken into consideration: that socialized human beings, who have to orient their own behavior constantly to the behavior of others, automatically consider such behavior from the viewpoint of their own practical interest. Besides this, two other components also play a significant role. First, it must be taken into consideration that the schemata for the interpretation of actions of others always ultimately point back to inner experience, and because of this, the results of the interpretation will depend on the degree of essential affinity between the interpreter and the actor. However, in addition – and this point is especially important – on the one hand, the factual basis (symptoms) available to the interpreter vary greatly in kind and extent, and on the other hand, these symptoms are incorporated into different interpretive relationships depending on the specific goals of knowledge in each case. Here the circumstance already mentioned, and to be investigated more closely in what follows, is of a great significance: namely, that in the human sciences – in contrast to the natural sciences – there exists no special heuristically preferred basis of interpretation. Later, we shall have to analyze a series of important consequences which follow from this, and in particular we shall investigate the limits of the relativity of interpretations.[139]

[139] *See below.* Part Two, section "The 'Historical' in the Social Sciences".

In our general exposition of the meaning of human action we already mentioned the concept of 'speech act' as an example of *sign-setting*. Now we must analyze this latter concept more closely, and in so doing direct our attention above all to the confrontation of *meaning-establishment* and *meaning-interpretation* as well as *levels* of interpretation.

With the help of signs, rational beings want to communicate something to other rational beings, to enable them to take part in knowledge of a certain kind. The subjective meaning of setting forth of signs as a communicative act thus is the transmission of certain knowledge of one's own to the person receiving the sign. The purpose of letting the receiver know from whom the communication comes can also belong to the meaning of sign communication; however this need not always be so, and in certain circumstances, hiding this fact can be the goal. As far as the goal of sign communication is concerned, as we have already mentioned, only rarely is this confined to the transmission of knowledge: rather, as a rule, the intention is to produce a certain motivation in the receiving person by means of the knowledge transmitted to him. In the analysis of the meaning of imperative sentences in the next section, we will see that this intention of motivating will under certain circumstances overlay the content of communication to such an extent, that the isolation of this content, by means of rational reconstruction, can only be carried out with difficulty. In what follows, we shall call the above intention '*communicative goal*' and the above content '*communicative meaning*'. As far as the intention of the communication is concerned, the person sending the signs can intend either to let it be known to the receiving person or to hide it, or else may have neither the one purpose nor the other.

Let us turn now to the *interpretation of the sign by the person receiving it*: here five levels of interpretation can be distinguished, which, to be sure, can coexist in time in the actual interpretive process[140]:

1. A certain phenomenon is interpreted by the receiver as a sign, as expression of the intention to make a communication, the meaning of which is as yet unknown.
2. The phenomenon is interpreted by the receiver as a sign for the communication of the certain meaning – i.e., meaning grasped in the interpretation.
3. From the communicative meaning the receiver infers the communicative goal, i.e., an inference is made from what the sender of the sign wanted to say to what he intended to achieve by saying this on this occasion.
4. The receiver is able to, and not infrequently actually will infer from the communicative meaning and the communicative goal as grasped by him – perhaps in connection with other facts known to him – the circumstances which led the sender of the sign to the project of communicating the sign; he will recognize the 'because-motives' of the sender of the sign (Schütz).[141]
5. The receiver can, and not infrequently will, infer from the communication other aspects too, e.g., character traits, which belong neither to the meaning nor to the goal, nor to the 'because-motives' of the communication.

---

[140] For this compare the analysis carried out by Schütz in *Aufbau*, p. 120, of the phases of interpretation, exemplified in the interpretation of the activity of chopping wood.

[141] *Aufbau*, p. 99 ff.

We wish to distinguish these five levels of interpretation in terms of a simple example. On August 4, 1914, the captain of a British ship notices certain movements of a colored object and assumes that they are to be interpreted as flag signals, signs of the optical Morse code (1). He deciphers the words: "Neptun 68° 17′ North latitude, 12° 17′ West longitude, with 10 cm of water in the engine room," thus communicative meaning (2). He furthermore infers the communicative goal to be a request for aid by his ship (3). He interprets the 'because-motive' of 'Neptun's' request for aid to be occasioned by a collision with an iceberg, which he knew about from a previous signal (4); and finally he deduces from the appeal for aid, that on board the 'Neptun' which he knows is a German ship, nothing has been heard as yet of the outbreak of war, for otherwise they would not have requested aid from an enemy ship (5).

It is obvious that revealing or hiding the 'because-motives' of the communication, and also certain facts to be included under (5), is also part of the communicative goal and (or) can be interpreted by the receiver as included in it. Due to this, the relation 'sign setting forth – sign interpretation' becomes still more complicated.

Now as a major part of research in the social sciences consists in the interpretation of signs, or in the interpretation of the interpretation of signs and thus is subject to these complications, it is especially important to set up a *taxonomical schema* of interpretations, which would permit grouping them systematically according to the most general possible points of view. It would be appropriate here to assume five grounds for this division, which would be determined by the following questions – no longer entirely unfamiliar in their structure:

1. What facts are to be interpreted?
2. What facts may be drawn upon to aid in the interpretation, and what is their weight?
3. What schemata of interpretation are to be utilized, and from what experiences do they originate?
4. Under what circumstances is the interpretation to be completed (what is the interpretation to achieve)?
5. What status is to be attributed to the result of the interpretation?

By inserting in this taxonomic schema the theses and antitheses brought forth in the *Methodenstreit*, concerning the various problems of interpretation, it will either become clear at which points these arguments 'actually' diverge, i.e., with respect to procedure or else it will be seen that they are formulated in too confused a manner to be able to attribute a precise methodological meaning to them.

In the preceding analyses we have always operated with the term '*interpretation*'; while we have not used the word so frequently employed in the *Methodenstreit*, '*understanding*'. This is because the use of this word is hardly uniform. For sometimes (to be sure not frequently) it is employed synonymously with 'interpretation', 'interpretation of symptoms', as such; in what follows we will however not employ this usage. At times, in a manner yet less general, it is used synonymously with 'interpretation of symptoms of psychical facts' (meaningful phenomena); furthermore, on occasion, only the interpretation of the meaning that others bestow on their experiences is called 'understanding', although 'understanding another person' [*Fremdverstehen*] is not explicitly expressed; finally, not infrequently, only those interpretations are called 'understanding' in which the psychical facts inferred from the interpreted phenomena are comprehended as the real grounds (causes) of the interpreted phenomena.

Now we will have to add a series of fundamental remarks about understanding; they can be formulated very briefly, because to a large part they represent the evident consequences of analyses already carried out:

1. All understanding that is not understanding of one's own self implies the basic assumption of the existence of other persons; therefore, because of the character of these assumptions as presuppositions for every meaning-interpretation, I have called them '*fundamental interpretations*';[142] Schütz speaks of the "general thesis of the *alter ego*".[143]
2. Just as little as, from the viewpoint of the theory of science, one can speak correctly of *the* explanation of phenomena, as though there were only *one* kind of incorporation of a given fact with the general context of experience, so too the *understanding* [*Verstehen*] of an object or a process can hardly be conceived as an *unequivocally* determined cognitive process. For every explanation, the following questions have to be answered: (a) on what bases is the explanation to be made, i.e., by what data will the explanation be supported? (b) by what means is the explanation to be made, i.e., on what general laws of experience (assumptions) is the explanation to be based? (c) what is the explanation to accomplish (under what conditions will the explanation be considered achieved)? Thus in the case of understanding, the questions must also be asked (a) on the basis of what psycho-physical facts is the given phenomenon to be understood? (b) on what general laws of experience is the understanding to be based? (c) under what conditions is a fact (an object) considered to be understood? However, while in the case of explanation in natural science one certain kind of incorporation is predominant in intellectual practice, namely, the one carried out in physics, in the case of understanding, it is by no means always one specific kind of direction of research which prevails; instead, the direction of research depends to a large extent on the specific direction of interests. Insight into these relationships leads to the problem of 'value relation', which we shall have to examine more closely in the section after the next.
3. The 'specific evidence' of understanding [achieved by empathetic emotion] cannot be a criterion of truth. Here the results of the pertinent reflections in our general section find analogous application.[144]
4. The fact that understanding frequently takes place especially *rapidly*, is no argument against the known fact that the process of understanding presents itself to reflective analysis as *a highly complex thought-process*. It would be

[142] Compare Kaufmann, *Strafrechtsschuld*, p. 86.

[143] *Aufbau*, p. 106 ff.

[144] See above Part One, section "Basic Philosophical Considerations".

wholly erroneous to assume that the experiential simplicity of an insight, the speed of its comprehension, could be regarded as a measure of the simplicity of the material content of knowledge. This remark is also directed against the kind of argumentation, by means of which Scheler sought to oppose the analogical inference theory of understanding, by maintaining that understanding could not be conceived as an inference by analogy, because phenomena of understanding have been shown in infants and chimpanzees too, and thus the creatures who were doubtlessly incapable of inferential thought.[145] But it is by no means required that a psychical process, which upon reflective analysis is shown to be an inference, also present itself as such in the process of experiencing it. Instead the phylogenetic or ontogenetic automation of a psychic process can leave its content essentially untouched. Thus, what is central for a theory of understanding [*Verstehen*] and for a systematic classification of the different variants of understanding, is not a dissection of the phases of the process of understanding in terms of the experiential [inner] time (of the understanding subject), but the discovery of the elements of experience which are contained in this process and the manner of their synthesis.

Because they did not understand this epistemological fact, those who hold various theories of the understanding of other persons did not reach the core of the scientifically relevant range of problems. This holds in the same way for the theory of analogical inference which teaches that understanding lies in the *formation of an analogy* with the events of inner experience[146]; for the *empathy* theory[147] which, though it assumes a specific cognitive mode for the comprehension of other minds, attributes this to inner experience insofar as it presupposes an affinity between the person who understands and the person understood as a precognition for empathy. And finally, it holds for Scheler's *perception theory* of understanding,[148] which not only claims that understanding is a cognitive mode *sui generis*, but also that it has priority with respect to knowledge of the external world. The correct core of the analogy theory lies essentially in the fact that the psycho-physical (i.e., something like one's own ego) becomes thematic for understanding, and in this sense inner experience is prior to understanding. But this by no means excludes the assertion that more profound knowledge of oneself as a rule is linked to the results of intensive observation of others. (Compare Nietzsche's statement which is also cited by Scheler: "Everybody is the most remote person to himself.") Unfortunately in his analysis – rich in new insight in other respects – Scheler does not distinguish sharply enough between these two elements, and this has misled him.[149]

[145] Scheler, *Wesen und Formen der Sympathie*, Bonn 1923, p. 274 ff.

[146] This theory is the oldest theory of understanding [Verstehenstheorie].

[147] Its chief representative is Theodor Lipps.

[148] Scheler, op. cit., p. 273 ff.

[149] For he does not succeed in grasping adequately the relation of understanding to external and inner perception. The most significant progress beyond Scheler's work is represented, in my view, by the investigations of Schütz in *Aufbau*, p. 106 ff.

5. With respect to the accusation of the '*irrationality of the process of understanding*' which some like to confront with the '*rationality of explanation in natural science*' and use as an argument for the fundamental difference between method in natural science and in the human sciences, and also with respect to the inapplicability of 'the logic of natural science' to the social sciences, the following is to be said:

To begin with, we note that even the thinking of the great natural scientists, indeed even of mathematicians, is by no means as 'rational' as many theorists of the social sciences would like to believe. As evidence – to be sure, hardly always totally reliable – the reports of their own work by such scientists (e.g., Henri Poincaré[150]) and on the typical manner in which they arrived at their discoveries, can be adduced; thus the famous statement of Gauss may be mentioned: "In mathematics this happens to me not infrequently: I have the result; only I do not know yet how I am going to go about finding it." Yet in spite of this it is by no means our intention to deny that a special sensitivity [*Fingerspitzengefühl*] plays a much greater role in the understanding of other minds than in mathematics and natural science, and that in the former, rational reconstruction may also be much more difficult than in the latter. However, this state of affairs should not be interpreted in such a way, as though there were *irrational criteria* for the correctness of understanding, as though 'physiognomical tact' (Spengler) could be regarded as an autonomous source of truth. 'Irrational comprehension of meaning contents' is the comprehension of these in the mode of confusion, and even if the removal of confusion in thought – scientific thought, too – is an 'infinite idea'; still it is an *error* to regard *this stage of imperfection of knowledge, as knowledge sui generis*. It should hardly be necessary to point out, however, that this insight does not in any way put into question the importance of the fact of irrational thought, or of action motivated by such thought, in the social world.

The above remarks concerning the concept of understanding [*Verstehen*] also contain the key to the evaluation of the various types of understanding distinguished by Max Weber. It emerges, but cannot be presented in more detail here, that a sharp distinction between *immediately insightful* and *explanatory or motivational* understanding is not possible, that the different variants of *rational* understanding can be traced back to *purposive-rational* [instrumentally rational] understanding, and finally, that between *rational* and *irrational* understanding there are only gradual transitions.[151]

But what follows from all these findings for determining the relation between the social sciences and psychology? Our reflections about 'meaning' and 'understanding' have made it sufficiently clear, that meaning-interpretation of phenomena presents itself as the link between these phenomena and psychical facts. Therefore for

---

[150] Henri Poincarés 'Wissenschaft und Methode' (*Science and Method*), Berlin, 1914.

[151] For this, see Landshut, *Kritik der Soziologie*, Munich 1929, p. 34 ff. Good references to the literature of the work concerning the theory of understanding in Sombart, *Die drei Nationalökonomien*, Munich, 1930, p. 192. For the history of the doctrines see I. Wach, *Das Verstehen*, vol. I, Tübingen 1926, vol. II, 1929, vol. III 1933.

those who conceive the interpretation of meaning to be the essential task of the method of social science – and as we shall recognize in what follows, this conception is justified – the question about the relation of social science and psychology appears to be solved by subordination of the former to the latter. For descriptive analysis shows that psychical acts are acts of *meaning-establishment* and that therefore every *description* of these acts must contain *meaning-interpretations*. Similar considerations also lead us to subordinate social facts, the objects of the social sciences, under psychological facts, the objects of psychology. For – so it can be argued – what makes social facts *social* facts, and distinguishes them from the facts of nature, is precisely their meaning-content. This holds not only for social actions, but also for artifacts and social collectives (e.g., society, the state). Thus the dividing line from the natural sphere, and therefore the constitution of phenomena as social facts, is marked (so it is said) by reference to meaning; but reference to meaning is nothing else than reference to the sphere of psycho-physical facts.

With respect to this, the following should be noted: first, it is necessary to observe that the line of argument just presented in no way corresponds to that which was brought forth around the turn of the century, when the controversy concerning the independence of the social sciences from psychology was at its height, by those scholars who wanted the social sciences to be considered as psychological disciplines, For at that time Franz Brentano's analysis of intentionality was hardly known by most of the scholars to be considered here, not to speak of its significance being recognized. The method of *associationist psychology* was the dominant method in psychology then, and it was this above all which was supposed to be applied to the social sciences. This has changed completely since then and therefore it must be clearly understood that the meaning of our question has undergone a complete change during the last decade.

The psychologist who is asked today whether according to his conception of the method of the social sciences coincides with the method of psychology will first have to ascertain *which* psychological method is meant. He will have to see whether it is the method of cognitive psychology, of Gestalt psychology, the psychoanalytic, the characterological, the behaviorist method or the method of psychological anthropology that is to be compared with that (or those) of the social sciences; whether the pure description of psychical facts and the pure description of social facts, or the explanation of psychical facts and the explanation of social facts, are to be brought into correspondence; whether the person posing the question regards 'self-observation' or 'observation of others' or a combination of both procedures essential for psychology. From this we can see how loosely the question of the relation between the social sciences and psychology is formulated; yet in spite of this, the main conceptual themes which underlie this question can be isolated and clarified with little difficulty.

The fundamental problem is whether the comprehension of social facts is accomplished with the aid of a *specific source of knowledge*, or whether it flows exclusively from the sources of external and inner experience. As shall become fully clear from what is to follow, social facts are either human actions, yet to be characterized more precisely, or phenomena which point back symptomatically to such actions.

The core of knowledge of social facts – about the character of which the pertinent methodological controversies mainly resolve – accordingly is the interpretation of the meaning of the actions of other persons, or else, of the phenomena linked with these in the manner previously indicated. Now, as we have recognized, a specific source of knowledge is not contained in this interpretation of meaning. Rather, the interpretation is a synthesis of inner and external experience, and accordingly, the concepts of social facts can also be constructed by use of physical and psycho-physical facts. How this construction is carried out will be shown in our section "Fundamental Concepts of the Social Sciences"; but from what has been said up to this point, it can already be gathered that *social facts are not intra-mental*, and that therefore they are not psychological facts, insofar as the concepts 'psychological fact' and 'intra-mental fact' are equated. But dealing with a definition of psychology which partially includes the theory of human actions, it will not be possible to answer the question negatively from the outset. Instead we will have to examine carefully – and best in terms of the universal schema presented above – to what extent the goals and the approaches to knowledge set on both sides correspond. In so doing it will in particular be shown that in the social sciences the systematization of observational data and consequently the formation of concepts is carried out according to other points of view and that in them 'historicity' and 'value relation' have a different function than in psychology (or the psychologies) of human action. But in any case there are gradual transitions between the various specific approaches in psychology, and the various specific approaches in the social sciences. It must never be forgotten that the traditional delimitation between the sciences is linked on the one hand to specific stages in the development of scientific inquiry, and on the other hand that they are most frequently presented in a mode of considerable confusion. Confronted with this, it is then the task of rational reconstruction to isolate from each other the various aspects under which this delimitation is carried out, and to analyze each of them individually.

3. Value Problems in the Social Sciences

Our investigations concerning the concept of value have led to the result that we cannot speak of a specific 'knowledge of value', to be placed in partial or total independence beside the 'knowledge of what has Being'. This result is of fundamental significance for the problem of 'value neutrality in the social sciences', the (historical, juridical, and ethical) 'attribution of responsibility' and the 'value relation', with which we shall have to deal in this section, and also the next.

All the basic things we have said about 'values' can be applied to 'norms'; for a norm is nothing else than a statement that (future) behavior of a certain kind is of value (or correct). Thus the entire critique which we have applied to the idea of practical (axiological) correctness, can be transferred quite readily to the concept of norm, and this holds true especially for the insight that there is no 'correctness as such', but that the concept of correctness is a relational concept.

Accordingly, 'ought-statements' are only accessible to verification insofar as they contain specifications of goals. For the sake of clarification, we wish to give a few examples.

1. You ought to follow the marked path if you wish to get to N in an hour's time.
2. You ought to always speak the truth if you wish to keep the respect of your comrades.
3. You ought not to insult your opponent in a legal trial if you wish to avoid a severe punishment.

We have already discussed fully the complications which arise here due to the plurality of the goals involved, their simultaneity or succession, and the intermingling of heterogeneous goal systems.

What has just been established about 'ought' in the combination 'you ought to' also holds true for its occurrence in the context 'I ought to'. 'I ought to act in a certain manner' means 'it is correct that I act in this manner', and this 'correctness' requires completion of the specification of a referential goal system. Again, here we must not be misled by the fact that in the 'ought-[lived]-experience' these goals frequently are not explicitly contained and, that the functioning of psychophysical automatism is apt to induce the illusion of the evidence of absolute correctness. What is decisive for the meaning of 'ought', however, as emerges clearly from our general investigations, is not its lived coloration, but the system of truth criteria for ought-statements. A 'feeling of ought to' (feeling of obligation) is characterized by the – more or less strongly – emotionally colored conviction (thus the name 'feeling') that a certain conduct is correct, or also, that the contrary conduct is incorrect. The meaning of this conviction, however, only becomes clear when the goal system is indicated to which the correctness (incorrectness) refers.

The range of problems of 'ought to' is however further complicated by merging of this concept with that of the *imperative*, where the person giving the command can remain more or less *anonymous*. Here we speak of *heteronomous norms*, and consequently that such a norm exists would mean nothing else than that a command (or order) has been issued by a – more or less precisely specified – person. To be sure, a more careful analysis that the fact A has addressed an imperative to B is considered as a norm-establishment by B or a third party only if one assumes that B would do the right thing by obeying the command, so that here too an *element of correctness* appears. But in any case the question concerning the meaning of ought-statements is so closely linked with that of the meaning of imperatives, which also has greatly troubled logicians and philosophers, that we have to pursue these problems in this direction too, and seek to clarify the meaning of imperative statements. This is also important because these questions have played a great role in the theory of the law.

The problem of imperative sentences arises (in conformity with that of optative sentences and questions) from the following considerations: – if we have recognized the connection between the meaning of a sentence and the methods of its verification, then it is incomprehensible how a sentence could be possible for which the question of its truth or falsehood could not be posed at all. For instance, the question whether a command like, say, 'Come here!' is true or false appears to be entirely inappropriate; in spite of that, we do not hesitate to designate the imperatives as 'sentences'.

Here it must be observed, to begin with, that not much is gained by the customary distinction between judgments (declarative sentences) on the one hand and imperatives on the other, according to which we characterize *judgments* as *expressions of opinion* and *imperatives* as *expressions* of *willing*. For the same sentence can express different things, i.e., can be a symptom of different psychical facts. Thus we have to ask: are sentences of the kind just named, judgments at all? and (if the answer is to be affirmative) what is asserted in them? Furthermore: if these sentences are judgments and therefore subject to question as to their truth or falsehood, how can it be understood that this question is regarded to be inappropriate with respect to them?

Let us begin with the first question: here, to begin with, it is easy to realize that in case the imperatives are at all judgments, they must be judgments about *one's own inner conduct*, and indeed that is the case. More difficult, however, is to grasp precisely which inner conduct is actually meant.

We can hardly doubt that someone who issues a command to another is communicating something to that person; however, the task lies precisely in determining what the specific *communicative meaning of the command is*. The most immediate conception, and one that arises repeatedly, is that the imperative 'Come here!' is equivalent in meaning with the sentence: 'I want you to come here!' But that conception is not tenable, as will become immediately evident from the following example. We proceed here, for reasons that will be made clear at once, not from the imperative in the narrow sense, but from the imperative in the form of request 'Please, come here!' For the point decisive for our reflections, however, nothing is changed in the slightest by this.

Now we can say the following: 'I wish that you would do this, but I don't beg you to do it.' This sentence obviously has a perfectly good meaning and corresponds to a readily understandable attitude of the speaker toward the person he is addressing. It does not contain any inner contradiction, and from this it follows that the second part of the sentence cannot be a negation of the first. That we started with a request and not with a command, has, as its reason, that we wanted to exclude the problems of the 'willing of somebody else's action', which has nothing to do with the question that occupies us for the present. But our example, which till now has only served to remove an erroneous opinion with respect to the meaning of an imperative request and thus of the imperative in the narrower sense, also leads us to understand this meaning by rational reconstruction. For we can replace the statement: 'I wish that you do this, but I don't beg you to do it' by the following sentence of equivalent meaning: 'I wish that you would do this, but I do not wish to *induce* you to do it'. Thus it emerges, conversely, that the *request of A* directed toward B is nothing else than the *communication to B of A's wish to induce B to do something*.

The same holds true with respect to the command in the narrower sense. Here too we have a communication to another person that we wish *to induce* this person to engage in a certain behavior. The difference from a request here lies only in that in each case the inducement intends to appeal to a different *motive*. The person who begs another for something expresses thereby that he presupposes the other person's friendly attitude or goodness of heart, or his desire to observe social forms, to be the decisive motive for the fulfillment of the request. A person giving a command, however, appeals to the other's obedience, which can in turn spring from a variety of submotives. These can be described by the limiting words 'awe' and 'fear'.

Thus command (order), offer, request, imploring (appeal to pity)[152] are distinguished by the variety of the motives to which an appeal is made. But the meaning common to all these cases is the communication of one's own wish to induce the person receiving the communication to engage in a certain behavior. That we do not speak of 'wish' but of 'will' in the case of command, is only related to the fact that as a rule in this case the chance of fulfillment is so great, or is assumed to be, so that the person issuing the command has the belief that, through his command, he is *causing* the action of the other, thus determining it in a similar manner to his will determining his own action. This is expressed with special clarity in the sharpest form of the command 'You will do such and such a thing' ('You will come here.') Here the belief in the obedience of the person addressed to the command is so strong that the fulfillment is unquestionably anticipated.

Now in order to comprehend the difficulties confronting the understanding of the meaning of imperatives, we have to recall the *distinction* between *communicative meaning and the communicative goal* made in the previous section. The difficulties lie in the apparent correspondence of the sentence's meaning with the typical motivation for uttering the sentence by way of address to certain persons in certain situations. Because the *content* of the imperative is: 'I want to induce you to do this', while the typical motivation is 'By communicating to you that I want to induce you to do this, I want to induce you to do this.' The danger of confusing these is especially great because in statements of this kind, the *main attention* passes *through their meaning* and is focused directly on the *purpose* they are intended to serve, which obscures the clear comprehension of the sentence's meaning. The question which points to the specification of the meaning of a command reads: What, by issuing the command, does the person issuing the command want to let the receiver *know*? The answer to this question is obviously that he wants to let the receiver of the command thereby know his resolve to induce him to engage in a certain behavior.

We want to call a sentence, in which the purpose of motivation in the manner just described stands in the foreground a *communication* [*Kundgabe*]. The boundaries are of course fluid, for a communicative goal or, stated more correctly, an intention extending beyond the mere fact of uttering that sentence underlies just about every utterance of a sentence.[153]

Now we can conduct a test of the result of our analysis by asking ourselves how things stand with the *truth* of a command, a request, etc. For with the isolation of the sentence-meaning, we should also have specified a method of verification. Now this is actually the case. Though one does not say a command is 'true' or

[152] For this compare Sander, *Allgemeine Gesellschaftslehre,* Jena 1930, pp. 381 ff.

[153] See above, section "The Social Sciences and Psychology".

'false' – which is related to the circumstance just mentioned that the main stress is placed on the motives of the command – we can, however, say that someone has *deceived* another by means of a command, or that someone has been *deceived* by a command. Such deception is at hand when the alleged wish to induce the other to engage in a certain behavior does not exist. This is the case, for example, when by means of the particular communication we wish to achieve just the opposite outcome from the one pretended – say, to induce the person receiving the command to disobey it, so that one can take corresponding measures against him. In this case the command is *untrue*, for the alleged will to induce the other to engage in a certain behavior is not present.

After this excursion about imperatives, let us return to the concept of norm. In the analysis of the value concepts above, we have established that their meaning varies with the stress placed on the aspect of correctness. This is true especially for the concept of norm. In the borderline cases on the one side, the aspect of correctness is hardly considered at all; 'norm' is identified with 'command'; in boundary cases on the other side, however, this aspect obviously forms the *core meaning* of the norm concept. On the one side stands the commandment which is blindly obeyed, and on the other side the guideline for one's own action established after the fullest consideration,[154] i.e., the resolve based on weighing the practical correctness of keeping one's own future action in harmony with certain principles. Here – in contract to command by others, to *heteronomous* norms, – we speak of *autonomous* norms.

But even within heteronomous norms, there is still an important distinction to be made; its basis is the significance to be attributed to *the fact of issuing orders*. What is involved here is whether the issuing of the order is regarded as being *constitutive* or *declarative* with respect to its character as norm. The first case is such that, acting according to the guideline presented in the order only (at least predominantly), appears correct because this order was issued by certain persons or, possibly, under certain conditions; here the belief in its correctness often (but not always) springs from the assumption that whoever issues the order will, directly or indirectly, do harm to those to whom it is addressed if they fail to obey the command. In cases where such harm is already threatened in connection with issuing the order, one speaks of *sanctioned* norms. In other cases, however, obedience of a command is viewed as correct because we have confidence in the person issuing the order, that (as a consequence of his wisdom, wealth of experience and benevolent attitude) he will command the *correct* thing. Thus here the issuing of the order is a *symptom* of its correctness. Comparing a police order with a doctor's order can serve as an example of the distinction thus made, and makes the ambiguity of the concepts of 'authority' and 'competence' emerge fully. As far as the reference system for practical correctness is concerned, norms of the first group always require the complement: 'If you want to obey the order (or, act according to the will of the person issuing the order),' where in most cases the goal of acting in accordance with the order will be incorporated within a further complex of goals (avoidance of harm,

[154] [Missing in original edition – Ed.].

achieving rewards, showing loyalty); in the second case, however, no uniform goal can be indicated. It has to be inferred from the context of each case; for example, with respect to the correctness of a doctor's orders: preservation of life, preservation or restoration of health, avoidance or termination of bodily pain.

From these considerations, combined with the results of our general analysis of value concepts, it follows directly that *we cannot speak* of a *unique normative method* which would justify a rigid separation between the '*normative sciences*' and the *sciences of Being*'. In normative considerations, two kinds of ascertainments must be distinguished; first, ascertainment of goals, second, ascertainment of the conditions of practical correctness with respect to these goals. Once the goals have been set, then the question of the efficacy of a certain action with respect to the goals (correspondence to norms) is a question of fact. Frequently, however, this is not clearly understood – especially in the analysis of heteronomous norms – and most generally because here too, empirical statements are confounded with the results of rational reconstruction. For most frequently the rational reconstruction of the content of the norm plays a decisive role in the subsumption of actions under norms, so that the actual ascertainment of fact, that a certain personally-temporally fixed action satisfies certain conditions, recedes into the background. But the rational reconstruction itself is not infrequently composed of heterogeneous levels of knowledge. Here again it is a question of distinguishing between the meaning of a sign (the 'content' of a communication) and the purpose – or purposes – of issuing the sign. In the interpretation of the meaning of heteronomous norms, the content of the order is often separated only with the greatest difficulty from the purposes of the order: the question of which purposes can be assumed to underlie the content of the command can at times hardly be answered. The investigative situation becomes still more unintelligible due to the fact that the meaning of the norm also has to be inferred from symptoms (usually linguistic signs), and that this interpretation of symptoms is not properly distinguished from the rational reconstruction. When finally the erroneous Platonic conception of '*ideal objects*', to which the norms belong as '*meaning contents*', is added, the result is a confusion of concepts almost impossible to disentangle.

An argument brought up again and again in favor of an alleged *dualism of spheres* between the 'ought' and the 'is', is that from an 'is' no inference of an 'ought' can be made, and from an 'ought' no inference of an 'is'; it is this thesis that we now have to examine:

Let us first establish that the thesis has dual meaning insofar as on the one hand it can be interpreted as a statement about the relation between 'ought' and 'being-there' [*Da-Sein*] (realization), and on the other hand a statement about the relation between 'ought' and 'being-thus' [*So-sein*] (general determinateness). According to the first of these two interpretations our statement says: From the fact that the carrying out of an action forms the content of a norm, past or future realization of such behavior cannot be inferred, nor is the inverse inference possible.

According to the second interpretation, on the other hand, the statement contains the assertion: From the fact that an action displays certain general features it cannot be inferred (deductively) that the action conforms to a norm and from the statement that an action, generally not precisely defined, conforms to norms, it cannot be inferred that the action possesses certain features or that it does not possess such features.

With respect to the first of these theses, the following is to be said: As we have recognized, the meaning of the assertion that an action of a certain kind *ought* to be performed, is that an action of this kind is *practically correct*, i.e., with reference to a goal to be specified it is appropriate to that goal. (The alternate meaning: 'norm = imperative', according to which 'ought to do something' has the same meaning as 'having received a command with respect to this' can be disregarded in the analysis of the relation of 'is' and 'ought'). Obviously, from this general characterization, the reality of that which is characterized can be deduced just as little as from any other general characterization. On the other hand, from the assertion that behavior of a certain kind was actually performed, it cannot be deduced that it is practically correct – even in the case of pre-established goals – because from the determination of a position in terms of which the statement of facts is presented, no general determination of features can be derived – or as it is frequently put, no 'being-thus' assertion from a 'being-there' assertion.

However, – and now we come to the first part of the second thesis – that from the general characterization of an action (the specifying of its 'properties'), its 'practical correctness' cannot be deduced, follows from the circumstance that in this characterization no determination of the goal system, to which practical correctness is referred, is contained. Finally, the fact that from the judgment of an action being practically correct – with reference to a given goal system – no further features of this action can be deduced, is due to the fact that from a relation the features of one of its terms can never be derived. But it must always be taken into consideration that in each of the four cases discussed, by 'inferring' only deductive inference is to be understood; induction is not only possible in all these cases, it forms a core of the meaning-interpretation. Failure to recognize this circumstance also has contributed considerably to increasing the confusion.

These results of reflection, as well as the relevant ones of Part One, section "The Concept of Value", make it possible for us to take a position with respect to the *postulate* (subject to especially heated controversy in the *Methodenstreit*) of *value-freedom* in the social sciences.

For from these results it follows that this postulate must not be interpreted as though there [actually] were knowledge of values, but that the investigator in the social sciences had to forego making use of that knowledge in the same way that the geometer foregoes the use of compasses as soon as he has agreed to the postulate of making certain geometrical construction only by use of the ruler. Instead our postulate has to be understood as indicating, that the investigator would have to realize clearly that we cannot speak of 'absolute values' at all in a meaningful way, and accordingly, insofar as he operates with value concepts, he would be obligated to specify the *criteria of evaluation*. As soon as he is aware of the *relational character* of value statements, he will also comprehend their *relativity*, i.e., grasp the possibility of different goal systems.

What is most important for the methodology of the social sciences in this relation is already to be found in Max Weber's methodological essays, especially in

Der Sinn der Wertfreiheit der Sociologischen und Ökonomischen Wissenschaften [*The Meaning of 'Ethical Neutrality' in Sociology and Economics*, tr. Shils and Finch, in Max Weber *The Methodology of the Social Sciences*. – Ed.] There the chief misunderstandings, to which the postulate of value-freedom has been subject, are thoroughly discussed – for example, in pointing to the important role of human valuations (this intended as an objection) – in the formation of the social world. In their main thrust we agree with Max Weber's results; but the only difficulty is that he lacks a clearly formulated insight that the alleged dualism of the spheres of being and value (validity) does not exist. This may well be explained as due to his partial dependence on the current of thought of the Southwest German school of Neo-Kantians especially of Rickert). We want to quote an especially characteristic passage from Weber's essay:

> It may be asserted without the possibility of a doubt that as soon as one seeks to derive concrete directives from practical political (particularly economic and social-political) evaluations, (1) the indispensable means, and (2) the inevitable repercussions, and (3) the thus conditioned competition of numerous possible evaluations in their *practical* consequences, are all that an *empirical* discipline can demonstrate with the means at its disposal. *Philosophical* disciplines can go further and lay bare the 'meaning' of evaluations, i.e., their ultimate *meaningful* structure and their meaningful consequences, in other words, they can indicate their 'place' within the totality of all possible 'ultimate' evaluations and delimit their spheres of meaningful validity. Even such simple questions as the extent to which an end should sanction unavoidable means, or the extent to which undesired repercussions should be taken into consideration, or how conflicts between several concretely conflicting ends are to be arbitrated, are entirely matters of choice or compromise. There is no (rational or empirical) scientific procedure of any kind whatsoever which can provide us with a decision here. The social sciences, which are strictly empirical sciences, are the least fitted to presume to save the individual the difficulty of making a choice, and they should therefore not create the impression that they can do.[155]

With respect to this exposition it must be remarked, that the function assigned to the philosophic disciplines, i.e., to explore the ultimate meaningful structure of valuations and their meaningful consequences, is nothing else than the task of rational reconstruction of the 'essentially' intended meaning of the valuations, where consideration must also be accorded to the implicitly intended relationships, along with other valuations.

Thus we can formulate the core concept contained in the postulate of value-freedom to mean that from thought in general, and from thought in the social sciences in particular, 'ultimate', 'absolute' ends or goals can never be derived. In all investigations concerned with valuations these must be assumed to be already set (implicitly or explicitly) and thus must be *presupposed*.

From this insight the task now emerges, that wherever human action is qualified as 'correct' in social scientific investigations, we must clarify, with respect to which goal system this 'correctness' is to be understood. This task has been neglected only too frequently by the proponents of the postulate of value-freedom, and thereby they can easily be shown to be wrong by their opponents, by exposing the

[155] Max Weber, *Gesammelte Aufsätze zur Wissenschaftslehre*, Tübingen 1922, p. 470.

underlying (implicitly presupposed) goals. But we must not overlook the fact that value-freedom in the sense just clarified, and 'reference to goals' by no means are opposed. Incompatibility results only when it is erroneously assumed that some sort of (scientific, philosophical or metaphysical) knowledge makes it possible to comprehend absolute values, or if it is believed that the examination of practical correctness with reference to given goals represents a special method of gaining knowledge. Only insofar as the opponents of value-freedom raise one of these claims – which in addition are sometimes alleged to be supported by misconstrued 'ideal objects' and 'meaning contexts' – is their judgment false.

In the pertinent controversies, the problem discussed at great length in our first part, about the *objectivity of science*, very often also plays a role. Scientific judgments – so it is argued on the one side – are theoretically correct insofar as they correspond to the facts; the criteria for this correctness are thus 'objective', i.e., conditioned by the nature of the objects themselves; subjective *attitudes or position-taking acts* play no role here. Values, on the other hand, are characterized precisely by taking positions in this very way, and that is why there is such an unbridgeable *chasm* between *knowing* and *valuing*.

This thesis can, however, be invalidated by the opponents by questioning the supposed objectivity of this knowledge: Every assumption that goes beyond the 'immediate given' (and this cannot be isolated at all by means of judgments) – contains presuppositions – above all of a certain uniformity of world events – which cannot be objectively justified, as they are based on spontaneous acts of acceptance, of 'belief' [Kaufmann's English – Ed.], which must be characterized as altogether '*subjective*'. This characterization is justified not only because these acts of acceptance can find no ultimate justification 'in the thing itself', but also because the motives for these decisions are at least partially conditioned by affect, and thus to be found in the subjectivity of instinctual life. It is no different with valuations and therefore a sharp delimitation between knowledge and valuing, between *acquiring knowledge and taking positions* cannot be drawn.

With respect to the arguments just presented, the following is to be remarked, to begin with: our investigations, concerning the general theory of science have shown that the conception of knowledge as pure receptivity is not tenable, so that invoking the spontaneity of knowledge is justified. Furthermore, it must be admitted that the syntheses carried out in the pre-predicative and predicative experience can be interpreted as *goal-directed acts*, which serve the orientation of human beings in the world, and therefore the preservation of their vital existence in it. Finally, it is undeniable that the same *data* can be systematically *grouped in different ways*, and that the manner in which this grouping is carried out is to a large extent determined by affect-guided, practical interests.

But does an invalidation of the entire set of conceptual motifs contained in the postulate of the value-freedom of science indeed follow from this? That is by no means the case, even if from what has been said it emerges that the justification we have cited cannot stand up under criticism. We now wish to extract the correct core of the postulate.

To be sure, the processes of knowledge are not to be confronted with purposive practical behavior as if they were theoretical 'activity free from purpose'; however, the *purposes of knowledge* can very readily be delimited with respect to other purposes, and here the fact that as a rule these purposes of knowledge are linked to 'practical' purposes does not form a counter argument. What is decisive is that the goals of knowledge can be *thematically isolated*, and therefore that the 'correctness' of human behavior can be ascertained (exclusively) with reference to them. The common (essential) aspect of all goals of knowledge, in every-day thought as well as in scientific thought, however, is forming judgments that prove valid. This validation – which, as we have shown, can never be final – mainly consists of more or less direct [coherent] agreement with the observation of data of external or inner experience.

Accordingly we can say that, correctly understood, the postulate of the value-freedom of science is a combination of the two following results of reflection:

1. *Scientific thought*, qua scientific, *refers solely to the goals of knowledge*: whatever one may 'actually' mean by 'correctness' or 'incorrectness' ('truth' or 'falsity') of scientific judgments is to be determined exclusively with reference to the goals of knowledge.
2. The assertion of *absolute* values, or also *absolutely* correct goals, cannot enter into science, because it is contradictory. Meaningful value judgments – that is, assertions of practical correctness with reference to given goals – however are by all means subject to scientific evaluation and have their place in scientific thought.

That *valuations*, as historical or sociological facts, can form topics of the pertinent sciences is self-evident.

However, there are still several words to be said about the grounds, from the viewpoint of the *psychology of knowledge* for the *typical linking* of the *postulate of value-freedom with naturalistic doctrines*, and the typical linking of its rejection with the different variants of anti-naturalistic doctrine. Very frequently a merging of two conceptual motifs is found. On the one hand, the object of the knowledge of nature – which according to naturalistic doctrines must also embrace social facts – appears to be *given* without the reference to the setting of goals and thus in this sense appears to be *value-free*; on the other hand, knowledge of nature appears to possess a method prescribed 'by the thing itself', while in the human sciences, according to all appearances, the method – and this includes the posing of the problem as well as its treatment and its solution – are determined by goals that transcend knowledge ('values'). The latter conception, however, is again based on the known fact that in abstract natural science we grasp the entire sphere of nature – or, at least inanimate nature – by using a relatively uniform procedure, and we can establish universally valid principles for it; while in the human sciences a multiplicity of aspects coexist side by side, among which a selection according to purely theoretical considerations is not possible, so that extra-theoretical considerations must decide among them.

With that we stand at the threshold of the range of problems of 'value relations' in Rickert's sense. But this is so closely linked with the problems of the science of history that we will not deal with them until the next section in which the meaning of the 'historical' for the social sciences is to be discussed. On the other hand, we do want to examine the problem of so-called *historical attribution* [*Zurechnung*, also translated as 'imputation' in what follows – Ed.] at this point, as the general problems of attribution have been linked most closely, in the history of ideas, with the problems of value, by way of the questions of 'significance' on the one hand, and of 'responsibility' on the other. It is toward this, and toward the problems of the concept of *freedom* – to a great extent coinciding with it – that we now turn.

The core of the concept of *attribution* is the concept of *causal connection*. We wish to examine the inferences that flow from this first, before we turn our attention to the axiological connotations of the concept.

In the investigations of our first part, we have presented the main misunderstanding linked to the concept of causality. Especially, the conception that the cause is contained in the effect, and that accordingly it is objectively and uniquely determined whether a fact F_1 is the cause of a fact F_2 was shown to be erroneous. Further, in the analysis of the '*ceteris-paribus* clause' we pointed to the role of accompanying circumstances in the formation of experiential laws; and here it turned out that each law is linked to the presuppositions of a relatively unchanging background of 'normal events'. Distinguished from this background of constancy are the variations of certain factors F, and (according to the law) the corresponding covariations of other factors G. Therefore if we wish to ascertain the significance of the variations of F, we have to compare with each other the various constellations which are distinguished from each other solely by differences in the varying factors. From this, in itself, it follows that in examining the question, 'What effects would there have been if event E had not taken place?', we cannot simply assume a vacuum in the place of the event E which has been thought as not occurring, but have to set in its place what would 'normally' have occurred. Now, due to the relatively unambiguous characterization of the direction in which links between experiences are sought, a clear awareness of this known fact plays a much lesser role in the abstract natural sciences. However, in the human sciences it is of extraordinary importance, and failing to take it into consideration leads to the most serious difficulties, which can also have very disturbing effects on the problems of the concept of attribution.

In the social sciences, the fact that every attribution contains within it the thought of a substitution was only taken account of consistently in the case of attribution in economics, and this especially in the '*subjective theory of value*'.[156] Here, we examine within the framework of the investigation, what share which goods of a higher order (especially production goods) have of the value (price) of consumers' goods, whether an eliminated higher good could be replaced at all, and if so by what (replaced according to economic principles)? Accordingly the economic significance

[156] Compare below, section "Remarks on the Methodological Controversy *[Methodenstreit]* over the Theory of Marginal Utility".

of such a good (its value or price) can only be estimated by means of consideration of its position within the system of economic experience.[157]

From what has been said it follows that questions about the origin of a phenomenon have to be answered the more precisely, the more *abnormal* this phenomenon is. If, for example, we ask why a certain express train operates on the Vienna-Salzburg line, then we will be at some embarrassment to supply a brief answer, as we will be confronted by the wealth of conditions required for realization of this fact. If, however, we are asked why a certain express train on this line was derailed, then, if a switch was set wrong, there will be no hesitation in giving this as the answer; for by this, the derivation from the normal constellation of facts which appears to be linked with the diversion of the train from its normal direction and thus its derailment, appears to be adequately characterized.

These considerations enable us to take a position with respect to a series of controversial problems of attribution which are closely related to methodological questions. One of the most important questions for the science of history is the question *what share great men* have in the shaping of historical events,[158] and we now wish to examine this, as an example.

First, in conformity with the preceding considerations, we have to observe that the question 'What would have happened if … had not …?' points to the comparison of constellations of facts of a *certain kind*. Therefore the answers given to such questions are not statements about a singularly occurring situation, but about situations of the kind under discussion.

If now we ask what significance Bismarck had for the unification of Germany, then we must seek to ascertain which of his plans and measures that were decisive for this outcome bear the stamp of his personality to such a high degree that it cannot be assumed another person could have accomplished something similar in his place. To be sure, with respect to the *projection of the background of normality* from which the significant personality distinguishes itself, there remains a broad scope of possibilities; the question arises especially to what extent we want to avail ourselves of knowledge of the concrete historical situation – thus, above all, knowledge concerning the qualities of the men who could have carried out the conduct of Prussia's affairs in Bismarck's place. The historian will at times defend the view that it does not reduce the significance of a man, nor his achievements, to say that other men were available who could have exercised his function just as successfully. Thus, here in the determination of a person's historical significance, a measure of the average is made to serve as a basis. The case is quite different when the question asked is to what extent a person was *indispensable* in a certain historical situation; for in this case the person's capacities and achievements presumed to be pertinent are compared with those of the *élite* that would come under consideration as

[157] Compare below, section "Remarks on the Methodological Controversy *[Methodenstreit]* over the Theory of Marginal Utility".

[158] For this see also the detailed discussion of this topic in J. Burckhardt, *Weltgeschichtliche Betrachtungen*, chapter 5.

replacements, in which case then the evaluation of their 'significance' may produce a rather different result.

Now as far as the *axiological connotation* of the concept of 'significance' is concerned, only a few words are required. The 'rank' of the action (achievement) is determined according to the place which the imputed facts occupy as positive or negative goals within the goal system; and according to the degree of rarity (irreplaceability) of the action, and this in turn is co-determining for the 'rank' of the person having acted insofar as he is evaluated on the basis of achievements.

From historical attribution, we now want to distinguish that *kind of attribution* which is designated as specifically *normative*, and exemplify its nature in criminal accountability. We can formulate the general schema for commands [orders] of the (objective material) penal law in the following manner: "Under certain circumstances a judge ought to order something to be due, to be characterized more precisely (e.g., putting to death, deprivation of freedom, or of money) to a certain person.[159] According to this formulation, a certain behavior on the part of the person upon whom the judge orders a certain action in accord with the prescription of law, is apparently not presupposed at all; thus among primitive people, an order of law could state, say: 'If it has not rained during the summer for the duration of 4 weeks, then a priest, specified more precisely, ought to be put to death.' In spite of this, in such cases, too, it is frequently said that the failure of rain to occur is 'attributed' to the priest. How is this to be understood?

To begin with, two different things have to be observed here: (1) Our conception of the existence or non-existence of a causal link between the behavior of the priest and the failure of rain to occur must be distinguished from the conception of those people in whose social domain the commandment under discussion has validity. (2) Besides the 'verbal meaning' of the commandment, implicit presuppositions of meaning have to be taken into consideration. If killing the priest is conceived in his circle as 'punishment' (*malum passionis quod infligitur propter malum actionis*), then our commandment receives the following meaning: In case, during the time-span named, the priest has not produced rain by means of measures to be more precisely characterized, then at the expiration of this time period he is to be put to death. Here the failure of rain to occur is actually causally related to the behavior of the priest (failure to bring rain); therefore saying that he is made *accountable* for this behavior is quite proper. However, should putting the priest to death be conceived solely as a sacrifice without being accompanied by any thought of punishment (atonement), then this mode of expression is incompatible with the 'actual' meaning of our term, to be explicated in a rational reconstruction. But it is to be carefully noted – for failure to take this aspect into consideration has contributed much to the existing confusion – that even where putting the priest to death is conceived as punishment, the question whether he *could have brought about* the onset of rain need not to be raised. For he could have been punished for his incapacity; the statement '*ultra posse nemo obligatur*' is not only not an analytic judgment,

[159] Compare for instance Hans Kelsen, Reine Rechtslehre, Vienna 1934, p. 25 ff. [Rev. Eng. Tr. in *General Theory of Law and State*, Cambridge, Mass, 1946].

resulting from the analysis of the concept '*obligatio*'; but even in historically existing legal orders, it holds only to a very restricted degree.

We had to stay with this point longer, because here lies one of the chief sources of the erroneous conception that a 'normative accountability' would have to be distinguished from 'causal attribution' in which then again an argument for the existence of specifically normative methods was seen. Here normative accountability was such as would follow on the grounds of norms – for example, legal norms – and not on the grounds of the recognition of causal relationship.

With respect to this it can be said that, the existence of a causal relation between the conduct C_p of a person P (as cause) and the fact F (as effect) is asserted in a judgment if the judgment contains the thesis that F would not have taken place had P conducted himself otherwise (in a manner to be further characterized). Now the command, P is to be put to death, can be supported just as well by a false as by a true judgment of this kind, and for the question of the legal validity of this command, it is irrelevant whether the judgment which supports it is true or false, and thus whether the person is correctly or incorrectly held accountable. But the inference cannot be drawn from this, that a normative element is contained in the concept of such accountability nor can it be inferred that the command that upon the failure of F to take place, P is to be put to death, contains as such the accountability characterized.

The difference – to be sure, not a sharp one – between this kind of accountability and historical imputation in the narrower sense, lies chiefly in the different *purposes* for which the *selection of the 'essential' causes* from the multiplicity of *possible* causal imputations, is carried out. While in historical imputation, *knowledge goals* stand in the foreground, on the other hand, in legal accountability, on which imposition of painful consequences depends, considerations about the chances of *achieving practical goals* by influencing people are decisive. Therefore the main goal of penal law is to induce human beings to avoid certain actions; and the means for achieving this goal – aside from rendering the individuals incapable of doing harm – is the psychological effect on the delinquents themselves, in order to deter them from further offenses (individual prevention), and the psychological effect on the other members of the community subject to the law (general prevention). Now the penalty is to be selected in such a way that it is appropriate to the purpose and accordingly, if the penalty appears to be linked to the condition of a certain accountability, then the kind of accountability must also take its orientation from those purposive viewpoints. That means, that there will be a tendency to hold a person accountable for a certain undesirable outcome, when there is a chance that his punishment, executed on the basis of this accountability, will lead to the avoidance of similar occurrences by the subject of the penalty, and in any case also within the remaining community that is subject to the law.

Being contrary to the criminological goals, and prohibited, the conduct which has been subjected to penalty, now receives the 'negative' value-index of 'illegal', and as the threat of punishment appears to be linked to the accountability, the false semblance arises as though that axiological moment were contained in the concept of accountability itself.

But we have not yet arrived at the end of our reflections. For the following aspect also has to be taken into consideration: from the viewpoint of criminal policy, the psychological context of the actions subject to penalty in given cases will play an important role, and especially the question whether the action was carried out with premeditation, whether the actor could have foreseen the outcome to be attributed to the action, whether he had known or 'could have known' that the action was in violation of the law. Now in case the assumption appears justified, that the person, who has carried out an action regularly subject to punishment was not capable, due to mental deficiency, of foreseeing the expected outcome (which is the reason such action was prohibited), or to clearly realize the illegal character of the action, then the interest in punishment from the viewpoint of criminal policy is a much smaller one, and therefore frequently there will be a tendency in such cases to exclude punishment. Viewed in the distorting mirror of the doctrines of absolute value, these factors lead to the following construction: The lack of certain intellectual faculties on the part of the actor in itself excludes his responsibility (guilt) and 'therefore' his punishment in such cases would simply be unjust (incorrect). It is this complex of conceptions which is typically linked with the term 'unsoundness of mind'. We cannot enter into discussion of the multitude of problems linked to the concepts of '[legal] responsibility' or 'irresponsibility for one's action', but have to be content with ascertaining that in them there is by no means any axiological element, any *qualitas occulta*, free of purposiveness.

We still have to pause at one especially interesting point, namely the analysis of the interconnections between (moral and legal) 'responsibility' and the '*freedom of the will*'.

That such an interconnection exists becomes clear when we consider that the question of accountability, which is to decide whether a person can be 'held responsible' for his actions, usually is identified with the question of whether this person is *capable* of voluntary action, or of *free will.*

Yet first we shall have to observe, that the concept of 'freedom', with which we operate in the social sciences, has a number of meanings. The fundamental level of the problems is characterized by the opposition 'causality' and 'freedom'. In analyzing it, it is best that we take our departure from the Kantian distinction between 'natural causality' and 'causality' [based on freedom].[160] A person acting with premeditation is conscious of the fact that the individual stages of his action are not entirely determined by the laws of external nature, and to him his own action appears to be free insofar as it takes its course undisturbed by external influences in the narrower sense, i.e., by natural events which inhibit the execution of his intent, further, by actions of others which oppose the realization of his own purpose, and finally, by inner distraction (passions, habits). Analogously, the conduct of another human being is considered to be free when we assume that it takes its course according to this person's intent, from which, incidentally, it follows that the entire range of problems of the interpretation of another person's subjective meaning enters into the

[160] For this see especially the exposition of the Third Antinomy in Kant's *Critique of Pure Reason.*

problem of freedom. Consequently 'freedom' does not mean 'unpredictability', but on the contrary 'predictability on the basis of knowledge of psychical data.'

It is comprehensible, that freedom understood in this way does not contrast with natural law in an opposition without transitional gradations. For even disregarding entirely that in every realizable intention, the lawfulness of nature must explicitly or implicitly be considered in the calculations (since it provides decisive points for the 'correct' way to attain the aspired goals) and the idea that we might not be hindered by external disturbances, and especially by intervening natural events, also permits gradations.

But the assumption of an opposition between freedom and causality contains yet another consideration, which we can formulate as follows: The problem of free will would not lie in whether one can do as one would wish, but whether one could will what one wants. The decisive question is whether the fact that one has a certain intention is itself determined by laws, or whether this intention has to be conceived as a *prima causa* in the true sense.[161]

But this way of posing the problem loses its sharpness – as do so many formulations of questions in terms of '*aut-aut*' [either/or] – as soon as we seek to account more precisely for the *criteria* on the basis of which judgment, in either of these two senses, is to be reached. For as soon as we have clarified the meaning of the causal relationship (as we have done in our investigations in the general [first] part of this work) we recognize that this relationship is not a pre-established one that can be regarded as linking individual things or events unequivocally, but that it represents nothing else than the hypostasized assumption of other facts. If we apply this insight to the problem of the determined or undetermined character of the will (intent), then we will recognize that this problem can be resolved on the one hand into questions of the *degree of predictability*, and on the other hand into questions of the *kind of phenomena* which are to form the *basis for prediction*.

The thesis that one can will whatever one wants to, would mean, according to the traditional conception, that will is undetermined, and therefore insofar as it is related to the willed action, as its cause, it must be conceived as the *prima causa* of this action, since every attempt to trace beyond it would mean a transgression into a sphere of complete indeterminacy. Therefore the assertion of free will, understood in this way, coincides – from the viewpoint of the history of doctrines – with the assertion that the *will* is the *prima causa* of action. However, in the application of this concept of freedom, there is an ambiguity. It results from the fact that the question of the freedom of the will – emerging in the framework of a particular treatment of problems aiming at an explanation of human action – has more than one kind of answer. The answers, as a rule, are made to depend upon whether it is *advisable*, in dealing with this investigation, to carry the explanation of action only as far as the pertinent intent, or whether it is wise to go back beyond that intent.

Under certain circumstances it is already to be regarded as an explanation of the behavior of a person, when we indicate that this action was intended; in this case the explanation can be formulated in terms of the following syllogism:

[161] Compare Kaufrmann, *Strafrechtsschuld*, p. 133 ff.

The person in question always carried out his intentions.
He has resolved on the behavior under discussion.
He has carried out this behavior.

Most frequently, however, more is demanded of an explanation of human behavior; to establish that this behavior was willed (intended) will not (by itself) be satisfactory but, in case the answer to that question is affirmative, we will further seek to incorporate this *intention itself* within a more general complex of experience. Now this endeavor can either have the result that this intention is understood as part of a more comprehensive intention, i.e., an intention whose partial realization is served by the intention to be explained; or, on the other hand, that the latter intention is traced back to causes which are *not* in their turn also intentions. And the investigation, under which circumstances the first or the second kind of explanation proves appropriate, leads us to implicit presuppositions which are contained in the problems relating to freedom of the will, and which have been exaggerated.

Now, as we have already made clear in formulating the syllogism, to trace back the conduct of a human being to his intention can only count as an explanation when, in the first place, his conduct is of such a kind that one can 'reasonably' assume it was intended by him, and in the second place, that *as a rule* he is capable of carrying out his intentions, i.e., his intentions of the kind in question. However, carrying them out can be frustrated by three kinds of factors, (1) by external obstacles in the narrower sense, i.e., such as cannot be attributed to one's fellow human beings; (2) by the behavior of other human beings; (3) by one's own irrational impulses. Accordingly, those are called free, who in shaping their lives uninfluenced by external facts, the intervention of other human beings, and their own passions, *live according to their plans*. At this point, we begin to see the contours of the interrelation between the concept of freedom just presented and a *rationalist* ethic; this relationship has led to the view that the concept of freedom is a central concept of ethics, understood as a system of propositions of practical reason.

What has just been said can be applied quite readily to the *explanation* of *intentions* themselves. For even a person's intention can only be explained by means of other intentions only insofar as this person lives a planned life, as his conduct is 'governed by reason'. From the viewpoint of cognitive practice, an important role will be played here if the intention to be explained proves to be a typical means for achieving typical goals, and thus whether we can incorporate this intention into a more or less specialized rational method, or not.

But here lies the second point of connection with ethics. For due to an easily understandable change in the meaning of the concept of the *normal*, and accordingly of what is understandable and thus rational conduct, it has come about that only such conduct is considered to be 'rational' ('reasonable') which serves the pursuit of '*well-understood interests*'. Accordingly, only those persons are designated as 'truly free' who act in their 'well-understood interest' ['enlightened self-interest'] and that characterizes – especially according to the conception of the ethics of antiquity, going back to Socrates – those who act morally.

Finally, the third point at which the concept of freedom merges with ethics is that of its link with *responsibility* (moral and legal accountability). Here the argument is that a human being can only be held accountable for conduct the causes of which lie within himself. However, most frequently the will alone is regarded as 'cause'.[162] Due to more profound insight into the psychology of the unconscious, this conception has recently been recognized as a prejudice. Indeed, it was never able to dominate in a field, where practical regulation of human society is concerned, i.e., in juridical legislation. Negligence as well as fraudulent damage was always subject to penalty. Initially, above all, this was according to the principle of liability for the results alone (without any principle of culpability); later, predominantly it was associated with the indication that 'negligence' also contained an element of culpability. Only in a very limited number of cases and predominantly when typical anomalies of the person's physical constitution or general capacity of understanding rendered the abnormality of the person being judged especially evident, was accountability, and therefore responsibility for action, considered non-existent. The insight that no sharp borderline could be drawn between normality and abnormality, as well as between conscious and unconscious endeavor, and that therefore a doctrine of accountability based on these distinctions could not draw a sharp borderline between responsibility and irresponsibility either, has led to the concept of 'diminished accountability' which contains in it *gradation of the freedom of the will*.[163]

The decisive insight that results from these considerations is that the problem of the freedom of the will, in the form in which it appears in the social sciences, is a *pseudo-problem*. For acceptance or rejection of the concept of freedom of the will is only another expression for the selection or rejection of certain research methods, which, under certain circumstances, turn out to be more or less appropriate to the goals with respect to given theoretical or practical aims.[164]

Finally, concerning the *stress on value* in the concept of freedom, it follows from our investigations of the concept of value and the problem of accountability, that this is methodologically irrelevant.

Through the preceding reflections the most important sources of error underlying the untenable assumption of specific axiological, or normative methods, have been uncovered. But the insight that, from this viewpoint the division of the sciences into sciences of being and sciences of value (normative sciences), proves to be inappropriate, must not lead us to overlook the significance of psychological-sociological relationships, in human valuations (a) as immediate topics of social science investigation and (b) as motives in the choice of method.

It must only be noted with care that relationships of this kind cannot be regarded as foundations of a *logic of value*. In such a misinterpretation, failure to recognize the relational character of the value concept again plays a contributing role. Anyone

---

[162] Recall Kant's famous dictum in his *Grundlegung zur Metaphysik der Sitten* "There is nothing anywhere in the world, nor indeed conceivable outside of it which could be considered good without reservation, except only good will."

[163] Compare Kaufmann, *Strafrechtsschuld*, p. 136 ff.

[164] [Missing in original edition – Ed.].

who is clear about his own universal or particular value system, and thus knows what he really wants, will (as we have observed), frequently recognize that some of his goals collide with certain others, and if he also desires to act with awareness of his goals and with consistency, he will have to decide to sacrifice some goals or other. With reference to those goals G_n which have been accorded precedence, the sacrificed goals will appear *incorrect*, because they are incompatible with the former, while on the other hand, the goals G_m whose realization appears bound to the realization of G_n, are *correct* goals with reference to G_n. As soon as the criticized absolutization of the sphere of values takes place, the false appearance arises that these incompatibilities or implications, were internal relations of value qualities and represent *a priori* norms for value judgments and actions, in the same manner as the principles of logic would represent *a priori* norms for thought.

Faced with this, in order to counteract this confusion, we want to call the valuations which reflect the real relationships of the goals '*axiologically consistent* valuations' (or 'actions'). Now it will readily be realized that in most cases only axiologically consistent conduct – conduct taking reality into account – will be successful, and that consequently only axiologically consistent valuation will form a suitable basis for the design for actions. Therefore it is not surprising that most of the automatic taking of positions and automatic impulses for action are to a great extent adapted to the principles of axiological consistency and thus – as rational reconstruction shows – possess a pervasive rationally-purposive structure. Action that is axiologically consistent in this way by no means has to be rational in the sense that in the design of the action itself the relationship of the goals has been grasped completely, or even only casually. As we wish to point out in advance, this is frequently overlooked in investigation of the character of the *ideal type*, which has created considerable confusion, especially in economic theory.

This error in thinking, which presents itself in the failure to recognize the meaning and significance of rational reconstruction, is also responsible for the erroneous belief that in pointing to the '*irrationality of value*' a counter-example has been found for the claim that a considerable proportion of human valuations is based upon strictly *formal* principles. Once we free ourselves from this prejudice, the correctness of this claim can be readily seen. By the postulate of axiological consistency, the orientation of action and valuation according to the formal viewpoints of consistency and of causal relationship of the set goals is required. Quite apart from this, a series of formal goals is empirically posited, simply with the general goal of making life in a community possible for human beings. These are, on the one hand, related to cooperation between the members of the community, and on the other, to the protection of the human beings living in the community, from each other, and against attacks from outside. Thus the negative valuation of lying (*aliud dicere aliud sentire*) to other members of the community is understandable because almost any cooperation is made impossible, when one can no longer trust the words of others. About the 'normal' principles of justice and its teleological background there are still several things to be said later.[165]

[165] See below, Part Two, section "The Concept of Positive Law and the Pure Theory of Law".

The relation of 'formal values' to their unspoken underlying purposes, however, cannot simply be conceived is such a way as though these purposes were the 'real meaning' of these values, for in the value judgments of human conduct it is not necessary to have recourse to them: it is sufficient to disclose that someone knowingly has communicated a falsehood in order to evaluate his conduct negatively as a 'lie' without first investigating whether by his doing so the goal of cooperative living together of the members of the community had been impaired. Rather, the goal of truthfulness is understood as an independent goal, as a reference system for practical correctness that can be isolated. But in order to *understand* this goal, it is important to take into consideration with what basic facts of communal life it appears to be linked. In this connection we refer to our considerations, in the first part of this book, concerning the relation of the categorical imperative to the conditions for the establishment, or maintenance, of human communities.

Finally, a few words of criticism must still be said about the erroneous conception characteristic of several variants of the *rational doctrine of value* (especially rational ethics) according to which formal values were to be seen as rationally evident 'absolute values'. This prejudice can be understood as arising out of a tendency which is closely linked to the striving for a unified orientation of action and valuation (axiological consistency), which seeks to deduce the totality of values from the smallest possible number of 'ultimate values'. To these latter, which are to serve as a basis for the entire system, absolute value has been attributed in the manner described above. The untenable character of this conception follows from our analysis of the value concept.

The fundamental task for axiology, which ought to form an important instrument for research in the social sciences would be to work out a *scheme of types of valuation*, which would contain the basic types of ultimate goals and on the other the main lines of the empirical connections between valuations. The following would be considered fundamental classes of goals: *personality goals*, in which the realization of a certain ideal personality is set up as the ultimate purpose; *happiness goals* by which the realization of a certain state or condition of human beings is set as the 'highest value'; and finally *esthetic goals* in which a certain state of the world (*harmony*) appears as the desired end, independently of any relationship with moral or eudaemonistic purposes, solely for the sake of its 'inner perfection'.[166] Then the most general possible schemata of connections between goals would have to be worked out, which would permit a comprehensive survey of the number of degrees of freedom to be considered on a certain level of abstraction. Finally one would have to coordinate the axiological types established in this manner and the characterological, sociological and historical (in the narrower sense) facts. Much preparatory work for this has already been done, but its systematic unification, free of pseudo-philosophical prejudices, is still required.

---

[166] See Kaufmann, 'Soziale Kollektiva', *Zeitschrift für Nationalökonomie*, vol. 1, p. 302 ff, (1925).

4. The 'Historical' in the Social Sciences

In our basic reflections in the first part of this book, we have placed great emphasis on the observation that even in pre-scientific concepts of facts, and all the more so in scientific concepts of facts, it is *not purely receptive findings* that are registered, but that they contain assumptions about general interrelations as well. Furthermore we have shown how, taking this fact into account, the demarcation between 'fact' and 'law' is to be formulated. These results must also be carefully taken into consideration in determining the concept of *historical* fact, as the topic of the historical sciences. The most thorough investigations, and the ones of the greatest importance for inquiry into theory of history during the last generations, were those of Rickert[167] based on Windelband.[168] He defines the logical place of the historical sciences in the narrower sense – i.e., the sciences of human history – with the aid of classification according to the paired oppositions: '*generalizing (nomothetic) sciences*' – '*individualizing (idiographic) sciences*' and '*value-free sciences*' – '*value-related sciences*'. For the value-free generalizing sciences, physics can be adduced as the model; for the value-related generalizing sciences, economics; for the value-free individuating sciences, natural history; and for the value-related individuating sciences, history in the narrower sense of historiography. Now, to be sure, we must not misunderstand Rickert by saying that he assumes history is concerned with the individual as such, in the sense that no general laws enters into it. There can be no question of that in a philosopher of Rickert's rank, who knows very well that all scientific inquiry is incorporation into general contexts. But in spite of this, his thesis, which recurs in diverse variants, that abstract natural science has as the goal of its inquiry the ascertaining of *relative uniformities*, while historical science, on the other hand, aims at ascertaining *significant singularity*, must be interpreted with great caution. For the concentration of attention on singular phenomena in history can be traced back – insofar as esthetic interest (in the broadest sense) in great personalities is disregarded – to the circumstance that in the case of the explanation of historical events, or also of historical prediction, it is possible to a much lesser extent to disregard *anomalies* (*singularities*) than it is in physics. The attempts to arrive at laws of historical process, such as have been undertaken ever anew for thousands of years, down to the present day, make it obvious that interest in the general traits of the historical process, and the desire to find as comprehensive historical laws as possible, are very much alive. Only a closer examination shows that the results of large scale historical constructions all too frequently are quite questionable, and that in most cases we can only arrive at explanations of historical processes which are considered adequate by basing ourselves to a very great extent on detailed descriptions of singular data, i.e., of significant personalities who 'intervene in his-

[167] See notes 123 and 124 of our first section. An interesting analysis of Rickert's theory is contained in Fritz Kaufmann's 'Geschichtsphilosophie der Gegenwart', *Philosophische Forschungsbriefe*, No. 10, Berlin 1931.

[168] See note 167 above

tory in a formative way' (i.e., change the horizon of prediction and of specific historical situations). Regarded from this aspect, the great interest of historians in singular facts is an *interest forced upon them*. The epistemological fact that certain of the data are singled out as distinct made the question acute, according to by what criteria this selection is made. The rational reconstruction of the viewpoints from which the historian makes his choices in this respect, finds expression in Rickert's theory of *value-relations*.

The core of his presentation, which is incontestable in its main points, is that in this selection the historian takes his orientation from *pre-existing values*, i.e., he sets as his goal the comprehension of those interrelations of facts that are generally regarded as essential (significant, relevant) within his – broader or narrower – circle, such as, for example, the development of political relations of power, of religion, of art, of civilization (related to technology) in the narrower sense. (The concept of *culture* employed by Rickert in this context is only a summarizing concept corresponding to certain maxims for the thematic selection of certain objects for historical investigation). Accordingly, the historical fact in Rickert's sense is a *value-related* fact – in the manner just characterized.

Now, the range of problems of the 'historical' manifests itself in the social sciences in various ways. The aim of the brief analysis that follows is first to isolate these various meaning strata of the concept of historicity, which are often not kept sufficiently separate, and then to treat one of these problem-complexes, which is of special significance for the theory of science, namely the problem of *historicism*, somewhat more fully.[169]

The thesis that the social sciences are historical sciences has been justified in various ways; the most important arguments are the following:

1. The *goal* of the social sciences is the *acquisition of historical knowledge*; for all social-scientific knowledge aims at determining the cause of human behavior at certain historical places and thus at the determination of historical behavior.
2. The *laws* of the social sciences are *historical* laws, because all general assumptions which are supposed to make long-range predictions of social events possible, can only be attained through insight into the principles of the *development* of mankind, into the 'meaning of history', and thus carry an *index of historical time*.
3. The *laws* of the social sciences are *historical* laws, because the *material* [*basis*] *of induction* from which they are drawn, are *entirely historically transmitted facts*.
4. The *laws* of the social sciences are *historical* laws, because it is not possible to establish general laws concerning the course of human behavior during a narrower or broader time span, if they are linked to a specific *historical condition of the data* as their factual basis.

[169] For the dogmatic history of historicism see Tröltsch, *Der Historismus und seine Probleme*, Tübingen 1922.

5. All *social scientific knowledge* is 'historicized' by the circumstance that it is *relative to the historical situation* of the subjects of knowledge; objective knowledge independent of the reference system of the concrete historical situation of the researcher is not possible (*fundamental thesis of historicism*).

Let us now turn briefly to the individual theses. Assertion (1) requires no further discussion, for it is doubtlessly correct insofar as every social fact is a historical fact. But in this sense the assertion becomes quite trivial, because what the question concerning the role of the historical in the social sciences really aims at is the decision to what extent *general* propositions about historical-social reality are possible. It is in this direction that the theses (2)–(5) cited above, point.

First, as far as thesis (2) is concerned, according to which lawfulness within the social world is determined exclusively by the predesigned line of development of historical events, it can probably be stated today that neither Hegel's[170] idealist constructions of history, nor the positivist historical constructions of Comte[171] and Spencer,[172] can be called correct in the sense that reliable predictions could be derived from them.[173] The same also holds true for the many earlier and later constructions, whether these are based on the optimistic notion of progress or the pessimistic notion of eternal return. On closer analysis we soon discover that their attempt to incorporate all the heterogeneous spheres of human behavior into the same schema of historical development in itself forms a sufficient ground for these speculative constructions to fail when they 'confront the facts'. To be sure, such a 'confrontation with the facts' can only furnish a test for the validity of the hypotheses, when their combination by the fulfillment of specific *predictions* is required; for otherwise it is frequently only too easily possible to rearrange *past* events adroit willfulness in order to fit the sense of the construction in question, and then to declare that what has happened had to happen.

Therefore in recent times efforts have been under way to treat the various spheres of human behavior separately with a view to their laws of development. Here we must point above all to Alfred Weber's distinction between the *social process*, the *process of civilization* and the *process of culture*.[174] Certain investigations of Max Scheler[175] on the sociology of knowledge also move in a similar direction. It seems determinative for deciding the question whether there is intellectual progress to consider, in particular, to what extent the intellectual situation at any time can be interpreted as a more or less constant accumulation of intellectual possessions.

---

[170] Compare for instance Hegel, *Die Vernunft in der Geschichte, Einleitung in die Philosophie der Weltgeschichte* [Eng.tr. Reason in History, tr. Robert S. Hartman, Indianapolis, 1953], ed. Georg Lasson, Leipzig 1920.

[171] Main work: *Cours de philosophie positive.*

[172] Herbert Spencer's main work: *System of Synthetic Philosophy*.

[173] About Marx's historical materialism, see below section "The Way to Overcome the *Methodenstreit*".

[174] Compare Alfred Weber, 'Prinzipielles zur Kultursoziologie', *Archiv für Sozialwissenschaft*, vol. 47, p. 1 ff. and *Ideen zur Staats- und Kultursoziologie,* Karlsruhe 1927.

[175] Compare Scheler *Wissensformen*, especially p. 64 ff.

In conceiving hypotheses about developmental tendencies in historical reality, the requirement of such considerations, oriented toward the diversity of content of different modes of human activities, clearly shows that the material for induction, from which these hypotheses are constructed, is not drawn exclusively from historical experience in the narrower sense. More profound reflection will soon show that the foundations for classifications of that kind are to be found predominantly in introspective considerations, which only afterwards are tested on the historical material (in the narrower sense).

From this follows our evaluation of thesis (3). It presents itself as an application of our general findings concerning the *theoretical content of facts*, and states that first, historical facts as such are impregnated with general insights, drawn to a large extent from inner experience; second, the grouping of these, from which general laws are then derived is to a large extent imposed on the material by inner experience as well. To determine this contribution of general assumptions – originating altogether from the 'contemporary experience' of the historian and the social scientist – to social knowledge will be one of the chief tasks of the *theory of induction in the social sciences*. Only such a theory – to be sure, to be worked out in rather broad outline – will be able to settle conclusively the controversy between the 'theorists' and the 'historians' in the social sciences. The goal of our considerations here is only to deprive the pertinent controversies of their sharpness by freeing the topic of discussion from exaggerated minor conceptions, and thereby removing the expectation of a decision which proves one of the parties to be entirely and solely in the right.[176]

A theory of induction in the social sciences will also have to speak the decisive word concerning the problem of *historicism*. It is to this that we now wish to turn, while we postpone the consideration of thesis (4) concerning the role of historical data in setting up laws of social sciences.

The fundamental historicist thesis, according to which social science research is conditioned to an equal degree with respect to the posing, treatment and solution of its problems by the '*social position of the investigator*' has appeared in various stages of the history of doctrines; we must refrain from even the sketchiest presentation of these.[177] Thus we cannot enter into either the doctrine of the 'fraudulent priests' nor that according to which a certain class of people falsify historical knowledge for the sake of their class interest, although undoubtedly the historical roots of modern historicism and sociologism are to be found in these. Discussion of these doctrines is not required in the context of our considerations, because the question whether certain groups of human beings sabotage, or do or did want to sabotage knowledge is irrelevant from the point of view of philosophy of science, while it may be very significant as a historical or political question. What concerns the

---

[176] See also below section "Remarks on the Methodological Controversy *[Methodenstreit]* over the Theory of Marginal Utility".

[177] An excellent summary is contained in Ernst Grünwald, *Das Problem der Soziologie des Wissens,* Vienna 1934.

theory of science in this context is the *possibility* of objective knowledge in the social sciences; it is this that we now have to examine.

It is best to take our departure from the investigations of Scheler[178] and Mannheim,[179] in which the problems of historicism (and sociologism) have received their sharpest formulation. The fundamental conception here is that human thought, and above all human thought about historical-social reality, is influenced to an incisive degree by *unconscious* elements, which in turn are influenced to a great extent by the place the human beings in question occupy within this reality, by the 'place upon which they stand'. Mannheim presents this relationship in such a way that due to the placement of a human being within a certain social space, a certain – largely unconscious – 'world-willing', [*Weltwollen*] a specific total world view, i.e., the taking up of a certain attitude (position) with respect to the world (and especially toward the historical-social world) is formed, which then in turn forms the thought in question according to its direction. This formation, according to the conceptions of Scheler and Mannheim, who are in agreement here, penetrates so deeply, that it even enters into the categorical apparatus of thought. In Scheler this conception receives precise expression in the definition of the concept of a *relatively natural world-view*, which we have already mentioned in another context.

> To this belongs all that, which is generally considered to be 'unquestioningly given' within a group, and every object and content of opinion in the structural forms of what is 'given' without special acts of spontaneity, and thus generally considered and felt not to require, or even be capable, of justification.[180]

These relatively natural world-views mark the limits of the universal validity of human knowledge; every attempt at installing an absolute world-view only means absolutizing one's own relatively natural world-view. The same basic conception underlies Mannheim's concept of 'total ideology'. Here in particular the conceptual motif articulated sharply by Dilthey[181] plays a role: that human thought cannot be isolated, and that in order to understand it, human beings have to be grasped in their totality, i.e., the strata of their nature conditioned by instinctual drives and even those inaccessible to their own reflection must be comprehended. Thus human thought is '*situation-determined*' and only offers a *particular aspect* of the world. A *synthesis* of the various particular aspects can only be carried out by a group of human beings who are bound by tradition only to a slight degree and therefore capable of revealing the implicit presuppositions contained in the various relatively natural world-views. Mannheim sees in the '*free-floating intelligentsia*' a group of

---

[178] Compare above all: *Die Wissensformen*.

[179] For Mannheim, see among others 'Historismus', *Archiv für Sozialwissenschaft,* vol. 52: '*Das Problem einer Soziologie des Wissens*', *Ibid.*, vol. 53, *Ideologie und Utopie*, [Eng.tr. *Ideology and Utopia*, London 1936] , 1929; Article: 'Wissenssoziologie' in *Handwörterbuch der Soziologie,* 1932.

[180] *Wissensformen*, p. 59.

[181] See for instance Dilthey's *Einleitung in die Geisteswissenschaften*, *Gesammelte Werke*, vol. 1, 1922.

human beings who stand above the particularism of the various aspects; it is possible for them to reveal the ideologies that dominate thought.

We do not wish to investigate which differences – in part quite considerable – there are between Scheler and Mannheim with respect to the kind and extent of those traits which the thought of all human beings has in common, nor to deal with other more recent work in sociology of knowledge. Instead, we confine ourselves, first, to bringing out the arguments (shared by) historicism and the sociology of knowledge which are relevant to the theory of science, and secondly, to reducing to its proper dimension the *exaggerated relativism* expressed in them, in order, finally, to reveal those aspects of research opened up by the – more carefully formulated – theses of the sociology of knowledge.

For this purpose we wish to distinguish two partial statements in the fundamental assumption of the sociology of knowledge concerning the conditioning of thought by the social position of the thinker; (1) Thought, and especially thought about social-historical reality, contains a number of implicit statements which determine its structure in an essential manner, and which cannot be justified cognitively. (2) These implicit presuppositions are common to the thought of human beings who find themselves in the same existential situation [*Seinslage*], and can be understood by recourse to that situation and to the 'world-willing' evoked by it.

Now, first, it can be stated that the methodological content, in the narrower sense, of the fundamental thesis under discussion, is *contained in toto* in the first of the two partial statements, for the question, whence the origin of certain implicit presuppositions – no matter how important under some circumstances it may be for their discovery – is as little a problem of concern for methodology, as the question of whence the origin of computation errors is for arithmetic. Therefore in what follows we only have to deal with the first part of the assertion.

Here the most important question for methodology is, to what extent it can be said that the implicit presuppositions in question are ideologies, or points of departure for *ideologies*, and as such deviations from objective truth. This question, however, leads us directly to the analysis of the role to be attributed to *conventional* elements in the knowledge of a social-historical reality, and thus to the most significant points of contact of the problems of the sociology of knowledge with methodological problems in the narrower sense.

Our question can also be formulated in the following manner: Does the prejudiced attitude of social researchers revealed by the sociology of knowledge merely have the consequence that they *opt* for certain among other possible ways of posing and dealing with problems by (perhaps unconscious) choice (convention) or is reality actually *distorted* by these prejudices; and how does this come about?

Now it will be recognized that this question, so fundamental for the range of problems in the sociology of knowledge, can only be answered insofar as the *truth-conditions* for assertions about social-historical reality can be stated. Without such indication is would not be possible to ascertain at all what the synthesis of the particular situation-determined perspectives into a total conception was supposed to look like, as it would not be possible to know which aspects were to be posited as invariant. But once the truth-conditions have been determined, the implicit presuppositions contained in the various particular aspects can be tested to see whether they are false assumptions in the narrower sense, and thus to be removed unconditionally during the process of revealing them; or whether they merely involve a certain one-sidedness of viewpoint which had its rights alongside of others; or finally whether they point in the right (optimal) direction with respect to the cognitive goal. *For the fact that a presupposition is implicit is in no way evidence against its adequacy*. Let me refer back to our fundamental remarks in the first part of this work, about the implicit presuppositions underlying induction; the degree of generality of the preceding considerations follows from them.

But two complications, above all, make the pertinent problems in the social sciences much more difficult than the corresponding ones in the abstract natural sciences; the first concerns the problem of the interpretation of subjective meaning by an *alter ego* (treated in the second section of this part of our study) with respect to the question, which moments of the 'subjective meaning' of the object of knowledge can be considered invariant as against the plurality and heterogeneity of its interpretations. It has shown that even in the interpretation of the subjective meaning of an action by the actor himself, differences emerge according to the relative temporal relation between the act of interpretation and the interpreted action, and according to the stratum [of meaning] the interpretation is to be referred to. In the case of interpretation by another person variable factors were added, those of the relative temporal and spatial location of the interpreting person and the person whose behavior is interpreted. Taking these into consideration proves very important in the comparative evaluation of a plurality of available schemata of interpretation. A more profound analysis of this epistemological situation will lead to the result that a simple distinction between fact-bound and conventional aspects will not suffice, but that in an adequate description of the problem situation we will have to operate with '*conventions of the various orders*'.

The second main difficulty which social theory confronts in this respect lies in the epistemological fact, repeatedly stressed by us, that an unambiguous explanatory direction is not prescribed in the social sphere in the same manner as in the natural sphere, due to which the questions of topic selection (value-relation) and attribution play an incomparably larger role here. Consequently, in order to examine the question of the 'objective truth' of statements about social-historical reality, these statements will have to be grouped in a corresponding manner. It will then be seen that the significance of conventional aspects is different for the different individual groups. The most important division in this sense may prove to be the distinction of *statements of fact* from *interpretations of fact*, and finally from *attributions*, the transitions between these groups being gradual.

Concerning *statements of fact*, we will mainly have to examine the perspectives from which sources are selected and their relative weight estimated. We will have to examine especially (in the sense of the sociology of knowledge) to what extent communications of persons about their own behavior are to be regarded as credible. This investigation will, on the one hand, aim at judging to what extent it can be assumed that the person reporting about his own behavior was conscious of the decisive motives of his behavior, and on the other hand to ascertain whether it might not have been in his interest to disguise the 'true motives' of his behavior. From this will result a division of historical facts according to the *degree to which they are burdened with interpretations* and this division will serve as an *index* of the *degree* of their *objectivity*.

We mentioned, as a second group of problems, the problems of *interpretation*, of understanding in the narrower sense, though, as we have seen, every statement of historical facts already contains an interpretation. We have to distinguish between historical facts as such and their incorporation into more comprehensive contexts, in the same way as the distinction between laws and facts can and should be made in spite of the theoretical content of facts. Within the latter group, 'subjective' aspects are more important than in the case of mere statements of fact.

This share is still more considerable, finally, in the third group of the above-mentioned problems, that of the *problems of attribution and imputation*; here as has been shown, the conventional aspect of the selection dominates.

However – and with this we come to a decisive point for judging the philosophical significance of the fundamental thesis of historicism and sociologism – it holds true for all three groups of problems, that it is possible in principle for the methodologist to *maintain the pure attitude of the philosophy of science* in posing the problems. The explications (rational reconstructions) arrived at in the execution of this attitude make it possible to 'make one's own' the various aspects of research, and with that to overcome the limits of 'aspect particularism' rooted in social existence. However, every investigator becomes a theorist of science as soon as he clarifies for himself his scientific activity, his goals and his methods of knowledge[182]; therefore it is not possible to limit him within those barriers which are doubtlessly open for the philosopher. In the theories of history and of the sociology of knowledge, the distinction has also not been pointed out sufficiently between the *independent selection of topics and methods* on the one hand and the *acceptance* of problems clearly delineated by another person [*alter ego*]. Even if a scientist, due to being captive within a certain existential sphere, may not be capable of gaining sight of certain relationships by himself, or of freeing himself of certain presuppositions, nevertheless, when the way has been prescribed for him step by step, he will be capable of following it, insofar as he possesses the qualification of a thinker at all.

In summary it can be said: the real problem of philosophy of science contained in the theory of historicism and the sociology of knowledge is that of the 'objective' and 'subjective' components of the knowledge of social-historical reality. This problem can be dissected into three levels: (1) the relationship of these factors in all knowledge of reality; (2) their relationship within the problems of understanding another person, i.e., the 'objective knowledge of subjective meaning' (Schütz); (3) their relationship in the problems of attribution and accountability. The two last points are specific to the methodology of the human sciences, as they are related to the range of topics of the psycho-physical sphere and to the lack of an unequivocal universal order of experience that is characteristic of these.

---

[182] See above Part One section "Basic Philosophical Considerations".

The dependence of the considerations of the sociology of knowledge on insights of the philosophy of science is revealed most clearly as soon as attempt is made within the framework of the sociology of knowledge to salvage the idea of the objectivity of knowledge, by means of a synthesis, an *interpenetration of perspectives*. In Mannheim, pragmative motives appear in the foreground in this context. Objectively correct knowledge, according to this, would ultimately be that which proves itself practically, which makes possible an orientation in the world. But if we pursue the question as to what conditions this orientation is linked, then it will turn out quite readily, that it depends upon the ability to make correct predictions about future events. With that, the analyses of the sociology of knowledge – insofar as they are not regarded as merely limited to special fields – merge with the fundamental problem of the methodology of social science, the problem of social laws.

5. Fundamental Concepts of the Social Sciences

The main problem from which the analysis of social scientific concepts has been taking its departure for more than two thousand years is that of the relation between social collectives – especially society and the state – and the human beings who form them. The following considerations sharpen this problem: On the one hand it is clear that society is composed of a number of human beings, and that there is no society where there are no human beings. We can also readily see that it is human beings who act or who suffer when it is said that society acts or that harm has been done to society. On the other hand there is the insight, no less clear, that a society can continue to exist even when many, indeed all, of its former members have been eliminated, or wholly or partially replaced by others. One can point especially to the circumstance that the *societal spirit* [*gesellschaftliche Geist*], as it shows itself in the language or in the manifold of societal customs, as a rule survives changes in the members of the society. Thus we arrive at the antithetically sharpened issue of contention – and as we shall recognize, one that is not unambiguously formulated – whether society as a 'social totality' has priority of the individual human beings conjoined in it (and interchangeable), or whether, on the contrary, we must speak of a primacy of the individuals with respect to the society they form. Recently this controversy has been designated as one between a *universalist* and an *individualist* conception of society.

The essential arguments of the universalist doctrine concerning the essential nature of society[183] can be summarized as follows: *Society* is *not* a *sum* of *individuals* and the laws of society cannot be discovered by a simple sum of the actions of individuals. The individualist thesis according to which the individual human beings exist to begin with, and only subsequently form society by a certain

[183] Othmar Spann will most likely have to be recognized as its most significant representative in the German-speaking world today. [1936 – Ed.] see for example his 'Gesellschaftsphilosophie' in the *Handbuch der Philosophie,* Munich 1928.

kind of living together, is false. Instead, the individual human being is born into society, and we can actually say that he is only a human being insofar as he is, in fact, born into society. Aristotle's thesis that by his essence man is a social creature (ξωου πολιτικου) is not to be regarded merely as an empirical finding but as an *a priori* truth. This can be understood most simply when we consider that the nature of man by no means lies in a special property of his body – as a hypernaturalism blind to problems and ideas would assert – and thus, for instance, his upright posture, but in his participation (understood in the precise meaning that Plato gives to this word) in mind [*Geist*]. This participation, which can be exemplified most enlighteningly in a language community, shows that the individual only becomes a spiritual being, and thus a human being, through belonging (his membership) to the sphere of spiritual values.

The *individualistic* theses opposed by the above arguments, base themselves, as has already been noted, above all on the circumstances that society is nothing else than an association of human beings, that thus it is formed by human beings, due to which the concept of society presupposes that of the human being; in their basic conception, these theses are so simple that they do not require further explanation. We can therefore at once specify the position we ourselves take in this conflict of views; it follows from our general reflections about abstraction and ideal objects as such, as well as, especially, about those processes of abstraction which lead to the *meaning concept*. These in turn allow us to recognize quite readily that the philosophical basis of universalist doctrines is *Platonism* and that the objections to individualism are quite in line with Plato's polemics against empiricism. Consequently the resolution of this controversy also presents itself as an application of the results of our reflections on the *debate about universals*. For a careful analysis of the concepts of social science shows, that while the assumptions that there are social realities, or ideal social essences, which are independent in the sense that to them correspond specific sources of knowledge transcending physical or psycho-physical experience, is entirely erroneous, yet such independence must be attributed to them as is due a product of abstraction, in contrast to the exemplary experience which forms the basis for the abstraction. In designing those empirical points of departure, and indicating which aspects of the pertinent experiences are retained as products of abstraction, that is, as invariants, accordingly lies the fundamental task of an analysis of *concept formation in the social sciences*. It is to this that we want to turn, in order to come back again to the controversy between the universalist and the individualist conceptions of society but only after these questions of principle have been clarified for then we will also be able to shed light on the ambiguity of the basic question of the 'priority' of either individual or society.

This task was clearly grasped by Max Weber and attacked by him with admirable intellectual energy. His fundamental investigations, published posthumously in *Wirtschaft und Gesellschaft* [*Economy and Society*] have been carried forward by a series of scholars among whom above all Fritz Sander[184] and Alfred Schütz[185] were

[184] See especially 'Der Gegenstand der reinen Gesellschaftslehre', *Archiv für Sozialwissenschaft*, vol. 54, p. 329 ff.

[185] *Aufbau*, p. 161 ff.

able to considerably improve several of Weber's formulations. In the following investigation, we want to take up our departure from Max Weber's definitions and thereupon take up the reformulations of Sander and Schütz.

At the center of Weber's relevant analyses stands his real definition of *social action*, which reads: "Action is social in so far as, by virtue of the subjective meaning attached to it by the acting individual (or individuals), it takes account of the behavior of others and is thereby oriented in its course."[186] Now, Sander[187] has shown when compared to the use which Max Weber has made of the concept of social action – in conformity with scientific usage, that this definition is too broad; according to it, even every *perception* of another's action would be defined as such; Weber certainly did not want to subsume all this under the concept of social action. For the concept designated by the above definition of Weber, Schütz suggested the term 'other-orientation' [*Fremdeinstellung*], in order to examine then which of these orientations Weber actually wanted to have designated as 'social action'.

> … There does seem to be hovering before his mind a specific type of connection between the social action and a piece of behavior on the part of the other person. Social action no doubt exists for him only in the two types of situation: either (a) where the social actor intends by means of his action to induce the other person to behave in a particular way – if, that is, the goal of his action is to produce a certain effect on the other person's consciousness; or (b) if this same social action was induced by the other person's behavior – if, in other words, the perception and interpretation of the already enacted behavior of the other person is the genuine because-motive of the social actor. Weber's concept of social action covers both these cases. Social action can, as he puts it, be oriented to the past, present, or expected future behavior of others.[188]

The cases designated under (a) Schütz subsumes under the name 'affecting-the-other' [*Fremdwirkung*] or also 'social affecting' [*Soziales Wirken*]; the cases designated under (b), he subsumes under the name of 'action affected by others' [*fremdbewirktes Handeln*].

According to this in the action called '*affecting the other*' in the projection of the act future conscious experiences of others – of our fellow man or fellow men who will be affected by the action – are anticipated. The orientation of one's own action to the behavior of others thus in this case presents itself as including the expected behavior of others in the actor's project. The action *affected by another* on the other hand takes its orientation from the behavior of others that has already actually taken place – or is assumed to have actually taken place – in such a way that the attention to conscious experiences of the other human beings in question forms a 'because-motive' (see above) of the action, so that the project of the action as a whole is motivated by that attention, and that thereby appears to be causally determined.[189]

The distinction thus made between affecting-the-other and action-affected-by-another, and any such orientation towards others that is neither of these is also

---

[186] 'Wirtschaft und Gesellschaft', in *Grundriss der Sozialökonomie*, Tübingen 1925, p. 1. [*The Theory of Social and Economic Organization,* p. 88].

[187] Sander, *op. cit*, p. 335.

[188] *Aufbau*, p. 164.

[189] *Ibid.*, p. 167.

important for the analysis of the second (more complex) fundamental concept of social science, that of *social relation*. The 'social relation' is defined by Max Weber as "the behavior of a plurality of actors in so far as in its meaningful content, the action of each takes account of that of the others, and is oriented in these terms"; and this definition is further clarified by the following statement:

> The social relationship thus *consists* entirely and exclusively in the existence of a *probability* that there will be, in some meaningfully understandable sense, a course of social action. For purposes of definition there is no attempt to specify the basis of this probability.[190]

That such a chance exists, however, means nothing else then that under the assumption of the existence of such a mutual orientation, the actions in question can appropriately be interpreted, where the most important criterion of an appropriate interpretation lies in the confirmation of the prediction of the course of future actions based on this interpretation. Now it can be seen, as Schütz has said that

> … any social relationship within which a case of affecting-the-other occurs can be identified as such with greater confidence than a social relationship in which no more takes place than intentional acts of other-orientation. The first type of social relationship we shall from now on call "social interaction" (*Wirkensbeziehung*), and the second, "orientation relationship" (*Einstellungsbeziehung*). It is easier to observe the effect that the action of one person has on that of another than it is to observe the attitudes they may have toward each other, e.g., sympathy or antipathy. In other words, it is easier for me to state with objective probability that two people are socially interacting than it is for me to say that they are merely oriented toward each other in a certain way.[191]

After the definition of the concept of social relation, the *concept of society* understood in that broadest sense, which is applicable solely for delimiting the spheres of the social sciences – offers no fundamental difficulties any more.

If between a majority of human beings, social relationships of a certain kind – let us call them S_k relationships – exist and the interpretation of such relationships can be carried out by means of a uniform schema of interpretation, then we want to say that these people are members of a K-society. The requirement of S_k-relationships between individual persons is not to be understood in such a manner, that direct S_k-relationships must exist between any two of them, but already the incorporation within a context of S_k-relationships – A stands in an S_k-relation with B, B with C, D and E, C with F, G and so forth – can be regarded as sufficient. The possible range of 'indirectness' and the required minimum of the extension, intensity and duration of S_k-relationships cannot be specified precisely, and also varies with the kind of S_k-relationships. If we speak simply of 'society' without explicitly indicating or implicitly presupposing its kind (e.g., recreational society, economic society, religious society) we have in mind an indefinite multiplicity of relatively intensive social relations between the persons forming the society.

If we now recall our consideration about empirical laws and their field of application, then we can say that *society is a field of application of certain schemata of interpretation for social relationships*.

---

[190] 'Wirtschaft und Gesellschaft', p. 13.

[191] Schütz, op. cit. p. 172.

This insight applies to deciding the question, whether or in how far we can speak of a *uniform* method of social science which would encompass the interpretive sciences in the narrower sense, such as jurisprudence, philology, etc., the historical sciences, and the theoretical [law-like] sciences [*Gesetzeswissenschaften*] of the social world (e.g., economics). It will be seen that an essential affinity actually exists between these groups of sciences to such a degree that bringing them under a common name appears to be justified. This affinity lies in the task common to all three groups of sciences, that of the *interpretation of the social behavior* of other people. The apparently fundamental difference between meaning-interpretation and the empirical investigations of social collectives (society, state, etc.) will be recognized as not too profound as soon as it is realized that the constitutive principle of unity of social collectives is the unity of a complex of *meaning* [*Sinngehalte*]. It therefore appears to be justified to call all those sciences '*social sciences*' which set as their goal the interpretation of social behavior of other people – or the symptoms of such behavior. Superordinated to this concept is that of the '*human sciences*', whose topic is the interpretation of human behavior as such, and thus also that of non-social behavior.

We wish to clarify this quite abstract exposition by means of a simple example, which is also used by Max Weber, that of the game skat. A game of skat is a game society of three persons who have agreed to engage in a social behavior which follows certain rules, the rules of skat. Within the framework of these rules every player pursues his own ends which run counter the ends of at least one of the other players. Therefore the pertinent action of the players is also *understandable* for an observer in terms of the interpretive schema of the rules of skat, and these rules themselves are nothing else than potential maxims for game-playing behavior, or a potential interpretive schema for the understanding of this behavior.

A certain game of skat *begins* as soon as the behavior of the three persons, oriented to and therefore interpretable in terms of the rules of skat is initiated, and it *ends* as soon as this behavior is terminated, usually, but not always, in conformity with the rules of the game. The three persons are *actually* skat players in so far and only insofar as their behavior can be interpreted in terms of the rules of skat. The latter are by definition invariant with respect to all variations of the persons of the players, and place and time of play. But this must not be interpreted in such a way as though the rules of skat were transtemporal and transpersonal ideal essences, which are realized on occasion by the particular games of skat, without thereby being affected in their ideal being-for-themselves [*Fürsichsein*]. Rather, the epitome of these rules, as the epitome of meaning contents is – as is every meaning – *intended* meaning, only that in their application as schemata of interpretation, the occasional data of this intention remain open.

These considerations can be applied to any social collectives whatsoever. Thereby the speculative pseudo-problems are deprived of their basis, which – closely linked to empirical investigations – are treated under the title of the 'rise and fall of states' ('legal orders', 'languages', 'economic forms', etc.) and to a large part are rooted in an erroneous interpretation of the *concept of an organism*.

A brief analysis of the concept of *legal person*, which since immemorial has been regarded as very much in need of clarification by legal theory, can serve as example for further clarification.[192]

The problematic nature of this concept is rooted in the cognitive fact, that while every legal person is *represented* by 'physical persons', its representative 'organs', it remains the *same*, even when these organs change. For instance, whoever has a claim against a joint stock company, does not have to be concerned – from the legal point of view – with whether the chairmen or stockholders change; if the claim is not met on the due date, he will sue the 'company'; 'it' will be convicted, sentenced, and, if necessary penalties levied upon it. But what is this 'company' which is sued, sentenced, punished? If we refrain, in complete consistency, from all constructions that go beyond the recognizable state of affairs, and confine ourselves to a bare description of the facts, then it will be seen that these measures are always carried out against individuals, 'physical persons', only that these persons are *not individually* specified, but fixed through the specification of a law of formation, *i.e., a general designation*, namely, the statutes of the legal person in conjunction with the relevant legal order. Thus the legal person acts when persons act who satisfy certain relations, determined by statutes; and measures with respect to a legal person are taken by displaying a certain behavior towards these persons. Yet the legal person is by no means identical with those physical persons who are its organs; rather, these only 'represent' the legal person, and this only in so far as they exercise '*proper functions*' or, in so far as others treat them as *representatives*. However, both possibilities are clearly circumscribed by general rules – the statutes and pertinent general norms of the legal order, so that the *legal person* is to be defined as *the field of application* of certain *rules*, which determine the *representative organ functions*. These rules state that with respect to the occurrence of certain legal consequences, it is immaterial whether actions of a certain kind, e.g., the signing of certain documents, were carried out by person A or by person B, if person A as well as person B satisfies certain conditions (for example, appointed or elected by certain other persons to exercise certain functions). The designation '*persona incerta*' that was customary in Roman law for the 'legal person' is a suitable expression for these relationships. But it must not be forgotten that the criteria for the unity of a legal person ultimately always do point to certain 'physical persons' (human beings) and to certain points in time. We can speak of the 'rise' and 'fall' of a certain legal person, in the sense of our findings, when a *field of application* for such rules arises or disappears.

Taking up the reflections of our second section about interpretive *schemata* in general, important consequences result from what has been said with respect to the problem-determinateness or, more correctly, problem-relatedness of concept formation in the social sciences. In our analysis of the concept of 'legal person' we have just recognized that its unity is correlated with the invariance of certain legal consequences with respect to certain variations of the persons serving as

[192] See also Kaufmann, *Die Kriterien des Rechts*, Tübingen 1924.

representative organs. If we ask why precisely these variations in connection with this invariance have furnished the point of departure for juridical concept formation, then at first we would answer quite generally that this would happen because those aspects simply would be the '*juridically relevant*' ones. However, this answer receives a precise content, only when the reasons for this relevance have been determined. Obviously these lie in the fact that in the legal order to be interpreted a substantial extent of common legal consequences are attached to the various states of affairs within the range of variation, so that it proves useful to summarize, in a term, the states of affairs equivalent with respect to these legal consequences. Thus the law will generally protect a person's possession of a thing he has bought against the third party in the same way as it will a person's possession of a thing which did not have an owner. This similarity is taken account of in the formation of the concepts 'property' and 'kind of property acquisition'. Whether in such cases the concept was formed by the legislator himself, within the framework of juridical codification, or subsequently by the interpreting jurist plays no role in our consideration. Very important, however, is to point in this context to certain exaggerated interpretations of concepts of social science, whereby those concepts are comprehended as '*connotative*' in the sense that one attempts to derive 'from the essence of the concepts' absolutely valid criteria for the judgment of human conduct of a certain kind. Thus it was asserted, for example, that it followed from the *essential nature* of property that the proprietor of a thing could not be limited in its use; for property was the right of unlimited mastery over a physical thing. If the origin of such erroneous interpretations is sought, we arrive at the following result: The legislation which links to certain states of affairs, designated as kinds of property-acquisition, certain uniform legal consequences is based on the intention of the legislator to endow whoever gains possession of a thing with the most undisturbed possible enjoyment of it; and rather to protect him against being disturbed in this by a third party. However, insofar as these motives do not become thematic as such for the jurist, but only their reflection in legal propositions, the legal concepts too can only be formed immanently, i.e., can express only relations between states of affairs and legal consequences and nothing more. The interpretation of legal concepts, as containing purposes or ends, is therefore mistaken with respect to this juridical subject-matter, i.e., inadequate to the nature of the problem. To be sure, the situation appears different wherever decisions of judges are to be rendered in keeping with the 'intent of the legislator'. For here the task consists precisely in recognizing legal regulations of a certain kind, for example, the legal orders summarized under the title 'property', as *symptoms* for certain *legislative tendencies*, which then in turn are to be decisive for the case as yet undecided in the law. But justice is not done to even this problem-situation by the conception that the juridical concepts themselves contain *unequivocal* tendencies of purpose which can prescribe correct decisions '*praeter legem*' *a priori*. For as one can readily recognize, this conception is subject to the fundamental critique to which we subjected the *dogma of pre-stabilization*. In legal practice it leads, as a rule, to one-sided consideration of certain tendencies of legal policy, and neglect of others. In the case of property, this results, as already noted, in the most unlimited possible freedom of use of property, with neglect of the opposing interests of third parties.

But the elimination of misconceptions concerning the immanence of purpose in the concepts of the social sciences, must not lead to a failure to recognize the role of the goals of knowledge in that concept-formation – and therefore indirectly also of the role of practical goals that form the inducement for posing problems in a certain way. We have already stated what is fundamental about this in earlier sections and can confine ourselves to a brief recapitulation of the results.

Every empirical object – whether one speaks of objects in the narrower sense, or of facts – is determined by a context of experience, as will be recognized as soon as one asks for the truth-criteria of judgments in which the existence of objects is asserted. However, we can distinguish between contexts of experience of various levels – objects of various orders – where the objects of a higher order are based on the experiential foundations of objects of the first order. The manner in which these 'higher level syntheses' are carried out in scientific thought depends on which contexts the interest of the scientist is focused. This interest itself, however, is co-determined by objectively founded expectations with respect to the results of one or the other direction of research. That is why in abstract natural science, where one specific path of research has proved especially fruitful, concept formation displays a far greater uniformity than in the social sciences. Therefore in the latter the dangers of inadequate, or else ambiguous concept formation are essentially greater and therefore the need for a systematic critique of method all the more urgent. As for this critique, our concern is above all to describe the topic of inquiry precisely, it will be best for the critique to be guided by our *fundamental schema of the degrees of freedom of* [*research*] *topics* which we have designed. One of its most important tasks will be in each instance to differentiate the scientific concepts (system concepts) from the prescientific (extra-systematic) concepts. We will seek to clarify this task in the section on economic theory with a striking example (the concept of need).

When after all these considerations we return to the *controversy over universalism* characterized at the beginning of this section, we will recognize quite readily that the difficulties in resolving this controversy – disregarding the divergence of political tendencies that lie behind it – reside above all in the lack of precision with which the problem has been posed.[193]

If it is asserted that society has priority over the individual, or vice versa, then by this priority four different things can be understood, namely: (1) logical-ontological priority; (2) methodological-heuristic priority; (3) genetic-causal priority; (4) axiological priority. To be sure, the statement that these four meanings are different from each other does not contain the statement that they are pair wise independent of each other. Rather, we understand quite readily, that something that is prior

[193] For the following, compare Kaufmann, 'Soziale Kollektiva', op.cit. p. 299 ff. The problem of the ontology of social objects has recently been treated in detail by T. Otaka in *Grundlegung der Lehre vom Sozialen Verband,* Vienna 1932.

logically-ontologically cannot be posterior genetically. Furthermore we will see right away that even with this fourfold division the question of 'priority' has not yet been given an unambiguous meaning.

To begin with, as far as the '*logical-ontological priority*' of either individual or society is concerned, it must be pointed out above all, that we cannot speak of a logical relation in the narrower sense, that is of a relation between the general and the particular, as for instance that which holds between 'colored' and 'blue'. However, two kinds of other criteria for according priority do come under consideration, which, in any case, can be called as 'logical' in the broader sense, namely first those of *degrees of complexity*, where the more simple would be regarded as prior with respect to the more complex, and secondly that of *degrees of independence*, where the more independent object would be accorded priority over the less independent one. Now with respect to each of these two criteria, the universalist can maintain the logical priority of society. With respect to the first criterion this can be done by pointing out that the participation of the human being in the spirit which would constitute the essence of social being would form a common element of the members of society, who would have individual properties, and thus in addition to this commonality also would display their individual features. As far as the second criterion is concerned, the universalist can support his position by pointing out that a *foundational relationship* exists between the individual and society, so that the individual member of society is in need of complementation by the other members in an analogous sense to color needing completion of extension. But we have to be clear that these two concepts of priority are *incompatible* with each other, for the less independent object is always simpler than the more independent one in which it forms a 'part'. Indeed, when the relation between society and individual is subsumed under the general relation of 'whole-part', it must not be forgotten that here by parts one must not understand concrete pieces, but 'moments' (abstract, dependent parts in the sense of Husserl's *Logical Investigations*); the question, however, whether the whole is 'prior' to its parts, or inversely the parts 'prior' to the whole, will have to be decided in opposed ways according to whether by the 'prior' is understood 'priority' in the first or the second of the meanings just characterized, i.e., greater simplicity or greater independence.

In any adequate treatment of these two questions – after their separation from speculative intellectual constructions – the analysis of the 'share of the individual in the spirit of the society' will play an important role. It will be seen that this 'participation in' cannot be understood in such a way that thereby the belonging of a human being to a *specific* community (culture) could be *unambiguously* determined; rather it will have to be comprehended as a general capacity (disposition) to participate in the spirit of the society. From this, difficult problems emerge, such as, what role heredity and environment play in the origin and development of specific mental capacities, and these questions surely cannot be solved through ontological speculation.

The question of the logical-ontological primacy of the individual or of society becomes still more blurred because the problems just characterized, which relate directly to the structure of social scientific concepts, are overlain by another intellectual theme, as a consequence of which the community of *unconscious motives* among members of a society (which, as unconscious, appear to be alien to the ego) is used as an argument for the super-personal character of society. Frequently the social researcher can only classify the social behavior to be investigated empirically in the desired manner, by taking into account unconscious motives which are conditioned above all by inheritance, tradition and imitation, and thus historically conditioned. Accordingly, in his investigations he has to take *social givens* into consideration, thus, for example, to turn his attention to religious tradition when he wants to ascertain whether a newly introduced legal regulation will have a chance of being implemented within a certain social circle. Thus the (relatively) unreflective human being, who perceives his unconscious motives as fate, as well as the scientist who sees in them and their origins essential elements of the determinability of social processes, tend to hypostatize these, as a consequence of which the unconscious motives appear as developments or emanations of a collective spirit or a collective soul.

From such consideration, derives also the assertion of the *methodological priority* of the investigation of 'social wholes' over those which deal with individual human beings. For that reification appears to contain a justification of the 'macroscopic method' in the social sciences, a method urged upon the social scientist by other moments, indeed even forced upon him.[194] For he knows that his statements about social phenomena cannot count as exact, in the sense that they represent all the details of the behavior of socialized human beings. But he is also conscious that this does not represent a deficiency in social research which could be removed or reduced by more precise empirical observation, but rather an essential aspect of the social sciences. Further he recognizes, if not always with full clarity, that propositions about the behavior of social groups are not observations about the 'average' behavior of individual persons in this group, but that here we have the behavior of certain 'elites'[195] who play a disproportionally greater role than that of the rest of the members of the group in question. Finally, he will generally be clearly aware that this difference in significance cannot simply be dealt with by employing a 'weighted' average instead of operating with a purely 'numerical' average, but that the problem of the significance of the individual for the social group is an eminently *qualitative* problem. These considerations fall within the general range of problems of the value-relation and of attribution and imputation, with which we were concerned in the last two sections. In this context, the remark will suffice that the insight into the varied importance of the behavior of the individual within the group for the 'behavior of the group' forms a decisive conceptual theme in the universalist conception of the nature of the social.

In contrast to earlier periods, the question of *genetic priority* which involves whether human beings were animals form the outset, living in herds, or rather,

[194] For this see Simmel, *Die Probleme der Geschichtsphilosophie*, 4th ed., Munich 1922.

[195] Concerning the role of elites in social events see especially Pareto, *Trattato di soziologica generale*. 3 vols. 1923 and Ortega y Gasset, *Der Aufstand der Massen* [*The Revolt of the Masses*], trans. from the Spanish, Stuttgart.

hordes, or whether 'originally' they lived as isolated individuals, no longer plays any notable role in the discussions of today's philosophy of science. In particular, it has been recognized that no consequences with respect to axiological primacy can in reality be deduced from a decision in favor of one alternative or the other, as was believed formerly. We do not need to deal in greater detail with these considerations, though they are often of some historical interest.

There only remains the task of examining the question of *axiological priority* (axiological primacy). We know from our basic considerations, concerning the problems of values, of the impossibility of deducing the higher value of society as against its individual members, nor the opposite, relative ranking, from any *a priori* or empirical presuppositions, without referring to a pre-established reference system. Accordingly, the main task of a methodological analysis of this question will consist in grasping the *unspoken underlying goals* in each case by means of a rational reconstruction, and in demonstrating the relationship between these goals and the assertion of the axiological primacy of the individual or of society in terms of the psychology of knowledge. We can only sketch the main outlines of such an investigation.

We can readily realize that a question which aims at examining the significance of individual behavior for the social group will tempt one toward the interpretation to determine the value of the individual by his *contribution to the group*. Here the positive or negative evaluation of his achievement is made in relation to group goals – frequently power goals – and the rank accorded to the achievement is a result of more or less clearly conscious considerations of attribution, characterized in more detail in the section before last. From this it is typically inferred that the value of the group represents something that is prior with respect to the value of the individual, without analyzing adequately the concept of 'group goal', the pivotal point of the entire consideration, and without comprehending it in its relation to the 'goals of the members of the group'.

The further choice of the argument generally goes as follows: The achievements decisive for the value of the individual appear dependent in kind and extent on his capacities, knowledge, character traits and basic attitude, and accordingly these form the *actual* criteria of personal evaluation; from this then, as a rule, conclusions favoring an *aristocratic* social order are drawn. People generally operate with the vague concept 'value of society', only to stress the value of society in general, as against a state of affairs lacking social organization (usually erroneously equated with anarchy), or to simply identify the value of society with the value of a certain type of society – for instance, the existing order.

While, as we have just noted, the assertion of the axiological primacy of society over the individual typically leads to the preference of an *aristocratic order*, the opposite conception has links to the preference for *democracy*.[196] For if the achievements of the individual are not stressed, then under the influence of the visibly

[196] Kaufmann, 'Soziale Kollektiva', p. 303.

far-reaching similarities between human beings as such, or at least, between those belonging to a certain community (state, nation, religion, class) within the visual field of the observer, it is plausible to assume a 'natural' equality of value, the origin of which is seen in the 'equality of the nature' of human beings. But this equality, insofar as it appears to a considerable degree to be conditioned by the *somatic constitution*, consists above all in a far-reaching agreement of the *vital needs*, and the instinctual life associated with these. Therefore this approach leads to a strong emphasis on the value of conditions [of life]: the 'value of happiness' appears at the top of the scale of values. Typically, it is argued further that all human beings have an 'equal right' to happiness and as a rule from there the path leads to political doctrines of a democratic nature. This, however, does not always have to be the case, for, as the history of the eighteenth century shows, enlightened absolutism, and as more recent history shows, dictatorship, is quite compatible with the conception of a welfare state. Furthermore, it must also be taken into consideration that a spiritualized idea of community based not on a somatic conformity of human beings, related to basic needs, but on their spiritual (psychic) fellowship as, for example, of the Christian community of Faith, is quite compatible with the aristocratic political order.

This intermingling of levels of meaning, already quite complicated in itself, leads to further obscurities because *analogies* are adduced in order to support the various contending doctrinal views, without their *fundamentum analogiae* being clearly grasped. One of the most significant of these analogies in the history of doctrines is that with the human-animal *organism*.[197] The errors which the over-extension of this analogy has led to in the organismic doctrine of society, of the state and the economy, have already been adequately criticized in various places.[198] We can be content here with the observation that the basis of the organismic analogy in the social sciences lies in the reference and relatedness to ends, and that its methodological core lies in the postulate that investigations concerning the complex relation of achievement on the part of members of a community must occupy a central place in social research. This postulate must then be tested for its heuristic suitability, and the extent of its fruitfulness, and the result of such an examination must be independent of the speculative guise in which the postulate has been offered for examination. The result will indicate that the analysis of relations of achievement is of extraordinary importance for the social sciences, especially because a large part of social scientific laws can be derived from this relatedness to purpose. But of course it must not be concluded from this, that the organismic analogy is justified from the viewpoint of the practical pursuit of knowledge and therefore can be retained. Instead the only conclusion is justified that the approach, which was to be shown as the *only* appropriate one by the analogy, must be granted suitable scope in research.

[197] It goes back far into antiquity. Compare for instance the legend of the tale of Menenius Agrippa about the quarrel between the stomach and the remaining parts of the body.

[198] See for example Kelsen, *Hauptprobleme der Staatsrechtslehre entwickelt aus der Lehre vom Rechtssatz,* 2nd ed. , Tübingen 1923.

6. Social Laws and Ideal Types

Our basic considerations have clarified the essential misunderstandings in the *Methodenstreit* with respect to the character of laws in the social sciences, and with that especially the inappropriateness of sharply antithetical formulations of the relationship of the social sciences to the natural sciences. We wish to briefly summarize the most important results of these investigations:

1. Neither in the natural sciences nor in the social sciences there is an abrupt opposition without gradual transition between exact laws and mere rules or tendencies. Particularly as far as the mathematical method is concerned, its '*internal precision*' does not guarantee the '*empirical precision*' of mathematically formulated propositions in the experiential world.
2. The *laws of the social sciences*, just as those of the *human sciences* as a whole, are distinguished from the laws of the natural sciences by the circumstance that in them *inner experience* becomes *thematic*. The 'laws of understanding' of the social sciences, however, are not independent of the laws of events in the external world, as is shown especially clearly in our analysis of rational purposive understanding.
3. A *special status* of the laws in the social sciences which can be traced back to specific evidence of understanding as a source of knowledge *does not exist*.

   The thesis of the existence of *laws of value* or '*ought-laws*' which belong to an independent sphere, different from the sphere of Being, is not tenable.
4. After having revised the premature assumption that the laws of nature and social laws are different *toto coelo*, our perception is sharpened for ascertaining the *gradual* differences and we will now briefly characterize them. In so doing we wish to refer to our previous considerations about the various 'properties' of the natural laws, and to ask ourselves, Is there any prospect of establishing a system of propositions in the social sciences by means of some method, which in their totality, or at least to a large part, possess the heuristic merits of natural laws? Let us recapitulate the points toward which we have directed our attention in this context:

   I. *Unity of Laws.*

      (a) Unified basis of reference; (b) systematic (deductive) unity.

   II. *Simplicity of Equations.*

      (a) Small number of parameters; (b) simplicity of the relations between the parameters; (c) small number of constants; (d) functional continuity.

   III. *Domain of Validity of Laws.*

      (a) Temporal domain of validity; (b) spatial domain of validity; (c) density of validity; (d) material domain of validity.

   IV. *Precision of Laws.*

      (a) Degree of precision; (b) percentage of exceptions.

To what extent can these aspects be demonstrated for the laws of the social sciences?

To begin with, as far as the *unified basis of reference* is concerned, we have already established that such a basis does not exist for the whole of the social sciences. We have also pointed to the consequences which this circumstance entails for the problems of the value relation and of attribution and imputation. As they lack a unified basis of reference, we cannot speak of a systematic (deductive) unity of the social sciences.

Now we come to the question of the *simplicity* of social laws. Here we have to note first that today only a very small part of these have mathematical form; still an analogy with the laws of theoretical natural science can be noted as here, too, the classes of facts related to each other are frequently quite small in number, and the nature of their relationship relatively simple. Recall for instance the laws of classical economics, which deal with the relations between the interest of capital, wages and ground rent. Yet it must not be forgotten that the internal structure of the facts linked in social laws is much more complicated than that of the facts of nature linked in physical laws. Also, the nature of the interrelationship in social laws is frequently characterized very imprecisely. Thus when social facts of the kind C are designated as causes of other social facts of the kind F, it often remains quite obscure what kind of correlations hold between the variations of F and the variations of C.

If (thirdly) we compare social laws and natural laws with respect to their *domain of validity*, then it is especially important to clarify what *requirements* are posed for a social law, i.e., under what conditions it will be said that such a law exists. Above all the task is to ascertain what kind of *predictions* are made possible on the basis of the law.

Here the following emerges: if the required care is exercised in the formulation, then we may very well succeed in formulating rules, mainly on the basis of general considerations about the psycho-physical nature of human beings, their vital needs, their psychic reactions, their mental faculties. These rules can be tested historically over thousands of years and in the most diverse places, and may be capable of withstanding such tests. They will, perhaps, embrace the entire domain of human action, and their *density of application* will be high. But for the most part, future events will be too *weakly determined* by such laws to be able to offer suitable points of orientation for long-range rationally purposive action. We can give a striking illustration by comparing the predictability of astronomical events on the one hand and political events on the other.

Though the contours of *future* events can be predicted only very incompletely by means of the limits set by these general rules for variations in the course of social events, nevertheless, the contours of *past* events will demonstrate that these have taken place *within such limits*. Therefore in the social sciences, retrospective prophecy, the *vaticinium ex eventu*, which shows that what happened had to happen – where particular features of the events are isolated and incorporated within general complexes of relations – is so much more successful than is true prediction. These considerations as we wish to remark only in passing – also lead to an objective investigation of the questions to what extent one can *learn from history*.

Long-range concrete predictions of social facts – or, rather, the general assumptions on the basis of which such predictions can be derived – thus generally possess a small degree of precision and a high percentage of exceptions. On the other hand, *short-range predictions* for the various domains of social life can be carried out with the reliability required in practice. For otherwise every human community in our sense of the term, which after all presupposes the reciprocal orientation of human behavior, would be impossible. It is sufficient to reflect on how many presuppositions we make about the behavior of our fellow men – who to a large part are unknown to us – when we write a letter and mail it[199] or when we purchase a railroad ticket, in order to recognize clearly that the social world also knows laws which have a very small percentage of exceptions.

Especially important for grasping the differences in method between the natural and social sciences is the insight that the latter lack *universal laws* which would represent the highest principles of a hierarchy of laws embracing the entire social sphere. This forces a *methodological particularism* [upon the various fields] which has the consequence that *relevance* with respect to the specific goal of knowledge – which in turn may be suggested by practical goals – plays a much greater role in the social sciences for indicating the course of research than it does in the natural sciences. To be sure, the technician, in the narrower sense, has his specific aims as well, and accordingly turns his attention to certain special investigations, for instance, the determinations of material constants. Still, his path of research is determined in its basic principles to a large extent by the universal laws of nature, and his results, too, find their precisely prescribed place within the framework of research. In the social sciences, on the other hand, where universal principles of a similar fruitfulness and a unified explanatory basis are not available, we must seek to find *partial dominants*, to discover islands of firm causal relations in the sea of social events.

From a different angle, the investigations in the last sections have led us ever more close to the insight that the questions of philosophy of science at issue in the *Methodenstreit*, are almost all rooted in the epistemological situation just described. This insight is only obscured by the speculative prejudice that there is an intellectual authority that transcends empirical inquiry and could present one procedure as the only adequate in contrast to all others.

Due to the lack of universal laws which would permit application to the most varied constellations of data, the significance attributed to *certain historical situations* as *points of departure* for inquiry is considerably increased. In the first part of the book we observed that for the determination of events in the external world, knowledge of initial state conditions at least for one specific point of time is required. But – and here lies the decisive difference from the problem-situation in the social sciences – the selection of this point of time is in principle irrelevant: under the presupposition of a universal pervasive lawfulness of events in the external world, any past or future state can be calculated from the state of this system at any arbitrary point of time. Now to be sure this idealization which appears in the fictions of Laplace's demon does not do complete justice to the matter, but still, in its correct

[199] See Kaufmann 'Soziale Kollektiva' p. 299.

finite version a quite essential trait of the universal lawfulness of nature is clearly represented. In the social sciences, on the other hand, it can frequently be seen that even with a given research goal, the correct way for the explanation of social phenomena observed in a temporal cross-section can only be found by taking exact account of the historical context of the facts to be explained. Therefore, the *relatively constant factors* of the historical process already mentioned, e.g. the relatively constant structure of human drives, do provide guidelines for the direction of research by fixing the limits of the possible, but these can be regarded as *general sufficient conditions of prediction* much more rarely than in the case of the laws of the abstract natural sciences.

In this connection we cannot deal more fully with the plurality of areas of social lawfulness, examined by the various social sciences. Reliable predictions are, in particular, possible when the object of analysis is a systematically structured *organization*. In this case, all events to be treated occur according to certain plans, so that the direction to be followed in explaining individual events within the organization, i.e. the explanation of the events by relating them to the organizational plan, is unambiguously determined. If the behavior follows the plan, we will normally consider it explained if we have comprehended the plan. If the behavior does not follow the plan, we will investigate the *reasons for this deviation*. Constant incorporation of a great number of people within an organizational plan occurs mainly where the organization is, or is connected with, a *power* organization, so that behavior by a member directed against the organization results in the chance that sanctions will be applied to him. In order to best interpret the actions of the members of the organization described in the organizational plan, we will, accordingly, seek to grasp the *key goals* (dominants) of the organization, i.e., to ascertain which goals of which persons are decisive for the shaping of the organization. This *evaluation* of *dominance* is a *problem of imputation*. With that as the reference-point for the interpretation given, the further task then consists in investigating in what manner the actions of the 'subordinate' members of the organization are incorporated within the plan. The more completely the social events to be investigated appear incorporated into the framework of an organization, the more important the selection of such key goals will prove for their explanation. But the significance of posing the question in this way goes even beyond the framework of the investigation of organizational forms; for example, in the planned behavior of the individual, the same distinction can be made between *primary* goals, which then function as key goals for the inquiry, and *secondary* goals (auxiliary goals); thus the range of problems of the social sciences to a great extent can be comprehended from the perspective of this selection.

This is true for the methodologically extremely important theory of *ideal types* developed by Max Weber. Weber conceives of the construction of ideal types as the specific method of concept formation in the social sciences, and as such, in view of the correlation between concepts and laws in science, it must stand in the closest relation to the problems of discovering social laws. And indeed all the important pertinent questions do find their reflection in the theory of ideal types. Still, in Weber's exposition – which is admirable in spite of this – the various aspects are not sufficiently differentiated so that here a quite difficult rational reconstruction still remains to be carried out, in order to be able to judge the meaning and scope of the results. It is to this that we will now turn. Let us begin by citing the most important formulations of Max Weber in his own words:

> [The ideal type] is formed by the one-sided *accentuation* of *one* or *more* viewpoints and by the synthesis of a great many diffuse and discrete more or less present and occasionally absent *concrete individual phenomena* which are arranged according to those one-sidedly emphasized viewpoints into a unified *analytical* construct. In its conceptual purity this construct cannot be found empirically anywhere in reality, it is a *Utopian vision*, and it is the task of historical work to ascertain in each particular case how near to, or how far from, this ideal picture reality actually is.[200]

The ideal type must be constructed *meaningfully adequate*, and *causally adequate*. The concepts of meaning adequacy and causal adequacy, as well as their unification in an 'understandable action type' are defined as follows:

> We apply the term '*adequacy on the level of meaning*' to the subjective interpretation of a coherent course of conduct when and in so far as, according to our habitual modes of thought and feeling, its component parts taken in their mutual relation are recognized to constitute a 'typical' complex of meaning. It is more common to say 'correct.' The interpretation of a sequence of events will on the other hand be called 'causally adequate' in so far as, according to established generalizations from *experience*, there is a probability that it will always actually occur in the same way. An example of adequacy on the level of meaning in this sense is what is, according to our current *norms* of calculation or thinking, the *correct* solution of an arithmetical problem. On the other hand, a *causally* adequate interpretation of the same phenomenon would concern the statistical probability that, according to verified generalizations from experience, there would be a 'correct' or an *erroneous* solution of the same problem. This also refers to currently accepted norms but includes taking account of typical errors in calculation or of typical confusions. Thus causal explanation depends on being able to determine that there is a probability, which in the rare ideal case this rule of probability can be numerically stated. But is always in some sense calculable, namely that a given observable event (overt or subjective) will be followed or accompanied by another event.[201]
>
> A *correct* causal *interpretation* of a concrete course of action is arrived at when the overt action and the motives have both been *correctly* apprehended and at the same time their relation has become *meaningfully* comprehensible. A correct causal interpretation of *typical* action (understandable action) means that the process which is claimed to be typical is shown to be both adequately grasped to some degree on the level of meaning and at the same time the interpretation is to some degree causally adequate. If adequacy on the level of meaning is lacking, then no matter how high the degree of uniformity and how precisely its probability can be numerically determined, it is still an *incomprehensible* (or not completely incomprehensible) *statistical* probability, whether dealing with external or subjective processes. On the other hand, even the most perfect adequacy on the level of meaning has *causal* significance from a sociological point of view only in so far as there is some kind of proof for the existence of a *chance* that action *in fact* normally *takes* the course which has been held to be meaningful. For this there must be (in the average or the pure case) either a determinable frequency or an approximation to it. Statistical uniformities constitute understandable types of action in the sense of this discussion, and thus constitute 'sociological generalizations' (rules), only when they can be regarded as manifestations of

[200] M. Weber, 'Die 'Objektivität' sozialwissenschaftlicher und sozialpolitischer Erkenntnis', in *Gesammelte Aufsätze zur Wissenschaftslehre*, pp. 146–214, p. 191.

[201] *Wirtschaft und Gesellschaft*, op.cit., p. 5.

the *understandable* subjective meaning of a course of social action. Conversely, rational constructions of subjectively understandable action constitute sociological types of real empirical processes only when they can be empirically observed with a significant degree of approximation. It is unfortunately by no means the case that the actual likelihood of the occurrence of a given course of overt action is *always* directly proportional to the clarity of subjective interpretation. Only external experiences show this to the case. Only external experience can show this to be the case. There are statistics of processes *devoid* of meaning such as death rates, phenomena of fatigue, the production rate of machines, the amount of rainfall, in exactly the same sense as there are statistics of meaningful phenomena. But only when the phenomena are meaningful is it convenient to speak of sociological statistics. Examples are such cases as crime rates, occupational statistics, price statistics, and *statistics* of crop acreage. Naturally there are many cases where both components are involved, as in crop statistics.[202]

For the question toward which *goals of knowledge* the concept formation of ideal types is oriented, what heuristic function thus is to be attributed to the ideal types, the following quotations are especially revealing:

The empirical *material* which underlies concept-formation in sociology consists to a very large extent, though by no means exclusively, of the same concrete processes of action which are dealt with by historians. Its concepts are formulated and its generalizations are worked out, mainly in an attempt to justify its important claim to be able to make a contribution to the causal explanation of historically and culturally important phenomena. As in the case of every generalizing science the abstract character of the concepts of sociology is responsible for the fact that, compared with actual historical reality, they are relatively *lacking* in fullness of concrete content. To compensate for this disadvantage sociological analysis can offer a greater *precision* of concepts. This precision is obtained by striving for the highest possible degree of adequacy on the level of *meaning* in accordance with the definition of that concept put forward above. It has already been repeatedly stressed that this aim can be realized in a particularly high degree in the case of concepts and generalizations which formulate *rational* processes. But sociological investigation also attempts to include in its scope various irrational phenomena, such as prophetic, mystic, and effectual modes of action, formulated in terms of theoretical concepts which are adequate on the level of *meaning*. In *all* cases, rational or irrational, sociological analysis both *abstracts* from reality and at the same time helps us to understand it, in that it shows with what degree of approximation a concrete historical phenomenon can be subsumed under one or more of these concepts. For example, the same historical phenomenon may be in one aspect 'feudal', in another 'patrimonial,' in another 'bureaucratic,' and in still another 'charismatic.' In order to give a *precise* meaning to these terms, it is necessary for the sociologist to formulate pure (ideal) types of the corresponding forms of action which in each case involve the highest possible degree of logical integration by virtue of their complete adequacy on the level of meaning. But precisely because this is true, it is probably seldom if ever that a real phenomenon can be found which corresponds exactly to one of these ideally constructed pure types. The case is similar to physical reaction which has been calculated on the assumption of an absolute vacuum.[203]

The ideal types of social action which for instance are used in economic theory are thus 'unrealistic' or abstract in that they always ask what course of action would take place if it were purely rational and oriented to economic ends alone. But this construction can be used to aid in this understanding of action not purely economically determined but which involve deviations arising from traditional restraints, affects, errors, and the intrusion of other than

[202] *Ibid., p. 5 f.*

[203] *Ibid., p. 9 f.*

> economic purposes or considerations. This can take place in two ways. First, in analyzing the extent to which in the concrete case, or on the average for a class of cases, the action was in part economically determined along with the other factors. Secondly, by throwing the discrepancy between the actual course of events and the ideal type into relief, the analysis of the non-economic motives actually involved is facilitated.[204]

So much for Max Weber. Now his investigations have not infrequently been interpreted as though by creating the concept of ideal type he had provided an *instrument for social science research* which *formed a counterpart to the conceptual apparatus of the natural sciences.*[205] The interpretation contains various elements that require correction. To begin with, the designation of concepts as 'instruments of research' is misleading; it springs from the untenable conception that the world is given a chaotic manifold and reshaped by reason into the cosmos, in which process the conceptual apparatus provided by reason represents the tools for this formation. In truth, however, as we have recognized, concept formation is as such nothing else than *definition*, which, indeed can be regarded as a symptom for the chosen method of inquiry, insofar as specific procedural concepts are involved. Max Weber's achievement lies, however – and with that we come to a second error of the above interpretation – not in the creation (or discovery) of a new procedure, but in clarifying by means of rational reconstruction a certain type of procedure, characteristic of the human sciences. Finally, a third error of interpretation lies in the exaggeration of the oppositions between the natural sciences and the human sciences (social sciences). To be sure, Max Weber himself is not entirely innocent of contributing to this, as he did not work out the logical structure of the ideal type, which permits recognition of the degree of common characteristics nor did he sufficiently take into consideration the inner connection between meaning adequacy and causal adequacy.

To the following additions to and corrections of Weber's analyses; must be prefaced the following statements of principle: The evaluation of the knowledge content of ideal types presupposes *insight into the function* that is to be accorded to them in the procedure of the social sciences; however, this function is – an undoubtedly emerges clearly in the passages from Max Weber cited above – that of *schemata of interpretation*. The behavior to be investigated – whether this may involve 'historical' behavior that has taken place in the past, or predicted future behavior – is to be totally or partially *understood* by means of subsumption under an ideal type. Taking this into consideration, we want to examine criteria of meaning and causal adequacy.

As can be gathered from the quotations, Max Weber conceives 'causal adequacy' and 'meaning adequacy' as logically independent of each other, and sets up as the criterion for correct causal interpretation of human action that it satisfy the requirement of meaning adequacy as well as that of causal adequacy. The contrast of these two corresponds is part to the current distinction between 'statistical results' and

[204] *Ibid., p. 10.*

[205] See for instance Pfister, *Entwicklung zum Idealtypus*, Tübingen 1928.

'theoretical results' in the social sciences, and for this reason their critical analysis is significant for the *Methodenstreit*. This analysis has the following result:

What quite generally distinguishes 'theoretically secured' lawfulness from mere statistical probability, the regularities of succession or coexistence, is its incorporation within a *multi-stratified context* of experience which permits confirmations in various directions and – combined with that – permits predictions which go beyond the repeatability of directly observed statistical regularity. The statistical result that is unfounded theoretically is *inductio per enumerationem simplicem*, thus induction of a certain, in this case of the first level. But it is required of a law of experience that it be incorporated unequivocally into *multi-stratified* context of experience. Incidentally, this is also the sole legitimate sense of the dictum that in the causal relationship a '*proper hoc*' beyond the '*post hoc*' is asserted or established. It can hardly be doubted – after examining the examples of causal adequacy given by Max Weber, that the conceptual motif to distinguish between statistics and theory, in the manner just sketched, played a role in his considerations. But this motif gains its specific coloration for the human sciences as the lawfulness, which is crucial, for the sociological rules, a *lawfulness of understanding*. The connection between the postulates of meaning adequacy and of causal adequacy for the ideal type is then to be interpreted in the following manner:

As quite generally, laws are nothing else than general assumptions – thus 'rational constructions' – which have been set up on the basis of prior experience and have to continue to be confirmed by the facts, so ideal typical interpretive schemes are 'rational constructions of meaningfully understandable' action. The relation to prior experience lies in that it is established according to our 'average models of thought and feeling' and the confirmation in 'that they can be empirically observed with a significant degree of approximation'. Thus within the framework of Weber's theory of ideal types, the postulate of causal adequacy – and thus also statistical observation – does not have the function of a point of departure for induction, as is the case in the statistics of processes which have no meaning, but merely as (accessory) *control function*. It has to be pointed out, however, that in the actual process of pursuing knowledge these two functions cannot be sharply separated from each other.

As a concept that is constructed to be meaningfully adequate, the ideal type appears to be tied to interpretative methods. Consequently, all the general problems of meaning interpretation enter into its discussion, which we presented in the second section of this part, referring to the analyses of Schütz. This is especially true for the questions of the dependency of schemes of interpretation on the position of the interpreting subject, relative to the object of interpretation, on the one hand, and to the specific problem on the other. In his work which we have cited repeatedly, Schütz analyzed these relationships with great care, and it will be sufficient here to point to these investigations.[206]

However, it is incumbent upon us to follow further those conceptual motifs of Weber's presentation which seek to elaborate the specific character of the method of

[206] Compare *Aufbau*, p. 202.

ideal types within the framework of the interpretive method, i.e., at ascertaining the *distinctive character* of the *ideal typical* interpretive schemes. Max Weber's exposition, which we have cited above, concerning the function of ideal types within the framework of sociological procedure, will serve us best as a guideline. We want to depart from the analogy with physical processes in a vacuum, which is most revealing for Weber's train of thought; to such a vacuum is to correspond 'the consistent unity of the greatest possible meaning adequacy' in the formation of ideal types. Now the basis of this analogy is that in both instances we are confronted with a case of the so-called *isolating abstraction*. Just as in the formulation of the law of falling bodies, only the rime of fall is taken into consideration as an independent variable while friction and air resistance are neglected, so in the case of ideal typical interpretive schemata certain groups of motivations are not taken into consideration for the sake of the unity of perspective.

Such an isolation will be heuristically significant in the first place, where the isolated factor is a *dominant* feature of the processes to be explained – thus a *decisive motive* in the case of meaningful processes; in the second place, where *complementary rules* are possible. Thus on the one hand, we can conduct experiments with falling bodies under conditions in which friction and air resistance hardly play any role; on the other hand, it is also possible to establish friction and air resistance experimentally and by computation, and then to achieve a description of the process of fall that is faithful to reality by combining the various particular laws. With a view to explanations in the social sciences, it is absolutely requisite to carefully keep these two possibilities separate, if false (exaggerated) conceptions about validity are not to be aroused. Thus every time we operate with social laws, we have to be clear whether we want to apply them to reality *in isolation*, i.e., use them to make predictions, or whether we want them to be conceived as *partial laws* requiring supplementation. If the latter is the case, it must not be forgotten that the combination of partial schemata of interpretation, in order to form a unity faithful to reality, does not consist simply in an addition of partial schemata but will display a complicated structure, which has to be determined. Thus the motives to be taken into consideration in the explanation will frequently have to be comprehended as modifications of other basic motives that also enter into the explanation. How important insight into the way *partial interpretive schemata are linked structurally* is, will be seen from the circumstance that it is only this which makes it possible, to ascertain the degree of interdependence between the components of an action which correspond to the individual motives. Thus, for instance, in economic investigations the question may become urgent, to what extent a change in the economic data by which the economic subject has to orient himself (for example, an increase or decrease in his income) will displace the relative weight of 'economic' and 'non-economic motives' for his actions. It seems clear that the heuristic significance of a certain ideal type, in general will be the greater the more independent the influence is of the motives isolated in this ideal type on the course of the action to be interpreted, from the influence of the other factors which have to be taken into consideration in the interpretation.

The above considerations about ideal types have concerned themselves with the isolating abstraction of certain motivational units carried out in them. Now it is not too directly against language usage to designate such rational construction as '*idealization*'. For in cases that are formally similar we speak of an 'ideal gas' or an 'ideal fluid'. But when doing so we must be doubly careful to avoid confusing *this* aspect of idealization in the ideal type with the quite different kind of 'idealization' which lies in using, as a scheme of interpretation, behavior which is 'correct' behavior, 'appropriate to goals', with respect to certain given goals. We wish to separate these two aspects from each other terminologically as '*isolating abstraction*' and '*idealization*'. It is clear that the isolating abstraction is logically prior to idealization in the formation of ideal types, just as 'appropriateness to goals' as a relational concept presupposes 'given goals'.

This idealization we can designate as the '*assumption of objective instrumental rationality*', and with that express the fact that an action that evolves according to this rationality externally corresponds completely to an action which is carried out with full clarity as to its goal, and in conscious pursuit of the most efficacious path toward this goal. Yet with that, it is not stated that this clarity must actually exist for the actor. This is the difference Max Weber had in view when, in his analysis of ideal typical behavior, he pointed to the circumstance that the constructive concepts of sociology were ideal typical 'not only externally but also internally'.[207] However, it cannot be gathered unambiguously from his exposition whether he intended the criteria of objective and subjective instrumental rationality to be contained in the concept of meaning adequacy; his equating of 'typical' and 'correct' meaning contexts in the definition of 'meaning adequacy', and almost all his examples, appear to suggest this; but there are also counter arguments.

More important for us than this question, belonging to the history of doctrines, is that of the logical relation between the two concepts of objective and subjective instrumental rationality. In order to answer this, we want to become clearly aware of the *criteria of subjective instrumental rationality*.

In the assertion that a person P in the situation S acts *subjectively with instrumental rationality* with respect to the goal G if P undertakes temporally ordered action A_i, the following presuppositions are contained:

1. If in the situation S the actions A_i are undertaken by P, then the facts F_g will take place (the goal is to cause these facts).
2. The causing of F_g through the undertaking of A_i satisfies a series of secondary conditions, to be characterized more precisely, which are determined by the total goal system of P (say, bringing about F_g in the shortest possible time with the least possible effort, least possible costs, and without producing undesired secondary effects).
3. The person P has a *complete insight* into the circumstances mentioned under (1) and (2), and thus he is clearly aware of his actual goal as well as his total goal system. Further, he knows the causal nexus that leads from A_i to F_g and is also

[207] *Wirtschaft und Gesellschaft*, p. 10.

able to survey other ways leading from the same initial point to the same goal. He also knows that the way by means of A_i fits more completely into the total goal system than the other ways. On the basis of this insight he decided to undertake A_i and carries out this resolve.

The *criteria for objective instrumental rationality*, on the other hand, are already fulfilled by meeting the conditions named in (1) and (2). Thus it is to them that the judgment of the observer (social researcher) corresponds: "In the situation S actions of the kind A_i have the occurrence of F_g as a result. This causal sequence is distinguished from other causal sequences that lead from S to F_g, among other things by a series of 'properties' to be characterized more precisely (the causal sequence satisfies certain conditions C). Therefore a person who wishes to arrive at F_g from S, while at the same time satisfying the conditions C, will realize this purpose when he undertakes the action A_i." We can then characterize this behavior as 'correct', 'appropriate to the goal', or 'objectively instrumentally rational'. All these terms are synonyms.

From this it follows that the assertion of the objective instrumental rationality of an action also presupposes reference to the existence of a certain subjective meaning of the actor – a specifiable goal, consistent with the subject's own goal system. On the other hand, what is not conceptually presupposed, is (a) that the actor is fully conscious or aware of these goals, (b) that he has chosen the correct path with clear recognition of its correctness. The concept of objective instrumental rationality is thus contained in that of subjective instrumental rationality; and with respect to the latter it is logically prior.

Now, in evaluating the heuristic practicability of a given ideal type for a certain topic of investigation, it is important to keep in view that in the application of the ideal type it is not subjective instrumental rationality which is presupposed, but only – partial or total – objective instrumental rationality. The investigator will thus always be able to operate advantageously with the assumptions of behavior according to plan when – but not only when[208] – the actions under consideration *take their course on the average, as though they were according to plan.*

Now the insight into the relation of subjective instrumental rationality to objective instrumental rationality is obscured by various aspects. In the first place the objective instrumental rationality of a course of action can only be conceived to be fully understood (as meaning-adequate) when we have succeeded in tracing it back to a detailed plan, either of the actor himself or of some other initiator of the action; this is frequently seen as its 'derivation' from subjective instrumental rationality, which can then be easily misinterpreted as a logical derivation.

Secondly, the concept of subjective instrumental rationality is frequently employed equivocally, insofar as – together with the use of the term in the meaning just described – all goal-directed behavior is designated as 'subjective instrumental

[208] It is sufficient that the ideal type determines *one* motivational component of action, insofar as sufficient indications for the knowledge of the other components, and the manner of the linking among all the relevant components, can be gained from other considerations. See above, this section.

rational', in order then, in a further displacement of concepts to confront 'subjective instrumental rationality', as merely *alleged*, with 'objective instrumental rationality'. And with reference to this concept we will, finally, in a typical manner, arrive at the erroneous absolutization which we have characterized in our section on value in Part One: the action to be studied is judged with reference to goals which have been imposed by the investigator making the judgments, and these are frequently declared to be absolute goals. As what is most important for the critique of this exaggerated concept of the correctness (objective instrumental rationality) of action has already been shown in the passage just referred to, we cannot linger any longer at this point, and can be content with the remark that it was just Max Weber who most sharply rejected this erroneous conception – in accord with the postulate of value-freedom.[209]

However we have to turn our attention toward a statement which is important for the analysis of the ideal type, which concerns the legitimate concepts of objective instrumental rationality and subjective instrumental rationality, namely that the judging of the instrumental rationality of social action is linked most closely with the *problems of prediction in the social sciences*. For subjective instrumental rationality this seems evident without further proof, as the concept of a plan contains that of prediction; but that it also holds true for objective instrumental rationality will be recognized as soon as we ask at which *point of time* an assertion of the objective instrumental rationality of human action is decidable. For the affirmation or denial of the question whether with given goals (including secondary conditions) and a given initial situation the actions can be characterized as 'objectively instrumentally rational' – and that without these actions having been realized, and the success (attainment of the goal) not being ascertainable – obviously depends upon how the concept of correct prediction is defined.

Thus what is a correct prediction? Obviously a prediction – we will be inclined to say at first – which comes true (which is confirmed), and therefore its correctness can be determined only *after* the point of time toward which the prediction points. But still, under some circumstances, a prediction is said to be correct or incorrect directly after it has been stated, and this according to whether it has been made in conformity with, or in opposition to experience up to that point. In this case, by 'correctness' (or 'incorrectness') it is not the fulfillment (or failure) of the prediction that is understood, but the considerable (not considerable) chance of its fulfillment. Unfortunately, these two concepts are frequently and promiscuously used. Now if a prediction, which was carried out in conformity with experience referring to a certain state of the data, proves inappropriate then the 'explanation' will readily come to mind that the state of the data to which the prediction as well as an action in conformity with it was oriented, did not yet offer a sufficient basis for the predictions.[210] So in view of the problematic epistemological situation of the social

[209] See especially his essay, which has already been mentioned, 'Der Sinn der "Wertfreiheit" der soziologischen und ökonomischen Wissenschaften'.

[210] For this compare Oskar Morgenstern, 'Vollkommene Voraussicht und wirtschaftliches Gleichgewicht' (Complete prediction and economic balance), *Zeitschrift für Nationalökonomie*, vol. VI, pp. 337–357.

sciences – in comparison with the abstract natural sciences – with respect to the possibility of reliable predictions, it can be understood that the question, what *amount of data* is requisite for prediction in each case stands at the center of methodological controversies. But it is not only due to the elasticity of the concept of state of the data that the idea of objective instrumental rationality in the ideal type becomes ambiguous. This is also the case because the investigation of the *causal relationships* which is determining for its evaluation has to be *historically relativized to a certain state of knowledge*, so that actually it would be permissible only to speak of 'objective instrumental rationality according to the present state of knowledge' or 'according the state of knowledge of that time'.

But from this the hasty inference must not be drawn that the formation of ideal types would not be relevant to the comprehension of objective instrumental rationality at all, but only to the comprehension of 'subjective instrumental rationality'. For it is essential for the aspect of idealization in ideal types that the description of the 'correct' (instrumentally rational) courses of action contained in them has a *material basis*. In addition, it is only through this that the concept of 'deviation from the course prescribed by the ideal type' receives its specific meaning. Now in the sense of the observation just made, two things have to be noted: (1) that an objectively based assertion, one supported by 'objective knowledge', about causal relationships must not be given an exaggerated interpretation as necessarily and conclusively true; (2) that in most cases the idealization will be carried out relative to the knowledge available to the human community under consideration. Here we can speak of a *historization* of the element of correctness in the ideal type.

But with that, the 'historicity' of the ideal types is not yet exhausted, as we can comprehend with little difficulty if we keep in view how the formation of ideal types takes place *in stages*. To begin with, the social scientist dealing with a certain topic will have to seek to demonstrate *goals for action* within his domain of investigation, the knowledge of which can furnish key points for the interpretation of human action within that domain. In so doing, we must not leave out of consideration the fact that such goals have to be conceived within more comprehensive goal systems, as otherwise the secondary conditions for 'correct' action will be obscured. But in order to discern the contours of these goal systems with some degree of accuracy, it will not, as a rule, suffice to base ourselves on knowledge merely of 'the general nature of man', of his instinctual and mental constitution; instead we will have to utilize knowledge relating to a specific human type and specific social preconditions (such as, for example, the legal order).

The same is true for the *second* stage in the process of forming ideal types, namely determination of the *initial situation*, i.e., that state of the data at the point of time when the goal was set, as well as for the *third* stage, the construction of a *course of action that is appropriate for the goal*. Still the terms 'historical ideal type' or 'historicizing of the ideal type' must not be understood as though historical time entered explicitly into ideal types. An ideal type is a general concept and as such contains no element of individuation. The historization only consists in the formation of the ideal type assumptions as adapted to a specific historical epoch, and thus for the purpose of applying the ideal type, thus constructed, successfully to the events of this epoch. Still the possibility of application to other epochs remains open. A human being born in the twentieth century can also fall under the ideal type of Renaissance man.

The preceding exposition has made evident in what manner the fundamental problems debated in the theory of the social sciences, (which we treated in the second, third and fourth sections of this Part) enter into the theory of ideal types; it has also permitted us to recognize the kind of relation that holds between social laws and ideal types in the social sciences. In the section after next (on economic theory) we will have to deal again with certain pseudo-problems, which are related to the construction and application of ideal types.

7. The Way to Overcome the Methodenstreit

Our investigation up to this point has argued with ever new aspects and with ever increasing clarity, that overcoming the methodological controversy in the social sciences is not conceivable in such a manner that in each of the cases under discussion a philosophical authority beyond appeal could declare a specific method to be the only correct one, or even the best. Instead we must seek to gain empirical points of reference for the validation of one procedure or another. To be sure, reflection on successes attained by means of particular controversial methods will only, in relatively rare cases, have as a consequence the complete elimination of a procedure recognized by a wide circle of investigators. However, it will frequently lead to a determination – though never absolutely final – of the comparative practicability (effectiveness) of the method in question, and this in a variety of directions:

(a) After the determination or the rational reconstruction of the goals of inquiry and their rank-order (which in turn will be suggested by the '*praxis*' of life or the characteristic idiosyncrasy of the researcher) this reflection will show that the idea of a hierarchy of methods, encompassing all the social sciences, which would prescribe the research procedure in its main outlines, is for the foreseeable future impossible to realize. Rather, as a rule, when the choice is between two applicable methods, any 'advantages' obtained will have to be gained at the cost of 'disadvantages'.

(b) It will show that frequently the arguments with which a methodological decision is justified, that may in itself be completely acceptable, prove untenable, as achievements are attributed to the chosen method which it will not be able to attain. Not infrequently 'achievements' are involved here, which are not even formulated without contradiction (alleged apodictic character of empirical assertions).

(c) It will make evident that in many cases the degree of independence of controversial methods from each other has been overestimated. We have shown this in detail in the case of two especially important examples which are closely related, namely 'explaining – understanding' [*Erklären – Verstehen*] and 'causal method – teleological method', and in what follows we shall give further examples.

(d) The reflection will establish that the methods at issue are burdened with implicit presuppositions which were not taken into consideration in the arguments of the *Methodenstreit* – much to their detriment.
(e) Finally it will be shown to what a high degree the practicability, and with that the relative preferability, of a method of social science is situationally conditioned by the mere circumstance that to a greater or lesser extent it presupposes specific material – which is more or less readily available, or perhaps not at all.

It is psychologically understandable that this complicated epistemological situation, which is difficult to survey theoretically, again and again has evoked the wish to arrive at absolutely valid methods for the social sciences by means of an *a priori* principle – whether this may be declared to be logical or in the narrower sense, philosophical (metaphysical). One of the examples of such a 'logical principle', important in the history of doctrines, is the *postulate of purity of method.* It corresponds to the basic conception of this book, according to which the most urgent task for the theory of science, and one ripe for solution today, is to deflate the apparent *a priori* character of principles; and we start this section with a critical analysis of that postulate. Subsequently we wish to characterize, in all brevity, the actual methodological content of a series of controversies in the human sciences, which have become relevant in the history of doctrines, in order finally to demonstrate paradigmatically in terms of these the way to overcome the *Methodenstreit* – insofar as it can be overcome at all.

In judging the goal set by the postulate of purity of method – or negatively formulated – in the *rejection of the syncretism of method* – we must take special care to keep separate two aspects which have been confused only too often in the controversy over method. For the postulate has two meanings, insofar as usually, on the one hand, the logical requirement of the unambiguous use of scientific concepts is understood, on the other, the methodological requirement in the narrower sense of the unity (systematic closure) of the procedure.[211] This ambiguity can be explained in the following manner: In the first place, the two requirements are linked by the fact that closure of the procedure greatly facilitates avoidance of the logical error in question (failure to distinguish equivocal concepts). Secondly, and here lies the decisive point, as a rule a sharp distinction is not achieved, because *scientific construction and rational reconstruction* are not properly distinguished from each other. Thus, for example, the protagonists of pure economic science use quite promiscuously both the thesis that the immanent interpretation of economic facts – i.e., an interpretation that remains confined to the domain of economic phenomena – leads to optimal results, and the thesis that the founders of classical economics 'actually' always strove for the autonomy of their science, and therefore only the resolute pursuit of these efforts deserves the name 'economic science'. However, within the framework of rational reconstruction the discovery and removal of equivocation plays a pre-eminent role, and thus it falsely appears that questions of constructive procedure as well, which have as their object the systematic closure of a

[211] See above, Part two, Preparatory Remarks.

science, could be decided logically and unambiguously. In truth, however, in many cases it is not even possible to make an unambiguous empirical decision with respect to the preferability of the one method or the other.

This statement requires closer clarification. To begin with, as already noted, methodological stipulations of this kind depend to a high degree on the scientific and extrascientific interests of the investigator, as well as on his special talents. Insofar as *those* moments are decisive, a resolution of the controversy over method must confine itself to rejecting exaggerated claims of validity in favor of one method, generally based on speculative prejudices, in order to aid the other in attaining its relative rights. Included in this is the task of working out the meaning of the competing procedures with the greatest possible clarity by rational reconstruction.

Aside from these 'subjective elements', decisions over the choice of method – with respect to 'purity', closure or the opposite – depend on the state of knowledge at the time of the choice. The degree of this dependence is determined in the first place by the kind and extent of specific knowledge of data, or the chance of procuring such knowledge at a foreseeable point of time; secondly, by the knowledge of general relationships, or the chance of attaining this knowledge. There is a close mutual interrelation between the first and the second point.

In order to exemplify what has been said with respect to the state of the data we may point to the great influence which the 'fortuitous' knowledge of certain documents, and the lack of knowledge of others, has exercised on the method of many historical disciplines. Whether an independent, self-contained investigation, for instance, of the language or legal system of a certain people existing in the past can be carried out, or whether one is forced to confine oneself to general ethnology, which utilizes documents concerning this people from heterogeneous fields as sources, will evidently depend to a great extent on what documents are available, and a new discovery, say a successful excavation, can produce decisive changes here. That not infrequently, only the discovery of general relationships (laws) makes possible the creation of disciplines unified in subject matter, or renders them relevant, also requires no argument here. We need only consider the significance which the laws of consonant-shift had for the establishment of comparative linguistics as an independent method of inquiry. Similar significance can be attained when facts or laws which up to that time had been comprehended *obscurely* are brought into the full light of awareness.

But besides these factors, strongly conditioned by the situation, there appear also those for which this is true to a much lesser degree. For there are methods to which we will accord preference without much question, 'once and for all', preferring them to the others; such as, for instance, the methods of modern natural science as compared to mythical explanations of nature. The objection that the former methods are incompatible with the latter as the goals are different in each case will not hold up, for, in the one just as in the other, the main endeavor is *to make correct predictions possible*.

The methodological point, which is of fundamental importance, and which must never be left out of consideration in controversies about purity of methods as well, must receive special stress. Above all, it must be pointed out that our thesis also applies to the historical sciences, even where the concern is 'simply' to ascertain the facts, only that here we are not dealing with the prediction of future events, at least not directly, but instead with the *reconstruction* of *events which have already taken place*. For every ascertainment of facts is a contribution to the *understanding* of the relations between facts, and increases the chance of making correct assumptions about the formation of other facts in this interrelation as yet unknown or known only incompletely. The procedure of thought in which such assumptions are formed is based on similar inductions that are decisive for the prediction of future events and their confirmation (for instance, through finding a document that confirms the assumption) and is, from the formal viewpoint, of the same kind in both cases. This incorporation of each single historical assertion into a context of confirmation in terms of factual materials that are in principle open-ended, forms, incidentally – as we wish to note in passing –, the decisive difference between the *scientific* and the *artistic* (epic or dramatic) *representation of history*. The poet working on historical materials, just as the historical scholar, constructs a picture of historical personalities and historical events on the basis of the available material. But for '*artistic truth*' it is already sufficient that this construction has accomplished meaningful adequacy (believably); documentary proof that the artistic conception of a historical personality, or the artistic description of a historical event, does not correspond to the facts is not essentially damaging, but any claims to '*scientific truth*' would be totally refuted by a similar assertion.

The general criterion for scientific statements, which lies in their testability, remains untouched by all the differentiations which are significant for the choice of topic and procedure. In the latter, besides the various kinds of value relations which are characterized by specific material interests, divergence in 'formal interests', aiming at the achievement of the highest measure of unity, simplicity and precision of procedure, also play an important role. Such goals of inquiry find expression in the postulate of purity of method; most frequently they are only attainable at the cost of greater remoteness from reality, i.e., heightened indirectness of application, and the investigator has to make his choice between these alternatives – a choice which is frequently made, to be sure, with considerable confusion. The real core of the pertinent controversies lies therefore in the opposite answers given to the question, whether the purity of method thus understood is worth the sacrifices in 'truthfulness' that have to be made for its sake.

From what has been said already it follows with complete clarity that in controversies of this kind it is totally erroneous to seek to cut the highly involved knot of the problem with the aid of a 'logical' or 'philosophical' principle. Rather, constantly taking experience into consideration, we have to examine with all possible care whether the phenomena within the descriptively delimited partial domain of historical-social reality under discussion stand in such a relationship of interdependence that those facts within this domain, which are relatively easy to ascertain, will form a sufficient data base for the desired prediction of other facts within the domain. Now as can readily be seen, this will be the case above all where human beings orient their actions by the state of the data which appears to be determined by previous actions of the *same kind*. An especially important example of such orientation is the *economic behavior* of human beings. Here the postulate of the 'purity (closure) of method' has considerable prospect of asserting itself in scientific praxis.

However, we must consider that rational reconstruction of scientific thought reveals *various degrees* of fulfillment of this requirement. Consequently the case of complete purity would be characterized in the following manner: As soon as the object of the science has been descriptively specified, the *material content of all concepts* of the science must be *delimited* by that descriptive framework – all concepts must be '*system concepts*'- and accordingly all laws of the science must be presented as connections between facts which are describable solely by example of system concepts. Theoretical physics offers a paradigm example for the case of complete purity. But besides this, partial realizations of the postulate are also conceivable, and indeed important for methodology. It is of these we wish to speak whenever it is the case that, though all laws of the science contain *system concepts*, yet *extra-systematic* concepts, and *data* in the narrower sense, appear beside these as well. For example, if we conceive of economic science as the theory of the interdependence relation that holds between prices, the statement that – *ceteris paribus* – an increase in the quantity of money leads to a rise in prices, is a purely economic one; however, a statement that contains the concept of technical complementary is no longer completely pure.

The question that has already been noted: what can the correspondingly restricted conceptual apparatus 'achieve' remains heuristically decisive for the degree of purity sought for the procedure. But this achievement is determined by the degree of appropriateness (to the goals) of those laws which can be formed solely from these concepts. To make possible an overview in this direction, a *schematic survey* of the *system concepts*, which also displays the manner of their logical connections, presents an important aid. Such a schema is called the '*formal theory*' of a science.

Now since the attempts to set up such schemata generally are carried out within the framework of an endeavor to delimit a science in accord with the postulate of the purity of method, the *objection of exaggerated formalism*,[212] directed against methods orientated by this postulate, is usually also extended to such attempts, although these in no way prejudge the decision about choice of method.

With respect to the objection of exaggerated formalism, directed against a social scientific method, the following is to be remarked in principle: Most frequently the objection is directed, on the one hand, against the (allegedly) excessive *abstraction* from the historical situation, and the objection maintains that it will result in the *theoretical* barrenness of the procedure. On the other hand, it is directed against the *value-freedom* of the method, which lies in its neutrality with respect to the 'ultimate' goals of human actions, and, according to this objection,

[212] This objection, for instance, was raised again and again against Kelsen's pure theory of law. A brief reply is contained in his essay 'Juristischer Formalismus und reine Rechtslehre', vol. 23 of the *Juristische Wochenschrift,* Leipzig 1929. Compare also Felix Kaufmann, *Logik und Rechtswissenschaft* (*Logic and Legal Science*), Tübingen 1922.

brings about its *practical* barrenness. All that is fundamental for the critique of the second objection [of value-freedom] has already been stated.

It is more difficult to judge the objection of an excessive abstraction. In order to counter it, we must not try to appeal to logical or epistemological principles, for from these the preferability of one (contradiction-free) method as against others can never be derived. Instead we have to clarify the *teleological function* of formalism in the social sphere, in order to show that the action of people in social life is governed by an orientation provided by very abstract maxims, and that therefore the social science which isolates these aspects thematically has a rich and important field of work. With respect to this, in our analysis of the value concept, we have already pointed to the postulates of a social teleology underlying formalist ethics; furthermore, in the last section we have pointed out the possibility of reliable predictions in spite of far-reaching anonymization (thus formalization) for those cases where a pervasively social organization exists.

These two cases already make it evident that a sharp opposition cannot be constructed – as has so often been attempted – between formalist and teleological methods. These relations become especially clear when we ask – proceeding from the goal of providing for the future, for security of one's own life and that of others who are close and require protection – what conditions must be fulfilled in order to allow the striving for these goals to appear practically meaningful. For, contained in the idea of providing for the future is that of partial predictability; predictability in turn is linked to a certain conformity of occurrences and to the simplicity of the overview over the data that must be taken into consideration; due to this, conformity itself becomes a goal of human endeavor. We may think, for example, of the postulate of legal security in legal policy, which contains the idea of predictability as its formal core. Consistent analyses of concepts from the viewpoint just characterized can lead us to the center of comprehension of the relationships within the social sphere. Hence, these analyses offer insight into the social function of those unification rules of action which are called 'customs' and make comprehensible the reactions, which so frequently appear excessive at first glance, to the violation of these customs (social conventions). Whosoever does not conform to such a convention, say of fashion, proclaims thereby that either he does *not wish* or is *not able* to do so; both will make him appear 'eerie' to a community whose form of organization is linked to the presupposition of far-reaching uniformity (in certain respects) in the behavior of the members of this community.

Considerations of this sort also lead us to a sympathetic evaluation of Simmel's attempt to construct sociology as the theory of the *forms* of human cohabitation as a '*geometry of the historical world*'.[213] Although the philosophical foundations on which he bases his doctrines are open to various objections, it cannot be doubted that his analyses represent an important contribution to the knowledge of the social sciences, and the same is true of the scholars who have carried Simmel's investigations further in the one or the other direction. Beside Max Weber, Vierkandt,[214]

---

[213] Simmel, see his *Soziologie*, Leipzig 1908.

[214] Vierkandt, see his *Gesellschaftslehre*, 2nd. Ed., Stuttgart 1928.

Wiese,[215] and Sander[216] have to be mentioned here. However, as has emerged from our investigation, recognition of these achievements by no means presupposes the assumption of an *a priori* for the social sciences, given in advance of all historicity, and accordingly, the refutation of such an assumption does not in any way imply the rejection of a *formal theory of society*.

Whether, with Simmel, we define '*sociology*' as the '*theory of the forms of social relations*', whether we introduce this concept as an *encompassing concept* for *all the social sciences*, or whether we wish to use 'sociology' as the name for a *special science* to be characterized more precisely,[217] is relevant only insofar as in this, we should see an indication of the research method that is being proposed. For we cannot speak of any tradition in the history of social thought which is so unified that an unambiguous concept of 'sociology' could be extracted from it by rational reconstruction. Thus neither an *a priori* insight nor a legal title documented by the history of doctrines will prove decisive for the choice of the definition, but rather the weighing of the suitability of its method for specific goals of inquiry.

From the same point of view, the doctrine that most violently opposes formalism in social science will have to be judged which designates social collectives – society, state, economy, etc. – as more or less encompassing 'totalities' and rejects every attempt at a progressive construction of the social sciences by means of connecting elementary social relations as an '*atomization*' contrary to nature.[218] In the section before last, while discussion the 'controversy over universalism' we have already pointed out that the assertion of a methodological primacy of society with respect to the individual (thesis of the universalist doctrine of totality) aims at a method in which all actions of the members of a community are referred to community goals, and adjudged (valued) according to the degree of their relevance – as furthering or inhibiting of progress toward this goal. The decisive test for the methodological significance of the theory of totality within the framework of social science research will be furnished by the investigation as to what extent social actions can actually be understood by relating them to a more or less sharply delimited number of collective goals. The degree of comprehensibility will be determined here above all by the kind, extent, and reliability of the predictions that can be carried out with the help of the schema of interpretation used. We will undoubtedly find then that operating with social totalities as teleological entities is very useful for a large number of investigations in the social sciences. Indeed this will be the case above all where the understanding of relations of achievement, or cooperative efforts, in a narrower sense, are dealt with. Yet we must not be lured by this to entertain the assumption that these totalities represent 'ultimate' irreducible ontological entities.

---

[215] Wiese, *Allgemeine Soziologie*, I. Teil, Beziehungslehre, II. Teil, Gebildelehre (Theory of Cultural Formations), Munich 1929.

[216] See especially Sander's *Allgemeine Gesellschaftslehre*.

[217] A good description of this methodological dispute in F. Oppenheimer *System der Soziologie*, I. Bd., *Allgemeine Soziologie*, I. Halbband, Grundlegung, Jena 1922.

[218] See also below, subsection 8.

This erroneous assumption can best be avoided if in each case we ask about the *criteria for the realization of community goals*. To be sure, only rarely can these goals be traced back in a relatively simple manner to individual goals, such as is possible for the goal of increasing national prosperity. But in principle such a reduction must always be possible. Thus the '*external power*' of a community is essentially characterized by the circumstance that certain members of the community, as its representative 'organs', have the chance to impose their will on other persons not belonging to the community; this is so since they can point to the likelihood that any active or passive resistance will have as a consequence a deployment of the organized forces of the members of the community that would be dangerous for the resisters or the human group they represent.

A major difficulty in the controversies connected with the problems of holism consists in the profound misconceptions about the nature of concepts. A source for many other pseudo-problems in the social sciences is to be found in the *false conception of causal relations* which we have criticized already, according to which for every phenomenon there is an efficient cause or at least a single causal series, whereby one single mode of explanation is presented as being the only correct one for the social sciences. Not only more or less primitive doctrines which seek to derive the process of association from one single human drive, and from this genesis seek to derive the 'essential nature' of society, are influenced by this idea, but also carefully constructed social-philosophical or social-scientific systems, such as, above all, Marx's powerful system of *historical materialism* with its *principle of the economically determined character of all social phenomena*.[219]

The position we take toward this has been unequivocally defined by the results our reflections have attained up to this point.

The first step of the critical analysis will have to consist in replacing the concept of determination with that of *co-determination*, the second in characterizing the kind of dependency more precisely, by establishing series of correlations between the variations of the 'causes' and the corresponding variations of the 'effects', where the materials of induction will have to be taken from both historical experience in the narrower sense as well as from systematic observation of the present (statistics). In so doing, according to what we have ascertained above, we will have to test with special care to what extent the causal factor being investigated can be isolated, and thus to what extent the '*ceteris paribus*' clause can be applied empirically. We will recognize that this testing will be nothing else than the investigation of what prediction can be made on the basis of this causal correlation. Related to the fundamental thesis of historical materialism, this states that we must test what kind of predictions are possible and to what extent, given a knowledge of the economic state of the data at a certain point of time.

No one among the theorists of the social sciences has grasped more clearly the character of historical materialism as a partial aspect requiring supplementation than Max Weber, and therefore we wish to cite an especially significant passage of

[219] See, besides *Das Kapital*, especially his work *Zur Kritik der politischen Oekonomie* (*Critique of Political Economy*).

his programmatic essay '*Objectivity in Social Science and Social Policy*', where he argues against historical materialism, as follows:

> The explanation of everything by economic causes alone is never exhaustive in any sense whatsoever in any sphere of cultural phenomena, not even in the 'economic' sphere itself. In principle, a banking history of a nation which adduces only economic motives for explanatory purposes is naturally just as unacceptable as an 'explanation' of the Sistine Madonna as a consequence of the social-economic basis of the culture of the epoch in which it was created. It is no way more complete than, for instance, the explanation of capitalism by reference to certain shifts in the content of the religious ideas which played a role in the genesis of the capitalistic attitude; nor it is more exhaustive than the explanation of a political structure from its geographical background. In all of these cases, the degree of significance which we are to attribute to economic factors is decided by the class of causes to which we are to impute those specific elements of the phenomenon in question to which we attach significance in given cases and in which we are interested. The justification of the one-sided analysis of cultural reality from specific 'points of view' – in our case with respect to its economic conditioning – emerges purely as a technical expedient from the fact that training in the observation of the effects of qualitatively similar categories of causes and the repeated utilization of the same scheme of concepts and hypotheses [*begrifflich-methodischen Apparates*] offers all the advantages of the division of labor. It is free from the charge of 'arbitrariness' to the extent that it is successful in introducing insights into interconnections which have been shown to be valuable for the causal explanation of concrete historical events.[220]

The antithetical approaches of Hegel's idealism and Marx's materialism find expression in the question, which relative role is played on the one hand by factors of the instinctually conditioned reality and, on the other, by the ideal factors, (a) in the development of culture in the narrower sense (intellectual and spiritual life), and (b) in changes of the external conditions of life. Max Scheler, who has placed this question at the center of his sociological considerations comes to the following conclusion[221]:

The influence of the interests, in the narrower sense, which can ultimately be traced back to the structure of human instincts, is so strong that only those intellectual formations can be realized which do not conflict with such utterances. Thus though Mind does determine the scope of possible contents yet, by means of the reality factors (as negative factors of realization) that which is merely Utopian in these formations is separated out from that which can be realized practically. With respect to the external conditions of life, however, not only significance with respect to realization, but also the determinative significance must be attributed exclusively to factors of human drives. In this context, the function of mind is merely that it shows the instinct-directed will, those 'ideas' which appear enticing from the viewpoint of the instincts as well, and thereby diverts attention from other projects. In the long run the mind cannot force any renunciation on the instincts, but only open certain canal locks and close others.

In judging these theses we have to pursue similar considerations as above, in ascertaining the share of the *receptive element of experience*, the 'material' of the

---

220 N. Weber, Op. cit., p. 169 f.

221 Scheler, *Die Wissensformen und die Gesellschaft*, p. 6 ff.

senses, in the total content of experience. There we recognized that the *isolation of pure receptivity is impossible; just as little* is it possible to carry out the *isolation of pure instinctiveness.* Scheler's formulation has rendered considerable service – though, it halted at the initial effort – in that it described the invariants of human behavior, determined by the relatively constant structure of human drives, as limits for the possibility of variations in historical developments, without understanding the scope of variations determined by the mind. Nonetheless I can hardly see any direct application of this formulation to sociological and historical research that would be fruitful, with the exception of the rather obvious insight contained in it, that in the shaping of the mental sphere in the narrower sense, 'ideal factors' play a greater role than in the shaping of external conditions. In any case, on the basis of Scheler's exposition, the next task to be undertaken here is delineated at least with some clarity. It will consist in finding out what distinguishes those maxims of human action which we call 'ideas' from maxims we call 'determined by human drives'. For analysis shows that prescientific as well as scientific use of the term 'idea' is by no means uniform; considerable confusion arises especially due to the circumstance that in some cases 'idea' is understood as Being as such [Sein schlechthin] that is merely mentally represented (fantasized), and in others, however, the term receives that value connotation that is associated with the word 'ideal'. Thus the designation of people who make 'ideas' the maxims of their action vacillates between the – mainly negative – characterization as fantasts and – the mainly positive – characterizations as idealists. If we disregard this axiological coloration, then above all, the following points can be distinguished which are included in the intended meaning of the distinction between instinct-directed and idea-directed behavior: (1) Unconsidered as against well-considered behavior; (2) preference for goods serving vital needs as against spiritual goods in the narrower sense; (3) the uncompromising pursuit of vital self-interest, and at best those of one's immediate circle, as against the interests of the broader community.

From this it emerges that the question of the relative influence of reality factors and ideal factors has been posed much too vaguely to receive an answer that will provide an insight of value to the research interests of the social sciences, since in general a considerable part of the vagueness, which has so often been complained of, can be traced back to an imprecise manner of posing questions. One of the most important investigations to be carried out in connection with the problems just treated would have to seek to answer the question, in what way the readiness of human beings to take their orientation for their action from 'ideas' varies with the *degree of rationality* of thought, and how in the various levels of rationality *uncertainty about the future* is taken into account.

Reflections in conformity with the above considerations will lead to the rejection, or correction, of many other theories as well, that consider one single factor – whether this be heredity or environment or education – as dominant over the long range in the development of human communities. How the one-sidedness of methodological dogmatism can be overcome by rigorous empirical research, can be especially impressively exemplified in terms of the relative influence of *prenatal* and *postnatal* factors, as the well-known phenomenon of identical twins makes possible the isolation of postnatal influences.[222] The significance of such isolation for the progress of research hardly requires further elucidation; it is my conviction that the development of knowledge in the social sciences will depend essentially on the rigor with which these sources of inductive knowledge are pursued. From this it becomes comprehensible as well that *psychopathology* (including *psychoanalysis*) has great significance for psychology as a whole as well as for the social sciences, and that ethnological investigations of primitive people play an important role in the social sciences. This significance will increase once these achievements have been rigorously ordered with a view to their application to the methodological problems in the social sciences.[223] Above all, in all likelihood, important conclusions for the *degrees of freedom* of questions in the social sciences will result from them, i.e., for the number and the kind of factors which have to be established as independent variables in the treatment of given problems. For this, the universal schema developed in the last section of our Part One will prove a useful guide.

We now wish to test how an orientation according to this scheme would take shape in terms of a series of special questions of methodological importance, which in part have been closely linked in the history of doctrines with the postulate of methodological purity:

In the second section of this part of the work we already have given, in connection with our universal schema, the following schema of the degrees of freedom of interpretation:

1. What facts are interpreted?
2. What facts may be drawn upon as aids for the interpretation and what is their 'weight'?
3. What schemata of interpretation are to be used, and from what experience do they originate?
4. Under what conditions is the interpretation considered to be completed?
5. What status is accorded to the result of the interpretation?

This division can also serve as basis for further systematic classifications.

Thus with respect to the *first* point, interpretations can be divided according to the kind and quantity of the facts to be interpreted (of the material), where care must be taken to see to what extent a classification, apparently carried out according to purely descriptive moments, is already *burdened with interpretation*. Such a quasi-descriptive division is, for example, one into *indications* [*Anzeichen*] (which are not signs) and *signs* [*Zeichen*]. One of the methodologically important divisions of signs (but one which permits gradual transitions), further, is that between *conventional* and *non-conventional* signs. In the interpretation of the former, knowledge of

[222] Compare, for instance, H. Hartmann, 'Psychiatrische Zwillingsstudien', *Jahrbücher für Psychiatrie und Neurologie*, vol. 50 and 51 with numerous references to the literature.

[223] As important attempts in this direction for instance the works of Le Bon, Psychologie der Massen (trans. From the French) 2nd. Ed. 1923 and Freud, '*Massenpsychologie und Ich –Analyse*' (*Group psychology and Ego-Analysis*) 2nd ed. Vienna 1923, must be considered of value. See also R. Wälder, *L'esprit, l'ethique et la guerre*, Institut international de cooperation intellectuelle, Paris.

the communicative purpose linked to using the sign by the communicator is required to a much lesser degree than in the interpretation of the latter.

As far as the *second* and the *third* points are concerned, which relate to the sources of the interpretation – where the other *facts* drawn on for the interpretation of the material can be called sources in the narrower sense, and these same facts together with the *schemata of interpretation* can be called sources in a broader sense – here too a division according to kind and quality will have to be made. The more precise choice of the reasons for classification will, to a great degree, have to be determined by the goals of methodological analysis in each case.

In this connection a remark about the relation of '*material*' and '*sources*' is required. Material and sources can only be sharply distinguished thematically but not with respect to their heuristic function within the framework of a given thematics. For considered from the viewpoint of the choice of topic, the 'material' is defined as 'that which is to be interpreted', and thereby delimited *vis-a-vis* 'the sources that are not material'. Functionally considered, however, as a rule the individual parts of the material can be used as aids for their mutual interpretation, and thus in this respect cannot be distinguished from the sources that are not material. This consideration makes it understandable why those two terms are so rarely properly distinguished from each other.

The principle of the *degree of closure* [completion] proves to be especially important methodologically as a principle of classification – in relation to the material as well as the sources. To begin with, as far as the question of the closure of the material is concerned, the task may be set of interpreting material that is unambiguously specified historically (for example, a specific legal text); or all facts of a certain kind – even those to be discovered in the future – (for example, Etruscan grave inscriptions); or, finally, all those facts which promise insight into 'historically essential' events (thus all cultural documents relating to the Etruscans). An analogous division can be carried out for the means used to interpret the 'sources', and the combination of these two divisions leads to the customary distinctions in the theory of science. Thus interpretation in the *dogmatic* disciplines is characterized by the circumstance that the *material* is regarded as *completely self-contained* at every point of time. Whether it is a question of the exegesis of a biblical text, or a codification of law, or the interpretation of the grammatical rules of a certain language, the material is always regarded as completely given. Here, incidentally, lies, one of the chief roots of the doctrine of '*objective spirit*', which comes about because the *meaning content*, i.e., the result of interpretation, is attributed as a *property* to the facts to be interpreted, i.e., to be comprehended as symptoms for other facts.

Now with respect to *sources* there may be a rule that these are to remain confined to the domain of the material. This hold true, for example, for the '*objective interpretation*' of a legal order, which does not permit, as means of interpretation, either reports on motives or any other knowledge concerning the will of the legislator, insofar as it is not reflected in the laws themselves.[224] On the other hand, '*subjective*' interpretation of law is characterized, among other things, allowing other sources as

[224] The 'objectivation' results in anonymity.

well. But as a rule it will limit these in such a way that only those expressions of will on the part of the legislative organs are taken account of which are directly related to the content of the legal sentences to be interpreted (as for example reports of motives), But it is also entirely conceivable, that for the interpretation of the meaning of a law, which is comprehended as the expression of a certain will of the legislator, all facts are adduced which can be assumed to be symptoms of the latter. To be sure, such an approach corresponds less to the jurist's attitude, than to the historian's. The latter will be inclined to draw on every fact that can shed any light at all on his investigation, and in the hands of great historians a seemingly irrelevant state of affairs frequently proves to be an important symptom for the relationships under investigation. This is indeed the reason for the 'hunger for material' of great historians, which thus is by no means to be declared as 'pedantry', as superficial critics at times assume.

In any case, even among historians a considerable diversity of views with respect to the selection of materials and sources has emerged. They find expression above all in the alternatives '*political history or cultural history*'[225] and '*history of problems or intellectual history*'.[226] As far as an evaluation of the first controversy is concerned, superficial consideration would incline us toward the assertion, that we are confronted here with a pure 'question of taste'; the historian who has more inclination and talent for one of the two modes of inquiry should turn toward that one, and let other scholars whose condition is the opposite, work in their manner. However, the state of the problem is not really that simple, even if 'subjective moments' of various kinds come into the matter, as they do in most other controversial questions of procedure. For the actual significance of the question at issue is whether *general* historiography should place political, or, in a narrower sense, cultural facts in the foreground. For it is necessary to have some guideline – not only for the sake of description but also for material reasons – in grouping the events to be represented, and chronological ordering alone cannot furnish such a guideline. Consequently the task arises of finding certain *dominant features of life style* which would furnish the most universal schema of interpretation (even if not a complete one) for the *historically essential* behavior of the human beings within a certain circle at a certain epoch. Thus the question arises, to what extent (or also, under what circumstances) such dominant features of action are to be found in political goals or in cultural goals in a narrow sense. It is advisable, as soon as an orientation has been established in this respect, to make the *dominant sphere of the material the main topic* and for the time-being – though by no means exclusively – to incorporate the other domains of action from the viewpoint of their dependence of the constellations to be found in this dominant sphere. (Thus, for instance, a history of Italy taking its orientation from political events would presumably seek to relate the rise of Renaissance art to the multitude of small rival city-states.) To be sure, the decision to make the political

[225] See for a first overview the presentation in Rothacker, 'Logik und Systematik der Geisteswissenschaften' in *Handbuch der Philosophie*, Munich 1928.

[226] See note 225.

the main topic by no means entails in principle that the domains of life, lying beyond the political, will have to be interpreted solely in terms of the political, but in scientific practice these two aspects will generally be joined.

The controversy *the method of the 'history of problems' or the method of the 'human sciences'*, concerns the decision whether the sources for the interpretation of cultural documents (especially works of art) are to be confined to the *domain of the material to be interpreted*. Gothic architecture, for instance, runs the argument of the proponents of the method of the history of problems, is in the first place *architecture*; we must therefore proceed from general questions of architecture, and especially its technical problems, in order then to ask from what circumstances – above all knowledge and special skills and the goals, conditioned in part by these master architects and their clients – the specific elaboration of the general idea of architecture in the epoch in question is to be derived. Against this the proponents of the method of the human sciences conceive the nature of the problem in the following manner: Gothic architecture is a *creation of* '*Gothic man*.' In order to understand it we must therefore above all understand Gothic man. Therefore, we must not stop at the specific manifestations of his nature as they are displayed in architecture, but must in principle draw on accessible manifestations of every kind, in order to penetrate to the core of his personality, which will form the point of departure for the deeper understanding of all the expressions of his life.[227]

Finally, in an especially important example of the middle level of closure, or also openness of the sources, the so-called 'comparative method' must be mentioned, insofar as the comparison is only carried out within a uniform field of study (for example, comparative linguistics). Here research has had tremendous successes – especially during the last century and a half – and has developed procedures of extraordinary refinement.[228] A typological analysis of inductive thought in the human sciences (for which this book seeks to lay out a path) will find a wealth of material here.

From the above considerations we can already gather that it is possible to provide a clear classification of procedures in the social sciences [geisteswissenschaftlicher Verfahren] from relatively simple viewpoints. However, in this connection we must point to the not inessential complication that schemata of interpretation in turn point back to facts – as their inductive base. Therefore, not infrequently, a wealth of factual experience is contained in them as implicit presuppositions, to which, due to methodological prescription, explicit reference may not be made. (We can recognize quite readily here a specifically methodological variant of the problems of knowledge without presuppositions treated in the first section of our Part One.) Thus every

[227] Compare for instance for the field of art history, Dvorak, *Kunstgeschichte als Geistesgeschichte* (*Art History as intellectual history*), Munich 1924 and Worringer, *Formprobleme der Gotik* (*Problems of Form of the Gothic*), Munich 1922.

[228] From these investigations those which aim at the comparison of the methods of various sciences must certainly be distinguished. Among the latter the book of P. Oppenheim, *Die natürliche Ordnung der Wissenschaften* (*The Natural Order of the Sciences*, Jena 1926) should be pointed out.

objective interpretation of signs is already subjective in a certain sense, due to the fact that it is an 'interpretation of signs'; for signs are means of communication, and the interpretation of signs therefore includes (contains) the assumption of a communicative purpose on the part of the person communicating the signs. The difference in procedure between objective and subjective interpretations of signs, however, is the following: in objective interpretation of signs, general semiological knowledge supplemented by a general characterization of the time and the milieu from which the signs originate function *exclusively* as schemata of interpretation. Subjective interpretation also utilizes such insights, but in addition supplements them by means of data of a *lesser degree of anonymity*.

Now whether we decide in favor of the one or the other method will depend upon the aim of the interpretation, and the conditions under which it will be regarded as complete (successful). Thus for instance the procedure of the subjective interpretation of a legal text will frequently be indicated when the endeavor of the interpretation extends beyond the comprehension of the ostensive *meaning* of the publication to the comprehension of the *purpose* of the publication. Such a *goal for inquiry* may in turn be *suggested by practical goals*, but must in principle be sharply distinguished from these. That the pragmatist theory of science has most frequently failed to take this into consideration has been a serious obstacle for its efforts to grasp the full nature of methodological problems. To characterize a method as 'suitable for given practical ends' does not state anything concerning its internal distinctness.

Accordingly we will have to conceive, as one of the most important maxims for the methodology of the social sciences, that the theoretical content of a method must be carefully distinguished from its practical significance. The relational character of the latter with respect to *defined* practical goals is in turn to be determined with full clarity. Applied to the problems of sign interpretation, this states that problems of interpretation as such must be isolated from the questions, by what practical goals the one or the other kind of interpretation is suggested. Such isolation also forms the necessary precondition for determining the degree of 'objectivity'; ('rational status') of the results of an interpretation.

Our investigations have taught us to correctly evaluate this status, and have permitted us to recognize that between the Scylla of an uncritical objectivism and the Charybdis of an uncritical subjectivism there lies a free passage for scientific research, which neither surrenders to hasty illusions, nor to hasty disillusionment, concerning the limits to knowledge in the social sciences. These investigations have also revealed where the chief sources of the erroneous conceptions that are to be avoided are to be sought, and thus they have made possible taking critical positions also toward those conflicting doctrines in the *Methodenstreit* which we have not treated more fully before.

One of the chief desiderata for social scientific inquiry, however, would be to require each scientist who deals with more general problems, and takes a certain position, differing from other conceptions with respect to the choice of method, write a *report on his motivation* for this choice. Such a report would have to include, in the first place, an immanent characterization of his method, and the kind and extent of the deviation from competing methods would have to emerge; secondly, it would have to emerge from the report what considerations had persuaded the researcher to prefer this method. Reports concerning motivation of this kind, which, to be sure, would frequently be possible only at a relatively late phase of the intellectual process, would not only be an important aid to help others evaluate a doctrine, but could also prove an excellent means for controlling one's own work. Their arrangement could be made uniform to a great extent, if they were structured according to a universal schema, such as we developed above. By this, the true divergences between the various methodological tendencies could be clearly revealed, and well-founded statements could be made concerning the aspects of determining decisions in favor of one or the other method. To be sure, controversies about the appropriate choice of research methods would not be excluded altogether this way as assumptions about the success to be expected from a certain procedure could diverge greatly; but to permanently overcome this kind of controversy about method should not, after all, represent the goal of well-considered scientific endeavor. What ought to be overcome, and can be overcome, is the exaggeration of methodological conflict due to pseudo-scientific argumentation.

Now we have concluded our investigations of questions of principle in the methodology of the social sciences. The two sections which follow contain applications of the results of our reflections to vehemently debated problems in economic theory, and in legal theory; their main aim is to show how such applications can be carried out. Still these are so numerous, and the fields of study among which they are distinguished so varied, that one individual is hardly capable of independently achieving a complete overview. Here planned intellectual collaboration would bear rich fruits. It is intensely to be desired that the next years will bring this about.

8. Remarks on the Methodological Controversy [Methodenstreit] on the Theory of Marginal Utility

It is characteristic of the methodological conflict concerning the marginal utility school, which has continued for five decades with undiminished vehemence, that not only do the *theses themselves* confront each other with apparently irreconcilable opposition, but even the conceptions about the external *success* of the embattled doctrines are just as far apart. On the one side, Gottl and other scholars speak of the 'dying value theory', while on the other side, Schumpeter, in unison with a considerable number of eminent economists declares that "the theory of marginal utility is not one of many competing doctrines, but at the moment it simply is the only theory."

This initially astonishing fact already leads us to suspect that here too the arguments and counter-arguments are directed at cross-purposes in many respects, because they do not point directly toward the procedural meaning of the controversial doctrines but instead take their orientation from more or less *inadequate interpretations* of this procedure, by the originators of the marginal utility theory. Now modern marginal utility theory has made great advances in sharpening the formulations that are mainly involved here, so that today the attacks of the opponents are mostly directed against already abandoned formulations. But the path of clarification is by no means completed, and the following considerations are conceived as further steps along this path.[229]

In so doing we wish to devote special attention to the main parts of the theory, and especially to examine the following aspects:

1. The relation between the principle of marginal utility on the one hand, and the theory of needs, and the theory of goods, on the other, in the theory of an economy without exchange.
2. The nature of the application of statements and concepts derived from the analysis of an economy without exchange to an exchange economy.
3. The nature of the application of statements and concepts derived from the analysis of a barter economy without money transactions to a money economy.

In these considerations we will seek to isolate the pure procedural core of the theory from its inadequate secondary conceptions, in order to show subsequently that most of the arguments against the theory did not penetrate to this core at all, and therefore become groundless as soon as the desired stage of clarification has been reached.

Let us take our departure from our basic methodological schema. Accordingly, in order to characterize the theory of marginal utility, we have to clarify the following points:

1. its topic (research goal, or direction of research);
2. the kind of facts on which it bases itself;
3. the general assumptions by means of which it establishes relations among the facts;
4. the significance and status of its results.

The goal of the theory – and about this there is now essential agreement – is *to explain the relations of exchange between goods*, and especially to explain the prices of goods, where the concepts of 'goods' and 'exchange' are to be regarded as fundamentally clarified. By a good is understood the alternative and also cumulative chances for disposition (possibilities for utilization), and the use of the good lies in the realization of one of these opportunities. One speaks of an exchange between two exchange partners, A and B, when A turns over to B certain goods g_i under the conditions that he receives from B certain other goods g_k.[230]

---

[229] As there is by no means unanimity among the theorists of marginal utility about all the points to be discussed in what follows, therefore certain critical remarks will have in part no validity. For the presentation of a critical history of doctrines of marginal utility theory, this circumstance would be most significant; for the presentation within the framework of a doctrine of method which must disregard the details of doctrinal history, it is not.

[230] See for this Strigl, *Die Oekonomischen Kategorien und die Organisation der Wirtschaft,* Jena 1923; furthermore Kaufmann, 'Die ökonomischen Grundbegriffe'. *Zeitschrift für Volkswirtschaft und Sozialpolitik,* 1923; 'Logik und Wirtschaftswissenschaft', *Archiv für Sozialwissenschaft*, vol. 54, 1925; ,On the subject-matter and method of economic science', *Economica*, 1933, pp. 381–401.

With respect to the concept of explanation we point to our reflections concerning the 'achievement' demanded of the explanation.[231] We have observed that this achievement ultimately will be evaluated according to its confirmation through the prediction of future events, or the reconstruction of past events, and that the assumption of a single explanatory direction, prescribed unequivocally by the things themselves, which is to lead to the *causa efficiens*, is untenable.

The goal of inquiry just described is that of economic theory as such[232]; the specific feature of the *theory of subjective value* in general and *marginal utility theory* in particular, lies in indicating a specific *path* toward this goal. Its basic conception is the following: the essential problems of an exchange economy can already be demonstrated in a society without exchange; and in the same way the manner of their treatment can be developed under these simplified conditions. The economic subject knows itself to be dependent on *needs of varying intensity* (*urgency*) and the *satisfaction* of these appears as *goals*, the *ranking* of which is prescribed by this *order of intensity*. Furthermore the subject knows the *means* suitable for the satisfaction of these goals: *goods*, and it is clear to him that these are not at his disposal in *sufficient quantity* in order to satisfy all his needs. Therefore he sees himself faced by the necessity to manage his economic affairs in a planned manner, which includes that, in the first place, he uses the goods in his possession in the *most efficient manner* (with respect to the given goals); secondly, that when choosing whether he should *deny himself* good g_1 for the sake of good g_2, his decision will be based on the result of the consideration as to which of the two goods he can expect a *higher utility* from – presupposing the most efficient utilization. Subjective value theory claims that this train of thought must be developed to the level of full *rationality* and its consequences for the social economy pursued; this way one would obtain the best possible insight into the lawfulness of economic events, which by and large[233] take their course as though the economic subject acted with consciousness of his goals, and in a manner appropriate to them.

Proceeding to a precise analysis of this method which up to this point we have only sketched roughly, we want to clarify, to begin with, the role *attributed to* the assumption of the relative *scarcity of goods* (*necessities of life*) with respect to the extent of needs. The assumption is to render it comprehensible that the economic subject is forced to make planned provision for the satisfaction of his needs, and thus to justify a method which seeks to understand economic processes as conditioned by plans. This general concept of *scarcity* – to be distinguished carefully from that of the *comparative rarity* of particular goods – thus is not a concept immanent to the analysis of economic plans. In Schütz's mode of expression mentioned above we would have to say that scarcity is understood as the '*because-motive*' *of planning*.

---

[231] See above, Part One, section "Fact and Law"; Part Two, section "Value Problem in the Social Sciences"; Part Two, "Social Laws and Ideal Types"; Part Two, section "The Way to Overcome the *Methodenstreit*".

[232] See the works listed in note 230.

[233] If we incorporated the idealization into the definition of economic activity, and Mises and the scholars close to him do so, then the limitation 'by and large' disappears.

The same adds for the *concept of need* – and this is of still greater importance for the understanding of the subjective theory of value. Needs, with their intensities, do not enter into the economic plans underlying the analysis; instead they – or more precisely the expectations that they will arise – form the 'because-motives' of goals or of the stipulation of hierarchies of goals. Accordingly, in all investigations directed toward the utilization and selection of goods, in which a given hierarchy of given goals is *presupposed*, the concept of need can be neglected. This holds most importantly for those considerations which lead to the formation of the concept of *marginal utility*, and to the setting up of the *principle of marginal utility*. It is to these we now turn.

The question under discussion is: According to which principles will an economic subject, acting in a purposively rational manner and clearly aware of his goals and their ranking, and about the possible uses of the goods required for achieving these goals, decide whether he would – of all the units of goods in his stock – rather dispose of a unit of the kind G_i – let us call this unit g_i – or a unit of the kind G_i – let us call it g_j. If we assume initially that all units of goods belong to different categories, then the argument will take the following form: Given n + 1 goals Z_o, Z_1, … Z_i, … Z_j …, Z_{n-1}, Z_n with the ranking $Z_o > Z_1$ …>… Z_n… (read: Z_o is preferred to Z_1, and so forth). Coordinated with them are n units of goods g_1, g_2, … g_I … g_j … g_{n-1}, g_n in the following manner: If the economic subject has all these units of goods at his disposal he can attain Z_o. If he has all the goods at his disposal with exception of g_1 he can no longer attain Z_o but can attain Z_1. If he has all these goods except g_2 at his disposal then he can only attain Z_2. Then for *any arbitrary unit of goods gi of the store, the utility depending upon it* is: "The attainment of Z_o instead of Z to the ranking of goals $Z_o > Z_I$ …>… $Z_{n-1} > Z_n$ corresponds the ordering of utility (Z_o instead of Z_n) > (Z_o instead of Z_{n-1}) …>… (Z_o instead of Z_2) > (Z_o instead of Z_1) and from this results a ranking of goods for the economic plan that is appropriate to the goals $Z_n > Z_{n-1}$ …>…$Z_2 > Z$. That means: the ranking (order of preference) of the units of goods corresponds to the ranking of the utilities dependent upon them, which latter are in turn determined by the ranking of the goals and the possible uses of the units of goods. Presupposed here is that, for every combination of goods the *optimal total disposition* (with respect to the ranking of goals) will result. That g_I stands higher in the ranking than g_k means nothing else, *according to the definition* than that the economic subjects intends – *ceteris paribus* – rather to forego g_k than g_i and thus would give up g_k for g_i but not g_i for g_k. The utility depending on a unit of goods under the conditions characterized is called its *marginal utility*, and the statement that every unit of goods is valued according to its marginal utility is called the *principle of marginal utility* (*in the broader sense*).[234] Now we want to consider the case where the disposable stock contains several – say three – units of the same category of goods – let us call them g_{i1}, g_{i2}, g_{i3} – each of which is used in a different way. As, according to our presupposition, each of three units can be substituted for

[234] Compare for instance Paul Rosenstein – Rodan, article 'Grenznutzen' (Marginal Utility) in the *Handbuch der Staatswissenschaften*, 4th ed. (1930), p. 1192, "It is not difficult to recognize that the utility directly dependent on a good is identical with its marginal utility."

each of the two others for any arbitrary utilization, then – under the presupposition that the utilization of units belonging to other kinds of goods than G_i will not be changed by the elimination of a g_i – the utility dependent of g_i will be the least important among the three utilizations, and therefore to each individual one of the three units the same relative rank will be attributed which corresponds to the least important utilization (on the basis of the hierarchy of goals) (*principle of marginal utility in the narrower sense*).

Thus quite generally the principle of marginal utility states that under the presupposition of a systematic purposively rational economic plan, the relative rank of a unit of goods within the rank-order of the units of goods at disposition within the economic plan is determined by the rank of the utilization, which would be deleted with the loss of this good. This statement seems to be less simple than the customary formulation of the principle of marginal utility, but we shall soon recognize that the latter easily leads to misleading interpretations.

However, first we have to examine the specific *validity* of the principle of marginal utility. In this examination, our statement, which apparently is a hypothetical judgment about the course of human actions – (or about its unequivocally determining plan) – will reveal itself to be a *descriptive analysis* (*logical dissection*) *of the definition of* '*instrumental rational evaluation of goods in the case of a given hierarchy of goals*'. For if we were really faced with a hypothetical judgment of the form 'if the economic subject acts in a purposive rational manner, then it acts in this and that way', then it would have to be possible to determine the antecedent and the consequence in a manner logically independent of each other; therefore the 'purposive [instrumental] rational actions of the economic subject' would have to be defined in such a way that this concept would not contain within itself the meaning that is expressed by the principle of marginal utility. In that case a *falsification* of the principle of marginal utility would also be possible. But the statement is obviously not meant in this way. For the principle of marginal utility is 'derived' from the concept of rational action; and that means that it is already implicitly contained in it. Thus *nothing is posited* in our statement and therefore it is not subject to empirical confirmation. In spite of that – as we have shown above – a knowledge content has to be accorded to it, namely the knowledge content contained *in its presuppositions. Thus the principle of marginal utility clarifies the concept of purposive rational evaluation of a unit of goods within a certain stock, but it does not actually state anything* [new] *about it. It is result of rational reconstruction, but not an empirical assumption*.

However, in spite of that it is claimed by the marginal utility school that the principle of marginal utility is suitable for the explanation of *real economic* actions of human beings, because to a large extent they actually behave according to this principle: however, this assertion is of an empirical nature and subject to test. Yet in this case we have to guard most carefully against the misunderstanding that this test could bring about a confirmation or refutation of the principle of marginal utility understood in the sense just characterized. It would only decide whether, or to what extent, human beings act in a *purposive rational* manner, but not whether in the case of purposive rational action they act according to the principle of marginal utility; what is involved here is *not* empirical *coexistence*, but conceptual *identity*. Thus we have the result: either, the principle of marginal utility can be comprehended as a conceptual analysis; in that case it is 'irrefutable' but only because it does not contain any assertion; or it can be understood as an assertion that to a large extent the *economic behavior* of human beings can be *explained* with the help of the construction of types of purposive rational action (ideal type). The latter assertion, however, which in typical manner will appear in the form of a heuristic postulate (procedural prescription), that this procedure (which promises success) should actually be practiced is subject to confirmation, and it can be formulated in the following manner: If you want to find out whether a certain economic subject E_s would be more readily prepared to forego a unit of goods g_1 than a unit of goods g_2, then try to find out the goals E_s wishes to attain by the utilization of all the goods of his stock and the rank-order of these goals. You must further clarify the best utilization with respect to the hierarchy of goals, and in doing so you must consider that every displacement in the stock of goods can involve corresponding changes in the overall disposition, and assume that E_s will decide on the choice between two units of goods in such a way as though he had a clear grasp of all these circumstances and took his orientation from them. Then you will determine his economic behavior correctly.

Now we wish to remove some remaining obscurities and misunderstandings linked directly to the principle of marginal utility.

The first point concerns the customary coupling of the principle of marginal utility with *Gossen's law of the satiation of needs*.[235] What is sought is the justification for the proposition that within the necessarily limited timespan of an economic plan there are uses of identical utility only for a limited number of goods of the same category. Thus the marginal utility of a unit declines when the number of units at disposal in the economic plan exceeds a certain limit. This '*law of decreasing marginal utility*' which is most frequently confused with the principle of marginal utility itself is obviously an experiential statement, indeed an empirical statement of extraordinary importance for research in economic science, and is supported by the introspectively demonstrable fact of the satiation of needs. But for the 'derivation' of the *principle* of marginal utility, Gossen's law is not required.

A further important observation is that with the determination of the marginal utility of the total units of a stock all the questions concerning the *comparative rank of these units of goods* have by no means been decided. It only determines, with respect to two units of goods, whether the one would be foregone in favor of

[235]The original version in Gossen (1854) reads: "The magnitude of one and the same enjoyment will decrease continually if we go on uninterruptedly preparing this enjoyment, until at last satiation sets in." Sharpened versions were carried out by Wieser in his *Theorie der gesellschaftlichen Wirtschaft* (Theory of Social Economy) and H. Mayer, "Untersuchungen zu dem Grundgesetz der wirtschaftlichen Wertrechnung" (Investigation of the fundamental Law of the economic computation of Value), *Zeitschrift für Volkswirtschaft*, vol. II 1922. See also the interesting remarks concerning this law in Mises, 'Vom Weg der subjektivistischen Wertlehre' (On the way of the subjectivistic value-theory) *in Grundprobleme der Nationalökonomie*, Jena 1933, pp. 137–155, p. 144 f.

the other or vice versa. But from the order of preference $g_n > g_{n-1} \ldots > \ldots g_2 > g_1$ one cannot derive whether, for example, the unit g_{n-1} is foregone in favor of the two units of goods g_2 and g_1, or whether perhaps the reverse may be true. Indeed, one cannot even derive from it that g_1 *and* g_2 are foregone rather than g_n *and* g_{n-1}, for the loss of two units of goods, can in principle displace the total disposition of goods in such a way that the rank-order will be changed completely. Thus in order to determine the acts of choice possible with respect to the goods of the stock, one would have to determine the marginal utility (mediatively dependent utility) for each partial stock separately, and then set up a rank-order that would encompass all these marginal utilities.

The great significance of this consideration for the understanding of the marginal utility theory lies above all in that by demonstrating the existing complications it works against an *exaggerated interpretation* of the achievements of marginal utility computations. For it can readily be seen that the procedure of economic science, as intended by marginal utility theory, by no means aims at the construction of exceedingly complicated scales of utility of this kind. In fact, in no single problem of economic science is the task set to work out completely this schema of dispositional possibilities, and the hierarchies belonging to them, by incorporating empirical data in order to explain concrete economic behavior in this way. Instead, its procedural function is to give a *survey of the factors that have to be considered* and certain indications about the manner in which they have to be considered. Thus one of the strongest impulses which marginal utility theory has given to scientific research in economics, is in pointing to the *dependence of the economic subject's evaluation* of a unit of goods on the kind and size of his *total holdings of goods*, an indication which the theory of needs and the theory of goods complete by an analysis of the phenomena of *psychological and/or technical complementarity*. But *to what extent* such influences are empirically significant in each case has to be established in terms of the actual state of the problem; it is from the latter that the investigator has to conclude for what phase of his analyses he should orient his work by idealizing theoretical schema. For clarification one might draw on the comparison – only a loose one, to be sure – with the mathematical physicist, who could work out certain magnitudes appearing in his calculations, for instance the number π to the thousandth decimal but who knows very well that only exactness to, say, the thirtieth decimal would have any scientific significance. This divergence, however, brings with it the danger of ambiguity in the formulations of economic science; for such ambiguity will arise when the question remains open, whether a certain assertion relates to the 'purely theoretical' level (the idealizing schema) or to simplified presuppositions closer to reality.[236]

We wish to summarize the main result of our investigations up to this point: The considerations leading to the principle of marginal utility can be carried out independent of any reference to the concept of needs. The principle of marginal utility

[236] A good example is offered by the comparison of Böhm-Bawerk's method and that of Wieser in determining the value of a store of goods of the same kind, it the marginal utilities of the individual units of this store are given.

as such is the result of the logical analysis of the concept 'purposive rational choice among goods' and therefore does *not* contain any *positing* (assumption) subject to confirmation. However, the procedural prescription which states that one ought to carry out computations of marginal utility in order to gain indications for the exchange behavior of economic subjects, can be confirmed (in varying degrees) or not confirmed.

Now it is fully justified to claim with respect to *this* part of the theory that it is accepted – more or less consciously – by the overwhelming majority of the investigators dealing with the general problems of economic science. For most investigations in economics are based on the construction of economic subjects acting with purposive [instrumental] rationality; still, in the theory of marginal utility this construction has probably been pursued to its conclusions with greater intellectual energy than in any other doctrine.

But also with respect to the less abstract parts of the doctrine, to which we now wish to turn, far-reaching agreement with apparently contradictory methods can be attained, once certain inappropriate interpretations of the theory – whether on the part of its adherents or its opponents – have been removed. This is especially true for *concept formation* in economics, as will be briefly illustrated by the concept of *economic values*, the central concept of the *Methodenstreit*. Here one must above all understand that 'value' in the theory of a society without exchange and 'exchange value' from which the road then leads to 'price' are disparate concepts. In order to realize this, we have to first show an ambiguity which appears linked to the value concept in the theory of a society without exchange. For on the one hand the words 'value' and 'marginal utility' are declared to be synonyms, but on the other hand, by the value of a unit of goods is understood the place it occupies in the ranking of the units of goods at disposal in the economic plan.[237] One has to be clear about the difference: the marginal utility of a unit of goods is defined as the utility that is to be expected of it in the case of the optimal utilization of the total stock and therefore the rank-order of the marginal utilities is *by definition* identical with the rank-order of the utilizations which, for purposive rational behavior, follows from the hierarchy of the goals. Furthermore, an unambiguous criterion is established for the economic subject acting with purposive rationality through this rank-order, as to whether given two units of goods g_1 and g_2 he will have to forego g_1 for the sake of g_2, or g_2 for the sake of g_1 in a given case, or whether he can behave indifferently with respect to these two units. But the circumstance that one of the two rank-orders is *determined* by the other must not lead to regarding them as identical. One can also demonstrate this difference in terms of the twofold possibility for interpreting the term 'use value'. In the first case, the use value of a unit of goods is understood as identical with the utility to be expected in its use; in the second case the 'use value' is defined as 'the significance which the economic subject attaches to the unit of goods with respect to this utility';[238] the 'significance', however, is obviously nothing else

[237] See Kaufmann 'On the subject-matter'.

[238] Compare "Kapital und Kapitalzins" (Capital and the interest on capital), op. cit. [see note 242], vol. I, p. 223: "Formally defined the value is thus the significance which a good, or complex of goods, has for the purposes of the welfare of the subject."

than the relative position of the unit within the rank-order of the units of goods (ranking of foregoing). Now it is easy to see that the conceptual transition leading from 'use value' to 'exchange value' and therewith to 'price' does so by way of the second meaning – the comparative evaluation of the unit of goods on the part of the person making the demand is supposed to be decisive for the highest offer – but 'value' understood in this way is only an *index of ranking*; it does not signify a specific utility dependent on the goods. Therefore the concept of the exchange value of a unit of goods, or also, its price (the exchange value represented in terms of money), cannot belong to the same category as the use value; for the two concepts possess a different *syntax*.[239] This state of affairs is obscured by the identification of the term 'value' ('use value') with the term 'marginal utility', for the concept of the utility of a unit of goods – within the framework of a stock with a given ranking of goals and optimal utilization of the total stock for the realization of these goals – has an isolated unequivocal meaning; but the aspect of choice understood as alternative relinquishment, which comprises social exchange as well, is not contained in this meaning, so that there is no logical bridge that leads from it to 'exchange value'. Taking this aspect into consideration, incidentally, I also consider that mode of expression more efficacious according to which there is not, as it is so frequently formulated, an identity between hierarchy of utility and hierarchy of value, but only a one-to-one correlation. But this question is of secondary importance. What is very important, however, is to avoid confusion of the two concepts of value.

We shall encounter a further obscurity, connected with the concept of value that is related to its *eudaemonistic interpretation* and also plays a role in the problems of *measuring value*, as we now turn toward the theory of needs and goods.

It is the function of the theory of *needs* and their intensity, within the framework of the theory of marginal utility to furnish points of reference for the determination of those goals of the economic subject – and the ranking of these goals –which can be regarded as *grounds for determining the utilization of, and the ranking among, the goods*. To this heuristic function corresponds the designation of need as a desire evoked by certain circumstances, i.e., a more or less clearly conscious setting of goals, as in Tiburtius's[240] definition, according to which need is "a desire growing out of the feeling or conception of a (psychic) disturbance of equilibrium and aiming at preservation or restoration of this equilibrium". Accordingly, in order to understand the theory of needs, one must ask oneself *in what manner* the goals essential for the economic subject's disposition of goods and their ranking is supposed to be determined by it.

Here we first have to take into consideration that in our reflections on the theory of rational action, no indication of any kind was given, as to what sort of goals are to be considered the ultimate goals (key goals) of the economic subject. Now it is in the decision of this question that the *eudaemonistic* coloration of marginal utility theory begins to emerge, according to which the ultimate goals of human beings

[239] The expression is to be understood in the sense in which it is used in Wittgenstein's *Tractatus* and Carnap's *Logical Syntax of Language*.

[240] 'Der Begriff des Bedürfnisses' The concept of need) *Jahrbuch für Nationaloekonomie,* III. Folge, Bd. 48.

always consist in *securing pleasure*, or also *avoiding pain* [*Unlust*]; and from the viewpoint of the theory of needs these goals represent the *satisfaction of needs*. These philosophical (pseudo-philosophical) background ideas, from which the theory can be separated without great difficulty, and indeed in part has been separated in the more recent work of the school, has had as a consequence not only innumerable superficial criticisms in which the marginal utility school was rejected on 'philosophical' ('ideological') grounds, but also pseudo-problems within the doctrine itself. The latter are rooted in the erroneous conception that all the conditions of man, toward the establishment or removal of which man's economic behaviour is ultimately directed, were from the outset arranged in such a way that the order of pleasure-pain resulted in a *natural ranking* among those conditions. The more primitive conception assigns to each of these conditions a certain *quantum* of pleasure or pain, the less primitive conceptions a certain *intensity* of pleasure or pain.

In tracing back the hierarchy of goals to the hierarchy of needs, one arrives at the following argument: The intensities of need are allegedly given, and through them one attains first the hierarchy of the ultimate goals of economic activity, conceived as satisfaction of needs. With the aid of this hierarchy, the utility ranking of the utilization of units of goods, in the case of optimal total disposition, are determined and finally the ranking of units of goods in '*internal exchange*'. Thus according to this line of argument, the guidelines for internal exchange are already unambiguously prescribed in the economic comprehension of the given situation of needs; in addition, only knowledge of the *technical* aspects is required in order to set up the corresponding plan.

In reflecting on our fundamental discussion of the concept of pleasure, we can comprehend now with little difficulty that we have here an especially striking example of the methodological error we have called the *dogma of preestablishment*. For the illusion is conveyed that the solution of the problem of making predictions about purposive rational human acts of preference is unequivocally determined by presupposing *grounds for the determination of preferences that are rooted in prior givens*.[241] But the situation in fact is that the assumption of a uniform quality of feelings which would permit the persuasive comparison of the intensities of psycho-physical states, even of the same human being, finds no support in experience. Through this it becomes explicable that seeking to maintain this assumption leads to a *petitio principii, to a reversal into a postulate* by declaring that one of two states as being the less pleasurable which is *deferred* in the choice between the two. In the same way in the theory of needs, among two needs one is frequently designated as the weaker (less important, less urgent), the satisfaction of which has been deferred in favor of the other. Thus the illusion is produced that acts of preference are explained by prior psychical givens, while actually only a *new term* for the decisions reached in acts of preference has been *introduced* by definition. But this misunderstanding becomes especially dangerous here, as in numerous similar cases, because in certain *partial domains* of the field under investigation, such levels (gradations) of intensity can actually be found. This is true,

[241] See above Part One section "The Concept of Value".

above all, for the feelings corresponding to the 'primitive' needs of food, drink and clothing, of hunger, thirst and cold. As furthermore the typical conditions are known under which these feelings grow stronger, or are weakened to extinction (satisfaction of needs), *types of urgency* can be constructed here, which can furnish important points of reference for the explanation of economic behavior. In the formation of types of this kind it is really not only observations of 'external' acts of preference – above all the behavior of persons on the market – that are drawn on as sources of knowledge, but also *analyses of motives*; so here 'going back behind the acts of preference' is heuristically justified. But the assumption that *all* the goals relevant for the shaping of the demand for goods – even if only of one and the same person – must analogously stand in a pre-established order of importance, is an untenable prejudice that leads to the confounding of authentic explanations with pseudo-explanations, that turn out to be definitions, as just described. We want to clarify this in terms of the concept of the '*least important need*', as it has been introduced in the customary formulation of the principle of marginal utility. Böhm-Bawerk formulates the principle of marginal utility in the following manner: "The magnitude of value of a good is measured according to the importance of that concrete need or partial need which is the least important among the needs covered by the disposable total stock of goods of this kind."[242] If one now wants to have this sentence understood as a statement about reality, then it must be possible to ascertain, independently of the valuation of goods, which need is the least important in a given situation. Under this presupposition the principal of marginal utility is a proposition which is subject to empirical confirmation, and thus can be falsified. However, if it is said that economic behaviour furnishes the ultimately decisive criterion for which need was least important, then the principle of marginal utility interpreted in this manner is indeed 'irrefutable', but only because it asserts nothing; for then it represents a definition of the concept of 'least important need'. The error of thought which emerges here (and the marginal utility school has not always managed to avoid it) consists in on the one hand operating with empirically established rankings of the importance of needs in the procedure itself, but on the other hand, regarding the principle as irrefutable which determines the 'magnitude of value', i.e., the relative economic rank of the unit of goods in question by means of the least important need. This error corresponds to the confusion of the '*material*' with the '*formal*' concept of need in such a manner that *introspective insights which can be gained predominantly with respect to certain physiologically conditioned desires are prematurely extended to the totality of the motives decisive for the formation of the demand for goods*.[243]

With that, the fundamental question concerning the character of the validity of the principle of marginal utility appears to be clarified, including those versions in which the concept of need figures. We now have to ask quite generally, just in what consists the achievement of the *theory of needs* within the framework of the theory

---

[242] Böhm-Bawerk, Kapital und Kapitalzins, 2. Abt. Theorie des Kapitals, 3rd ed. 1909 and 1912, p. 246.

[243] Kaufmann 'On the subject-matter'.

of marginal utility. With respect to this it is to be observed at the outset that the constitution of a specific theory of needs is objectively justified if, or insofar as, explanations of the formation of the demand for goods can be gained with the aid of systematically grouped psychological insights insofar as the observation of actual acts of choice (market statistics) does not constitute the sole basis for knowledge of the economically relevant setting of goals. Were the latter the case, an unnecessary duplication of the object of knowledge would be provided by the theory of needs. But it is not the case. Rather, investigations of the mutual dependency of needs (bound needs, complementary needs, supplementary needs) as well as certain 'formal properties' of needs or of particular groups of needs (e.g., divisibility, periodicity)[244] are actually autonomous sources of knowledge for the behavior of economic subjects. Whether the term 'need' is a fortunate choice for the designation of the entire range of economic 'because-motives' remains a comparatively secondary question.

The main application of the insights acquired in the theory of needs and the theory of goods lies in clearly presenting to the economic investigator which aspects he has to take into consideration (all things considered) when he undertakes to construct *ideal types* – more or less approaching reality – for economic behavior within a precisely characterized social economy. Thus certain relations between changes in the price of certain goods will only be understandable if and only if he properly considers the aspects called 'psychical complementarity' and 'technical complementarity'. But one must not interpret these insights in an exaggerated manner, as if rigorous laws of economic action were contained in them *in nuce*. Aside from the fact that the idea of an irrefutably valid experiential law is in itself a *contradictio in adjecto*, the elements which the analysis of an economy without exchange provides for the comprehension of the laws of an *economy of social exchange* are by no means adequate for the formulation of these laws, and – it is especially important to note this – the need for supplementing these does not only involve the requirement of *incorporating new data*, but *new general assumptions* must also be made. To be sure, these new assumptions are not unrelated to the results of the analysis of a society without exchange; they can be rendered comprehensible (suggested) through these but cannot be logically deduced from them. One must clearly realize the difference between this relationship and that which exists between the general and special laws of a systematically closed discipline of physics (one already axiomatized or ripe for axiomatization). In such a discipline, the latter can indeed be deduced from the former, i.e., can be gained by mere substitution; in our case, however, such a deduction cannot be carried out.

From this epistemological situation arises the task, for the theory of science, of working out the required *additional presuppositions* with explicit clarity and assigning them to their place within the system. When that has been done, the true procedural differences between the conflicting doctrines can be precisely stated, as we

---

[244] See Hans Mayer, *Untersuchung zu dem Grundgesetz der wirtschaftlichen Wertrechnung* (Investigations on the fundamental law of economic value computation).

wish to exemplify in terms of several further important points of controversy in the doctrinal history of [economic] theory.

But first we have to formulate correctly one further assumption belonging to the theory of a society without exchange, and characteristic of the subjective theory of value to a high degree; we then have to pursue its consequences. We have already pointed out that in the work of the marginal utility school dealing with questions of method (especially in the older publications) the question of the discovery of *key goals* for the explanation of economic behavior is supposed to be solved by the assumption that all human beings ultimately strive for pleasure. We have established that this assumption is untenable. But by this, the procedural prescription, which was to receive its philosophical justification from this assumption, is in no way affected as to its efficacy. This is the directive – also implicitly contained in the theory of needs – to *take one's departure from the goals linked to the consumption of goods*. Thus key goals are not characterized as 'ultimate goals' in the procedure of the subjective theory of value, but as goals which are attained directly by the *ultimate mode of utilization* (*use*, in the narrower sense, and *consumption*).

Thereby the first element of the series of *goods of various rank* – more correctly in the series of *utilizations of various degrees of indirectness* – is established, and the point of departure gained for the *theory of attribution* – the core (center piece) of the subjective theory of value. Certain fundamental traits for the theory of attribution can already be developed for a society without exchange – thus excluding the assumptions of goods changing hands and *a fortiori* that of uniform prices of goods; but its elaboration is only conceivable within the framework of a *theory of prices*. Here we have as guiding conception, in addition to the idea of rational action, the stipulation that the *valuation of goods depends on the estimate of their* (*intended*) *ultimate utilization*. From this presupposition, the principle of the theory of production can already be derived, i.e., that the *value of a production good* is determined by the *marginal utility of the marginal product*. The observations we have made about the principle of marginal utility find significant application in this connection. The procedural prescription, implicit in the statement just cited, indicates that in order to find a frame of reference for the valuation of a unit of goods that is to be ultimately utilized not in isolation, but in 'productive combinations', one must seek to ascertain in which combinations the goods will be procured for consumption – given optimal disposition with respect to a pre-established order of goals – and which rank is attributed to the lowest valued of these combinations in the economic subject's plan of consumption.

Here the actual problem of attribution arises, namely the task of making inferences from the evaluation of this combination for the evaluation of the element of goods contained in it, that is under investigation. It emerges, to begin with, that the determination of the marginal product, and its rank in the economic subject's plan of consumption, permits one to establish the *upper limit* for the evaluation of the unit of goods in question; its rank in the economic subject's total economic plan – encompassing production goods and consumption goods – cannot be higher than the rank of the marginal product. But for the rest, everything remains open for the time being, and the achievement of the theory of attribution only consists in giving a characterization of the factors to be taken into consideration in the treatment of this problem, and the manner of their influence on the computation, Thus above all the significance is pointed out which is to be attributed to the substitutability of the unit of goods, to be evaluated, by other units of the same or another category of goods; and the question of the remaining utilizations of all units of goods combined in the marginal product. But these important and successful analyses are frequently accompanied by exaggerated interpretations of the results actually attained, or attainable. To be sure the problems of attribution were carefully separated by the marginal utility school from the ethical-political postulates (especially to find a 'just' wage) related to them in the history of doctrines. However it too has linked exaggerated theoretical expectations to the method of attribution as it believed that from the value of the products given into consumption, the value of the production goods could be determined in an unequivocal manner, presupposing knowledge of all their 'technical properties'. Here too we are once more confronted with the *dogma of pre-establishment* in its pure form. Only in the most recent works of the school on the problem of attribution has the required reconsideration set in,[245] but it has not yet penetrated in a fully radical manner.

In this re-consideration it is above all important to see clearly that the problem of the division of the value of a combination of production goods into its productive elements can be posed only as a *problem of price*. The next step will then be formed by the insight that most frequently the unequivocal solution of problems of attribution can only be reached when a series of more or less casuistic additional assumptions is introduced. Finally it must be taken into consideration that by 'solution' only the solution for a certain purposive rational schema of ideal types is to the understood; to what extent this schema is then suited for the prediction of prices of production goods is again a different question.

In summary, we can characterize the essential achievement of the marginal utility school, within the framework of the problem of imputation and attribution as follows: by furnishing a precise specification of the concept 'goods of different orders', including the specification of consumption goods as 'goods of the first order', it provides a firm point of departure for research. Furthermore, through the procedural prescription that, for the determination of the prices of goods of higher order, the corresponding marginal products and their prices have to be established first, it secures a frame of reference for fixing the upper limits for the highest demand for units of these goods. Finally, it provides a combination of factors which are to be taken into consideration for the question, what further inferences can be drawn from the price of the marginal product for the prices of production goods. In order to do justice to this achievement, one has to consider that with this theory the possibility of fundamental theoretical contributions to these problems have probably been exhausted. But the correct evaluation of the limits of what is attainable also forbids an exaggerated valuation of what has been attained, and shows the untenable

[245] Compare F.A.Hayek, 'Bemerkungen zum Zurechnungsproblem' (Remarks on the problem of imputation), *Jahrbuch für Nationaloekonomie,* III. Folge, Bd. 69 and H. Mayer, article 'Zurechnung' (Imputation) and 'Produktion' in the *Handwörterbuch für Staatswissenschaften*, 4th ed.

character of the exaggerated expectation which have been linked to certain 'solutions of the problem of attribution' (for instance that of Wieser).[246] In the analysis of the treatment of the problem of attribution within the framework of the marginal utility school, one should recognize with special clarity the misunderstandings which arise from operating with aspects of an exchange economy within an investigation allegedly devoted purely to the economy of the individual.

This confusion is based, not least, on the false assumption we have criticized earlier, that the laws of an exchange economy would be contained in the laws of an economy without exchange in the same way that the concept of social exchange is contained in that of 'inner exchange', i.e., the concept of foregoing a commodity in favor of another commodity. In order to radically refute this assumption, it would be necessary to systematically explicate the additional assumptions on which the general theory of an exchange economy and the general theory of prices, and further the additional assumptions of the most important economic disciplines are based, e.g., the theory of cycles of the market economy. However we cannot present such an explication in the framework of this book, but have to restrict ourselves to a few exemplary remarks.

A characteristic of the subjective theory of value in general, and marginal utility theory in particular, consists in that it departs from the *economic plans of the consumer* in order to first determine the prices of consumer goods and then, with these as presuppositions, seek to establish the prices of goods of a higher order.[247] This direction of research has the appearance – and probably is so perceived by almost all marginal utility theorists – of being logically prescribed, unequivocally – by the assumption that the evaluation of the *ultimate* utilization of units of goods is decisive for their evaluation. However, this is by no means the case, as one will recognize at once when one considers that even when based on the principles of marginal utility theory, a theory of a monopolized or semi-monopolized economy will necessarily take its departure not from the plans of the consumers but from the plans of the monopolists. From the assumption that every economic subject takes the orientation for his economic plan from his consumption goals, and thus that, for example a merchant will seek to sell his commodities as early as possible, in order to be able to acquire as many consumer goods as possible from his returns, it cannot be concluded that the key for the price formation of a certain good is to be found in the economic plans of those economic subjects who consume this good – whether in isolation, or in technical combination with other goods, – or that thus consumers' demand is the dominant factor in the formation of prices.

But this methodological point of departure proves to be heuristically justified by another consideration, because for the most important consumer goods, quite useful *estimates of the quantitative demand of the economic subjects* can be established. Taking into consideration the purchasing power (income) of the presumptive

[246] Compare *Theorie der gesellschaftlichen Wirtschaft*, op. cit.

[247] Compare Mises, 'Bemerkungen zum Grundproblem der subjektivistischen Wertlehre' (Remarks on the fundamental problem of the subjectivistic value theory) in *Grundprobleme der Nationalökonomie*, pp. 165–169, p. 163 ff.

consumers, guidelines for price-setting on the part of the sellers and producers, who count on a certain stratum of buyers as their customers, can be derived. To be sure, the quantitative demand of the consumers adapts to price with a lesser or greater degree of elasticity, but the *estimate of the demand of the consumers* in any case provides a useful foundation for the further considerations of the part of the economic researcher. Only, this point of departure must not be declared the sole adequate one, with the justification that the price of goods has its *origin* in the significance that the economic subjects attribute to these goods as means of satisfying needs.

The *price theory* of the marginal utility school can be *derived* from their theory of value just as little as the key position of consumers' plans. Above all, the additional assumption of a *uniform market price* plays a role here. It is very enlightening to see clearly what procedural prescriptions are implied in the famous statement, that the price is determined by the highest amount which the *marginal buyer* is willing to pay for a good and the lowest amount which the *marginal seller* is willing to accept for this good.[248] This prescription aims at the construction of ideal typical economic plans on the part of marginal buyers and marginal sellers, where the principles developed in the theory of an economy without exchange are supposed to find application. But the important additional question is how the marginal buyer and the marginal seller are to be determined – obviously, one is dealing here, in general, not with particular persons to be designated individually, but with types, with *strata of competitive capacity*. Thus one might be tempted to ask, whether it is not precisely the price of a good that is decisive for which strata will possibly become marginal buyers, or marginal sellers; therefore, is one not moving in a circle when one seeks to base the determination of price on the latter facts? But the semblance of circularity only rises through the formulation, which requires revision, and it disappears if one keeps in view the methodological meaning of our procedural prescription. It signifies that the economic position of those persons who come under consideration as buyers or sellers of the goods in question, has to be examined. For the first selection of possible exchange partners, with respect to the buyer (especially insofar as he is the consumer of the goods in question) mainly the magnitude and urgency of the demand must be examined; with respect to the seller, however, the capacity must be determined. The examination of the economic position will consist, for consumer-buyers, in the first place, in an investigation of their income situation, and for sellers in an investigation of their proper costs for the goods. It is beyond doubt that these methodological prescriptions are efficacious to a high degree for a great number of investigations of price formation, but the guise in which they appear in most presentations of marginal utility theory has led all too readily to an exaggerated interpretation of their significance, and does not point out sufficiently the additional elements that are essential for the analysis of an exchange economy – as compared to the analysis of an economy without exchange.

The relation between the theory of an economy without exchange and the theory of an exchange economy is of different kind than that between the theory of a

[248] See for instance Böhm-Bawerk, *Kapital und Kapitalzins*, op.cit., p. 357 ff.

barter economy (without money) and the theory of a *money economy*. Here we are not confronted by complications, but rather the operation of the 'general denominator', money, simplifies economic life itself, and its interpretation by economic science.

For the theory of marginal utility, and for the subjective theory of value in general, the problem of the determination of price was summarized in the question of how prices could be derived from values. Now it emerges clearly from our preceding considerations that the 'derivation' can only be understood to indicate that the methodological prescription, to determine the utility indirectly dependent on a unit of goods (marginal utility), also finds application to the theory of prices. But we still have to deal with another consequence of this question, namely that 'problem of the *measurability of value*'. This problem occupies an important place in the methodological controversy over the marginal utility theory. Should – so it is argued on the one side – it be possible to determine the prices of goods unambiguously through their marginal utility, as marginal utility theory assumes, then they must already be contained in this utility; it must be possible to describe the order of marginal utilities (order of value) as measurable (quantitative, as an order of *extensive* magnitudes). But this is – so the opponents maintain – impossible for marginal utilities (values) are *intensive* magnitudes and thus according to their essential nature not measurable.

Now we have already established that the designation of utility or of values as 'intensive magnitudes' is incorrect; furthermore, in the first part of this work we have pointed out that numerical comparability (which is solely involved in this context) must not be identified with measurability. But these two errors were not of too great significance for the *Methodenstreit*. It has been decisive, however, for the manner in which these controversies have been conducted, that both parties were not clear enough about the meaning of *indirect measurements*, or *the 'transformation' of intensive magnitudes into extensive magnitudes*. Those who contest the possibility of such a 'transformation', can be confronted with the method of physics which shows how to measure temperatures, brightness and the intensity [volume] of sound; on the other hand, those who concede (or maintain) this, must by no means therefore make the erroneous assumption that the numbers of measurement to be established are already contained in the phenomena to be measured, as *qualitates occultae*. The fact of the matter can be shown quite simply by the example of the measurement of temperature. What is involved is the coordination of spatial magnitudes (for example, lengths of columns of mercury) with sensations of temperature, which is based on the knowledge that, as a rule, the increase or decrease of sensations of warmth and the expansions and contractions of bodies exist together. This is usually expressed by the formulation – to be sure a not quite correct one – that it is the same cause which produces the expansion of bodies and the sensation of warmth. Thus with greater or lesser reliability, on the one hand inferences can be made from one's own sensations of warmth to the thermometer reading, and on the other hand, from a thermometer reading to the sensations of warmth to be expected in the space concerned. But we have already said all that is necessary concerning the inexact nature of this coordination. Now whether one declares that in the process described it is not the temperature that is being measured but merely the length of columns of mercury, or whether one wishes to speak of *indirect measurement* – as we have done in keeping with common usage – is a secondary question. What remains decisive is the insight that for such (indirect) measurement no specific descriptive requirements (characteristics of magnitude) exist with respect to the object to be measured, but only rules of coordination to directly measurable (spatial) objects must be specified. What has just been said for the measurement of objects also holds quite generally for every numerical (mathematical) treatment of an object domain. What is required is a rule for coordinating objects under investigation with numbers; however, for the possibility of discovering such a rule, a specific internal structure of the object domain 'to be mathematized' is by no means a necessary condition; especially it is not required that this domain presents itself as a complex of relations of intensive magnitudes.[249]

From these consideration the following conclusions result with respect to our controversy: The scholars who deny the 'quantification' of utility (or economic values) are right insofar as they object to the conception that one can operate with utility directly on the basis of its internal structure, without resorting to the aid of further specifications. Their objections are therefore valid against all those formulations whereby one operates without further ado with concepts of the sum or the difference of utilities (marginal utilities), or even with reciprocal values of marginal utilities which are equated to prices. But such critics put themselves in the wrong as they seek to deny the possibility of an *indirect* quantification on 'ontological grounds'. They can only require that those who operate mathematically with 'utilities' and 'values' *explicitly* state the additional assumptions with which they operate, so that the appearance of a possibility of direct quantification will be removed, and the empirical pre-suppositions on which the quantification is based can be surveyed and tested for their validity and import. With that, we have indicated how this controversy could be resolved, insofar as 'matters of principles' are concerned.

Until now we have contrasted the thesis: 'One cannot quantify utility (values)', with its antithesis 'We have quantified them: *ab esse ad posse valet consequentia*'. Now we have to ask and answer the following questions: Just in what do the procedures designated as 'quantification of utility' ('values') consists? Facts of what kind, and general rules of connection of what sort, are utilized? What scope and status is accorded to the results? Once one seeks to answer these questions, the fundamental objection against this or that 'quantification of utility' can no longer be formulated as 'What you attempt is an impossible, logically contradictory undertaking', instead it has to be formulated as 'You do not understand clearly what you are doing; your interpretation of your own method is incorrect – (for example, you believe that you are operating with 'utilities', while in fact you are operating with price; from the beginning) and consequently you cannot judge their significance'. Thus the 'logical' objection, 'on principle' will not be directed against the procedure itself but against

[249] See for this Kaufmann, 'Was kann die mathematische Methode für die Nationalökonomie leisten?' (What can the mathematical method achieve for study of national economy?), *Zeitschrift für Nationalökonomie*, vol. II, pp. 754–779.

its *inadequate interpretation* and the consequences of this; and it will be possible to test this at each point, so that a discussion carried on with intellectual honesty will have to lead to consensus.[250] Things stand differently with respect to evaluation of the heuristic significance of the procedure itself. Here, as we have shown, a complete resolution of the conflict of the contradictory conceptions is not to be expected, because they often contain different estimates of the *chances of success of future research* in applying the methods, but here too the elimination of erroneous interpretations forms the most important step in reaching a consensus.

Almost all the methodological controversies, more or less closely related with the theory of marginal utility, will have to be considered from the same perspective. We wish to mention the most important among them and seek to remove the misunderstandings impeding their resolution. Let us begin with the question of the *applicability of the mathematical method* to classical economics, a closely related issue.

From the results of our reflections so far, the untenable character of the two contending conceptions follows directly: the mathematical method is not applicable at all to economic science (and especially to the subjective theory of value) because in principle its objects cannot be measured, or, it represents the sole precise (and therefore scientific) method of classical economics. The erroneous character of the first thesis has just been shown, while that of the second (opposing) thesis follows from our general findings about the 'exactness' of empirical statements and the entirely different exactness of logico-mathematical propositions. In particular, we pointed to the fact, that in a procedure of empirical science to which the mathematical method is applied, three stages must be distinguished, namely the *approach, the mathematical operation, and the evaluation of the results*; and that the internal exactness of the second stage furnishes no guarantee of any kind for the efficacy of the entire procedure with respect to the research goals. The fact that some representatives of the mathematical method in economics – fascinated by mathematical symbolism and the rigor of the calculus – have lost their critical sense for this distinction, has brought a great deal of confusion into the discussion; but when this error is revealed, it is not shown that the mathematical method is inadequate for economics, but only that the ground has been removed from certain exaggerated interpretations of this method.

The mistaken view that empirical knowledge can be derived from mathematical insights is nourished above all by the circumstance that certain more or less simple empirical assumptions are associated with mathematical procedure, and therefore are falsely attributed to it. We want to clarify this in terms of the proposition – which forms the core of *quantity theory* – that with a rise in the quantity of money, prices also rise. At first glance this propositions appears to provide a logical insight and yet to contain a statement about reality. However, on closer analysis of this state of

[250] The attainment of such a consensus would be significantly facilitated by the reports on motivation mentioned above (ch. 7). Especially important in this connection would also be a unification of the terminology. Initiatives in this direction are already in evidence. See for instance Fritz Machlup, 'Begriffliches und Terminologisches zur Kapitaltheorie', *Zeitschrift für Nationalökonomie*, Bd. 2, pp. 632–639.

affairs, the following will be the result: Logical insight is contained in the analytic judgment that in *dividing* a larger sum among the *same number of goods*, 'on average' more will be allotted to each unit, or more precisely formulated that the sum of the price rises exceeds that of the (possible) declines in price. But quantity theory obviously wants to assert something different, namely that increasing the quantity of money will lead *de facto* to *a rise in prices*. In order to understand the essential difference of this assertion from the analytic proposition just presented, we have to recall the distinction which we made between '*logical entailment*' and '*empirical implication*'.[251] From it results the following: The proposition according to which the division of a larger quantity of money among the same number of units of goods is 'linked to' a predominating rise in prices, is not an empirical statement at all; and thus it is not a hypothetical judgment, but the result of a conceptual analysis. It does not state: 'If that is the case, then this is also the case', but instead, 'In the assertion that that is the case, the assertion that this is the case is logically contained.' Thus it is not permissible to say that the division of a large sum of money among the same number of units of goods *conditions* a predominating rise in prices, but one must say that this is what it *means*. Therefore any talk of additional conditions for the truth of this proposition, especially the introduction of a *ceteris paribus* clause, is nonsense. However, the case is quite different with respect to the empirical validity of the quantity theory. This will depend, for example, on whether or not the money pumped into circulation during the economic period under observation is withdrawn again due to additional hoarding, and whether or not a counter effect – or even over-compensatory – effect is produced by some other reduction of the rate of circulation of money, or by a rise in the quantity of commodities. The number of conditions to be taken into consideration is even larger for the more primitive formulation of quantity theory, according to which the rise in prices of all classes of goods will by proportionate to the rise in the quantity of money.

In the confusion just characterized lies *one* of the conceptual motives for the assertion that economics contains propositions of a *tautological* character which in spite of this contain *knowledge* of *real economic* events. This confusion obscures the distinction between a *hypothetico-deductive system* at the head of which stand certain assumptions about the world, and a *calculus of proof* at the head of which stand definitions (tautologies). As this point has acquired an increasing significance in the *Methodenstreit* just in the last few years,[252] we have to occupy ourselves with it a bit longer.

Tautologies are, as we have observed,[253] declarations about the relations of meaning of various words, or combinations of words. They can therefore be transformed into definitions in which one word (or the one combination of words) forms the *definiens*, the other the *definiendum*. No knowledge of reality of any kind can ever flow from such a definition, and it therefore also provokes misunderstanding

---

[251] Compare above Part One, section "Logical-mathematical Thought".

[252] See for instance Lionel Robbins, *An Essay on the Nature and Significance of Economic Science,* London 1932, p. 82 and *Introduction to Mises' Theory of Money and Credit*, London 1934, p. 13.

[253] Compare above Part One, section "Logical-mathematical Thought".

to say a proposition is true by definition. The contrary appearance, however, arises due to faulty operations with so-called *real definitions* by which an empirical concept is supposed to be specified unequivocally by definition, on the one hand, while on the other it is still *linked to connotations which originate in 'pre-definitory' experience*.[254] What is overlooked in this is that a concept defined with the aid of other concepts is determined solely by *these* concepts, and cannot draw additional properties from other sources of knowledge. Thus when, for example, 'savings' is defined as 'unconsumed income', then no additional meaning of any kind, may be linked with the word 'savings', if the unambiguous character of scientific language is to be preserved. However, one can very well formulate empirical propositions of varying degrees of generality about savings, and combine these to form a systematically structured (deductive) theory of savings.[255] Then one has a deductive system with assumptions of a (relatively) high degree of generality at the top, from which different levels of specific assumptions will follow by means of logical analysis. Each of these propositions is an empirical, falsifiable statement, and by no means a tautology.

But – so it will be asked – is not the statement: 'If those general propositions are true, then the more specific propositions are also true', a tautology? The proper reply to this is: To express the relations in question in the form of this hypothetical judgment leads to a misunderstanding, because it creates the false impression that this relationship depends on the truth of the general statements, and thus on empirical criteria. Actually, it is completely independent of this, for the specific proposition is contained in the general propositions in the same way, whether it [the latter] may be true or false. The ascertainment of this logical relation is to be designated as a 'tautology', but with that, it is not stated that all deductively derived statements are tautologies. Whether that *is* the case, has to be ascertained internally in each instance, and this internal character is not affected by the incorporation of the proposition into a deductive system.

But it is possible that with scientific progress a *change of meaning* of the terms occurs, in such a way that their meaning by definition (their content) is either extended or narrowed. In the first case, tautologies can be formed that state the same thing as the assumptions in an earlier stage of research, and in the latter case the opposite is true. That the danger of mistaken interpretations is a definite one can readily be seen, and it is also easy to understand that it will encounter less resistance in an investigator, the more the wish to gain special status for his theses becomes the father of the scientist's thought. Add to this the obscurities discussed above about the relation between what is posited and what is presupposed, as well as about logical analysis and the subsumption of the individual under the general, then errors in thinking become almost unavoidable. But this entire veil that wraps scientific procedure in such a mysterious twilight is torn away as soon as one relinquishes faith

[254] For this, see the consideration above in Part One, section "Basic Philosophical Considerations" on pro-predicative experience.

[255] Compare T.W. Hutchison, 'A note on tautologies and the nature of economic theory', *Revue of Economic Studies*, vol. II, pp. 159–161.

in the *logical miracle*, in the *spontaneous generation of knowledge out of nothing*. By the result of these reflections, the discussions about the value of the deductive method in classical economics are shunted away from the side-track again, onto which they were pushed by being linked to the problems of tautologies.

After this digression let us return to the question of the application of the mathematical method to economic science; and we will arrive at the following observations:

After deflating, on the one hand the exaggerated expectations with respect to the achievements of the mathematical method, and on the other hand the untenable thesis that this method is in principle inapplicable to the problems of economics, examination of two [further] questions will prove to be decisive for our problem. The one asks, to what extent mathematically exploitable *approaches* can be gained from theory, i.e., from the formation of *schemata of purposive rational economic action*, and/or what additional assumptions are required in order to make such approaches possible; the second question aims at examining what *significance the results* attainable in this way have for the explanation of *economic reality*.

With respect to the first question, one may remark that every investigation in economic science that has the goal of determining certain prices from other prices will include mathematical considerations. To be sure, frequently it is merely a question of the calculations of elementary arithmetic, the inclusion of which in economic knowledge is hardly denied even by the 'anti-mathematicians'; but as we have recognized, from the purely logical point of view the whole of mathematics is contained in elementary mathematics, so that a sharp line of demarcation can hardly be drawn here. But even the procedure of so-called higher mathematics – which are frequently characterized by the infinitesimal calculus – can find direct use in economic theory without difficulty. A well-known example is the fundamental problem of Cournot's *monopoly theory*.[256]

Cournot exemplifies his pertinent considerations in terms of the owner of a mineral water spring and presupposes (a) that the proprietor has no competition, and (b) that he can increase his production arbitrarily. One will recognize then, that demand is a function $\phi(p)$ of the price P and, accordingly, also the production costs $\psi(p)$ of production adequate for the satisfaction of demand. Therefore, the monopolist acting in accordance with economic principles, will have to limit his production in such a way that the expression $\phi(p) - \psi(p)$ is maximized. This maximum, however, can be found with the aid of the differential calculus in the well-known way, if the two functions $\phi(p)$ and $\psi(p)$ are given.

The methodological discussion has concerned itself not so much with Cournot's monopoly theory as with the production equation of Walras[257] and especially with

[256] Cournot, *Untersuchungen über die mathematischen Grundlagen der Theorie des Reichtums* (Investigations concerning the mathematical foundations of the theory of wealth), trans. (from the French by W. G. Waffenschmidt in Wänting's *Sammlung sozialwissenschaftlicher Meister* (Collection of the masters of social science) vol. 24, Jena 1924, p. 47 ff.

[257] Walras, *Mathematische Theorie der Preisbildung der Wirtschaftlichen Güter* (*Mathematical theory of the price formation of economic goods),* Stuttgart 1881.

the simplified form given by Cassel.[258] With the aid of these equations the prices of the means of production and the quantities of the products to be produced can be determined, if the following elements are known: (1) The quantities of the means of production (labor, land and capital); (2) the kinds of productive applications involved of the means of production (technical possibilities); (3) the dependence of the prices of the products on the quantities of products produced (demand conditions). Recently, on the basis of suggestions by Karl Schlesinger,[259] the mathematician Wald[260] demonstrated in a very ingenious manner the necessary and sufficient conditions under which the Walras-Cassel equations would have unambiguous positive solutions; these conditions are quite compatible with general economic theory and especially with the subjective theory of value. It is to be hoped and expected that this example will form the basis of a new approach and will lead, for most equation-systems of mathematical economics in the near future, first to a clarification of the mathematical presuppositions for the existence of economically utilizable solutions, and second, to an understanding of the significance of these presuppositions and solutions for economics.

Quite generally it can be said: In view of our considerations above[261] it is not to be expected that economics will receive a significant impulse from mathematics as abstract science has; but it can hardly be doubted that it is still to play an important role in the fuller elaboration of economic theory. Of a significance not to be underestimated will be the *compulsion* to arrive at a *precise formulation* of the economic presuppositions in the formation of mathematical approaches. In this compulsion lies in a narrower sense the *methodological significance* of applying the mathematical method. It has already proved beneficial in many cases and will become more effective, the higher the degree of clarity about the nature of the mathematical calculus and about the presuppositions for the existence of unambiguous solutions for certain 'properties' that have been attained. And by this the erroneous, frequently exaggerated, interpretations of the achievements of the [mathematical] method will be held in check. With these pre-conditions an established and far reaching agreement concerning the place of the mathematical method within theory – i.e., the construction of purposive rational schemata of economic action – could be gained. It would then be possible to concentrate methodological interest almost exclusively on the second of the questions mentioned above: to what extent is pervasive mathematical structuring of theory relevant for the goal of economic science, i.e., for the explanation of real economic processes, and especially the prediction of economic developments of shorter or longer range. That this question has almost always been

[258] Compare Cassel, *Theoretische Nationalökonomie*, 3rd ed. Erlangen 1923.

[259] 'Ueber die Produktionsgleichungen der ökonomischen Wertlehre' (On the production equations of economic value theory) in *Resultate eines mathematischen Kolloquiums,* ed. Karl Menger, Heft 6 (1933–34) Vienna 1935, p. 10 f.

[260] *Ibid.*, p. 12 ff; compare also the remarks in the discussion of this by Schams and Menger, p. 18 ff.

[261] Part Two, section "Social Laws and Ideal Types".

confused with that of the place of the mathematical method *within* the theory has been most detrimental for the evaluation of the epistemological situation.

These considerations, however, will lead in all probability to the following result: mathematical schemata furnish a *survey* of the factors to be considered, and points of reference for the estimation of their significance with respect to magnitude. In this way, they allow an approximate evaluation of the significance of the incorporation of certain data into the research procedure and of certain interdependencies for the explanation of economic reality. Here it seems to me, lies a broad and fruitful field of work for classical mathematical economics, which will become more clearly visible, the more theory reducing the level of abstraction will strive to develop rational schemata, that are structurally more complicated and closer to economic reality. The investigations already carried out about 'comparatively static' and 'dynamic' economic processes, and the kind of pertinent mathematical approaches, permit even now a quite clear recognition of the direction of future research.[262]

However that may be, the following results of our methodological reflections can be stated without any doubt: (1) the applicability of mathematical methods in general, or of the methods of higher mathematics specifically to the problems of economics, can *not* be denied by pointing out that economic objects are not extensive magnitudes; (2) The mathematical method by itself cannot lead to propositions about the relevant experiences of economic science, and no *proposition of economics* can lay claim to *apodictic* validity just because it was gained by means of mathematical methods; (3) An essential aspect of the significance of the mathematical method for economics is to be seen in its compulsion to *render precisely the implicit presuppositions*: (4) After removing the misleading conceptions, the core of the controversy will be recognized to consist in the difference in views concerning the *degree of computability* of economic reality, and with that, concerning the capacity for achievement of economic theory.

This question also lies at the root of the controversy about '*deductive* method or *inductive* method'; the debate between the theorists and the historians (or institutionalists), that has now continued for half a century since the famous polemic between Carl Menger[263] and Gustav Schmoller,[264] without the scientific opponents having brought any essentially new relevant arguments to bear. Our judgment of this controversy follows directly from our reflections about 'deduction and induction' and about the 'historical' in the social sciences and about 'social laws'. We conclude that the procedure of deduction as well as that of induction, and therefore also the relation between the two, has been inadequately interpreted by the theorists as well as by the historians (but especially by the latter), and that thus an unbridgeable opposition was mistakenly perceived, where actually a gradual transition

[262] For this, see Schams, 'Komparative Statik', *Zeitschrift für Nationalökonomie,* vol. II, pp. 27–61, 1930.

[263] Compare Carl Menger, *Die Irrtümer des Historismus in der deutschen National-Ökonomie* (The errors of historicism in German national economics), Vienna 1884.

[264] Compare Schmoller, 'Zur Methodenlehre der Staats- und Sozialwissenschaften', (On the methodology of the political and social sciences), *Jahrbuch f.G.V.* 1883.

exists. The theorists for the most part have exaggerated their conceptions of the significance, and the kind of validity, of the general principles that enter into their deduction as premises. The historians, on the other hand, typically fail to recognize that the utilization of general assumptions enters into the ascertaining of historical facts and relations of such facts, – i.e., the circumstance that in principle such ascertainment always was – even, if in differing degrees – based on rules of the interpretation of human behavior. After removal of these errors the debate between the theorists and the historians is revealed as a difference of views about the *level of abstraction* to which economic investigations should advance and – in connection with this – about the degree of generality of the laws to be set up, as well as about the *kind and extent of their control by the facts*. These are certainly methodological questions of the highest significance; but such problems cannot be solved uniformly *a priori* by philosophical reflection; instead a justification for reaching decisions with respect to the questions of procedure mentioned above, can only be arrived at in each instance on the basis of the most careful analysis of the particular distinctiveness of a given class of economic problems.

No less stubborn than this controversy is the one about the '*value-freedom of economics*', under which title, to be sure, quite diverse conceptual themes intermingle.[265] To begin with, it must be asked whether economic behavior is *value-bound* in the sense that *objective* values which are prior to knowledge are pre-given to be realized by it, so that one can speak of absolutely correct economic goals? A negative answer to this question – which was of significance especially for the economics of *scholasticism*[266] – results directly from our fundamental reflections about the concept of value, and the problem of value in the social sciences. However, this does not exclude, as we have also shown there, the fact that for all economically active human beings, certain general goals can be considered to be *unquestionably given*.[267] The second part of the question posed under the title of 'value-freedom' involves whether economics is *normative*, not with respect to the goals, but with respect to the means to be employed in attaining them; insofar as its task consists in indicating the correct means for given goals. These goals are generally characterized more precisely as the satisfaction of needs, and in this connection one points to the relative scarcity of goods [necessities of life] which makes planning necessary.

With respect to this, the following may be remarked: Whether one calls the topic of economics 'the determination of the *correct means* for attaining *given goals*' and thus wants to conceive the discovery of the purposive rational schemata as its ultimate goal, or whether one calls '*the explanation of economic reality*' its theme, is a question of *convention*. We can neither say that the tradition presented in the history of ideas speaks unequivocally in favor of the one, nor of the other definition. Where research was linked directly to the goals of economic policy, understandably the

[265] See the general discussion of the postulate of value freedom above, Part Two, section "Value Problem in the Social Sciences".

[266] The *Summa Theologiae* of Thomas Aquinas was authoritative here.

[267] Compare above Part Two, section "The 'Historical' in the Social Sciences".

first alternative dominated, and the same was the case where *pure theory*, thus the *establishing of ideal types*, and not their application to reality, commanded the center of attention. Still, today it can be said that the overwhelming majority of economists call '*the explanation of economic reality*' the theme of their science, and would grant to the purposive-rational schemata of theory a merely 'instrumental' character. However, this whole question losses much of its significance as soon as it is realized that one cannot speak of a specifically normative method which is essentially different from the empirical sciences concerned with what 'is', and as soon as it has been recognized that the criteria of the '*correctness*' of human actions, understood as purposive [instrumental] rationality with respect to given goals, are of a completely empirical nature.

The methodologically most important question treated under the title of 'value-freedom of economics', however, is the following: is it permissible to disregard a *concretization* of the pertinent economic goals as much as pure theory (the subjective theory of value) actually or allegedly does? Does not such a treatment of the problem have as its consequence, that either (in the case of actually performed abstractions) a series of highly significant concepts of economics, such as for example, national wealth are inadequately specified or that (in the case of merely alleged abstractions) these concepts appear to be burdened with implicit teleological presuppositions? (In the latter case, one easily arrives at an exaggerated interpretation of the *achievement* of the theory, because the fact that the results could only be attained with the aid of additional presuppositions of a teleological nature, not belonging to pure theory, remains hidden.) Would it therefore not be recommended to free oneself from the false semblance of a 'value-free' pure theory, *beyond the concretization of goals*, and to bring to explicit clarity the *value positions* (political postulates), contained in it as implicit presupposition? These questions have been discussed and answered affirmatively in recent years above all by Myrdal[268] and Souter[269] and Mackenroth[270]; in Myrdal's arguments, it is especially important for the thesis that the value judgments which are to be grasped explicitly, not only determine the setting of goals on the part of economic subjects, but also essentially influence the selections of the 'correct means'.

With respect to these assertions, the following is to be remarked: Without doubt a considerable series of general economic concepts which play a great role in economic policy cannot be made precise with the aid of so-called pure theory alone in a manner that does justice to pre-scientific and scientific language usage and traditional applications. This is true especially of the problems connected with '*welfare economics*',[271] for example, real wealth or real income (of the individual or the

[268] Compare Myrdal, 'Das Zweck-Mittel Denken in der Nationalökonomie' (Instrumental thoughts in national economics), *Zeitschrift für Nationalökonomie*, vol. IV, pp. 305–329. 1933.

[269] Souter, *Prolegomena to Relativity Economics*, New York 1933.

[270] Mackenroth, *Theoretische Grundlagen der Preisbildungsforschung und Preispolitik,* in *Sozialwissenschaftliche Studien*, Berlin 1933.

[271] For instance Pigou, *The Economics of Welfare*, London 1929 and Amonn, *Grundzüge der Volkswirtschaftslehre*, Jena 1926.

economic community), which are conceived to be quantitatively unambiguously determinable and therefore to be comparable, without adequate criteria for this being specified. Furthermore, it cannot be denied that in operations with these concepts by theoreticians, those gaps in definiteness frequently are filled by implicit presuppositions, the choice of which appears understandable only on the basis of specific – more or less conscious – valuations. Finally it is correct that these aspects are very significant for demands of economic policy – and these include the determination of goals as well as methods that are set up 'in the name of economic science' – for which reason their explicit realization is urgently recommended, not only from the standpoint of *pure* theory, but also from the standpoint of *applied* theory, of scientifically based praxis.

So far the arguments brought forward are pertinent and worthy of being encouraged. But they need to be weakened, or modified, in the following two points: First one has to be clear that they by no means apply to the entire theory – that is usually identified with subjective value theory in its diverse variants – but only to certain parts of the theory, mainly quite peripheral ones. This over-estimation of the polemical range of the arguments, to be sure, becomes understandable due to the circumstance that the overwhelming majority of the opponents (as well as of the adherents) – as we have ascertained above[272] – considered the contested theory to form a *unified deductive system.*

Of far greater importance for the theory of science, however, is the second point. The objections just outlined indicate, or at least suggest very strongly, the interpretation that a pure, thematically closed, economic science, as the theoretical foundation for a consciously purposeful economic policy, is not possible. Now, to be sure, the results of economics only form a guideline for economic practice insofar as the goals selected in the theory (including the secondary conditions due to which certain means of attaining these goals are eliminated) correspond to the goals and secondary conditions given in economic reality. But what is decisive is that economic goals can be *isolated.* That, for example, in this or that country, the creation of an autarchic heavy industry is attempted (and this disregarding the circumstance that considerations of *comparative costs* would seem to speak in favor of the import of the goods in question) may only be understandable when the defense policy of this country is taken into consideration; but as soon as this goal – and perhaps some secondary conditions which spring from the same motivation have been fixed – a problem of *economics* is formulated, and the motives which have led to this precise formulation no longer play a role. Here our remarks about the relations between the *sociology* of knowledge and the *theory* of science[273] as well as between theoretical goals and *practical* goals[274] can be applied by analogy.

Thus we arrive at the following result: clarification of presupposed values (political goals) which often exercise a long-range influence on topics and procedures within economics (because of the secondary conditions) can have great significance

[272] See this subsection.

[273] See above Part Two, section "The 'Historical' in the Social Sciences".

[274] See above Part Two, section "The Way to Overcome the *Methodenstreit*".

methodologically, especially by making possible a clear evaluation of the attained results. But it is not permissible to say, because of that, that these valuations *enter into the theory* and that therefore the idea of value-free economic science is impossible. For in the first place, as already mentioned, these observations pertain only to certain (special) parts of the theory; and secondly, the valuations as such never enter into the procedure, but only their 'objective reflection', in the form of certain presuppositions. Thus though the valuations are typically *inducements for posing certain problems*, because they condition the practical relevance of these problems, they are not *elements of the problems themselves*. The appearance to the contrary arises, however, due to the fact that scientific thought operating with incomplete clarity, in the first place fails to provide a sufficiently precise specification of the topic, and secondly, arrives at certain methodological decisions within the procedure, which cannot be justified objectively, but instead become comprehensible only through the uncovering of 'subjective backgrounds'. But it is just because of this that a clarification of the latter grounds must go hand in hand with their separation from the thematic content of the problem, if the entire task of methodological work is to be met. G. Haberler's analysis of the problems of *index numbers*[275] furnishes a good example of how such clarification should be carried out.

These results of our reflections also permit us to recognize the double meaning of the reproaches through which a doctrine, or its proponents, may be *ethically or politically discredited*, as is the case, for instance, when marginal utility theory is characterized by its opponents as a eudaemonistic or liberalistic doctrine. Usually such a designation is intended as an unmasking – generally considered to be defamatory – of the general attitude of the adherents of the doctrine being attacked, by pointing out that the adherence to such principles reveals a particular attitude, i.e., that the doctrine is *symptomatic* of such an attitude. So far, we are confronted with an *argumentum ad hominem*, which is not directed at all against the truth-content of the doctrine, and therefore is removed from the framework of objective discussion. But, with the *argumentum ad hominen* in most cases the thought is combined – more or less explicitly expressed – that, blinded by this general view, the proponents of the doctrine being attacked arrive at false results, or over-estimate the significance of their results in a biased manner. And those are material objections, if the asserted errors are more precisely designated. For in that case the *argumentum ad hominen* represents the attempt at an *analysis of motives*, which is supposed to explicate the occurrence of the main errors of thought in the doctrine being opposed, by pointing to the role played by immaterial motives, i.e., *motives alien to the research goals*. Accordingly, the task of replying to such objections in the guise of reproaches, will consist in extricating their material core, the assertion of certain errors of thought, and to refute these. In confronting the polemical objections to the marginal utility school mentioned above, this path has already been chosen by its adherents, but in order to pursue it to its end one requires the clearest insight into the theoretical structure of the doctrine.

[275] Haberler, *Der Sinn der Indexzahlen* (The meaning of the Indices), Tübingen 1927.

The same holds true with respect to invalidating the objections which seek to discredit the marginal utility theory intellectually by reproaching it for containing erroneous conceptions which were refuted conclusively long ago, and basing itself on outdated theories. The reproach of *psychologism* points in the first direction; the reproach that the psychological conceptions on which the doctrine is based are *antiquated*, in the second.

As far as the objection of 'psychologism' is concerned,[276] it is to be remarked in the first place, that it is raised usually without clear comprehension of its meaning. For usually it is the refutation of psychologism in logic by Edmund Husserl, in the first volume of his *Logische Untersuchungen* which functions as the paradigm example, and the usual train of thought is that psychologism is just as misapplied in economics as it is in logic; therefore, economics cannot be erected on a psychological foundation, as done by the marginal utility school. But whoever has correctly understood Husserl's critique[277] knows that it does not refer to the procedure of logic as such at all, but to the *interpretation* of that procedure – after all, the psychologists have the same logic as their opponents, only they misunderstand its meaning – while most opponents of the marginal utility theory mainly want to say, by the objection just mentioned, that the psychological method does not prove to be empirically valid in economics. But due to the failure to grasp this distinction, the two aspects are almost always confounded so that it remains undecided whether the reproach is intended to mean 'You are doing psychology in economics, but you should not be doing that' or instead 'You believe that you are doing psychology, but in fact you are not doing that at all, you are merely interpreting your procedure erroneously as psychological, which, just like ours, is based solely on the 'external' behavior of human beings on the market.'

As far as the question of the material justification of the two objections is concerned, the reply to them follows directly from the observations already made. The demand for the exclusion of psychology from economics is rooted either in the speculative error that conceives of an irreducible sphere of ideal social objects, which also includes a false conceptions of 'understanding' and 'meaning', or in the behavioristic misinterpretation of the allegedly solely 'external' facts of human behavior. If these two prejudices are overcome, then the discussion can only turn on the manner in which psychological knowledge is to be exploited in economic science, and which elements of psychological knowledge should be exploited, but not about its dispensability altogether. A main divergence in this connection will be whether *facts of the market* should be used exclusively as a data base for interpretation, or whether direct reference to *introspective* experience may also prove promising. The achievements of the marginal utility school seem to me to have shown the latter to be the case, even if the relevance which introspection has for this method has frequently been *over-estimated* by the school. Accordingly the reproach of psychologism understood in the second manner, is justified insofar, but only insofar, as it confines itself to rejecting that exaggerated interpretation.

[276] See above Part Two, section "The Social Sciences and Psychology".

[277] Husserl rejects this misunderstanding in his *Formale und transzendentale Logik*, p. 135 ff.

A clear understanding of the role which psychology plays within the framework of the marginal utility school also leads to invalidating, to a great extent, the objection that it operates with an *antiquated psychology*.[278] For what distinguishes the economic psychology, as it is pursued within the framework of the marginal utility school, above all from general psychology (or also, the schools of general psychology) is the fact that the selection and connection of facts is carried out from the outset from the perspective of its *relevance to economics*. It is never pure psychology that is practiced here, but even in the concept of needs, a link of biological-physiological with psychological aspects is established. But the pure theory of needs occupies a quite limited space within the subjective theory of value. Much greater space is occupied by those links between psychological and technological experience, which are analyzed in the theory of economic goods. However we by no means intend to deny that modern disciplines of psychology – especially depth psychology – can be successfully utilized in the treatment of numerous problems of economics.

Not infrequently, the marginal utility school is accused of psychologism and *subjectivism*; in this accusation the following 'subjective' elements within the theory are often not sufficiently distinguished:

1. The marginal utility school assumes that economic progress cannot be explained like facts of nature, but that the '*method of understanding*' is adequate for these processes.
2. The school assumes that economic processes (the formation of prices) cannot be adequately explained if an objective value, univocally determined by the process of production, is assigned to the individual goods, but that the *goals of the consumer* must be included in the analysis.
3. The school does not take its conceptual departure from intersubjective (social) facts (prices), but from the economy *without exchange* of a single human individual.

From the 'subjectivity' of the doctrine, the inference is falsely drawn that its propositions do not possess objective *validity*. However, a closer examination of this argument is hardly needed.

Related to the reproach of 'subjectivism' is that of (individualistic) '*atomism*'. It turns out to be a direct application of the fundamental universalist theses to the problems of economics; therefore we can content ourselves with pointing to the critique of this conception we have already carried out.

The above analyses have by no means provided complete insight into the controversies linked with marginal utility theory, but the theoretically most important arguments in the *Methodenstreit* were identified and critically evaluated. The conclusion of this section may be formulated in a few words concerning the question, linked in the history of doctrines with the problems of the marginal utility school, as to whether '*economic activity* is to be defined as '*social action*', and consequently whether '*economic science*' is to be defined as '*social science*'. In the extension of

[278] Thus recently for instance Mackenroth.

the concept of economic activity beyond the social sphere, the essential aspect of economic activity is regarded to be the disposition of goods according to plan, in which foregoing certain satisfactions of need in favor of others is included. From this point of view a theory of an economy without exchange has to be *coordinated* with a theory of an exchange economy. The other conception sees in the theory of economy without exchange merely an auxiliary construction to help explain processes of exchange, and it characterizes only *social* exchange, and not '*inner*' exchange, as 'economic activity', as it is actually understood within the problem range of economics.

The methodological background of this terminological question is obviously formed by the question, already investigated by us, of the significance of the theory of an economy without exchange for economic science as a whole, according to the present state of research. Still, considerations raised in the *history of theories* also play a role in the controversy. I myself am inclined to define 'economic science' as '*social science*', and to designate the theory of an economy without exchange as 'auxiliary science', as this terminology seems to me to have a better basis in the history of ideas[279]; but I can readily respect the arguments in favor of the alternative. However, it should be emphasized once more, that it is not the terminological question as such which has significance for the theory of science, but only the methodological attitudes which are documented by the opposing definitions. As soon as these are clarified and critically evaluated, the problem can also be considered to be solved, even if no arguments giving decisively greater weight to the one or the other definition, can be developed.

9. The Concept of Positive Law, and the Pure Theory of Law

Theoretical investigation that seeks to comprehend the concept of law in its full clarity is confronted by two problems of delimitation. It has to distinguish the law on the one side from the naked reality of domination, and on the other side from certain demands for justice, with validity based on rational insight, and independent of any codification: Thus the theory of law confronts legal validity as the '*ought*' in social existence, to the '*is*', and on the other side confronts it as *positive* validity to *natural law* validity, Especially the second of these two tasks has from time immemorial played a dominant role – even if in varied guises – in the philosophy of law. For by ascertaining the relation between codified law and natural law, it is to be shown to what limits power must subject itself in order to be able to justly claim obedience. In earlier times it was above all the legislator toward whom the demands of natural law were directed, in that it was declared that only that legislation which satisfies certain moral presuppositions could be called law [or right], while in the absence of such presuppositions, the assertion of laws by force was merely the

[279] Compare the detailed treatment of the problem in Amonn, *Objekt und Grundbegriff der theoretischen Nationalökonomie.*

exercise of arbitrary power. As late as the nineteenth century, Johann Jacob Fries wrote a textbook of natural law[280] which was explicitly conceived as instruction for the legislators, telling them what sort of principles they must follow, in order to confer the character of law upon their legislation and the *Philosophische Rechtslehre* of Leonard Nelson,[281] the most distinguished pupil of Fried in the twentieth century, which appeared in 1922 also serves the same goals.

In more recent times, however, much less thought has been devoted to the legislator than to the jurist who applies the law, especially the judge who is considered to be the implementer of the norms of natural law. This occurred in connection with the insight that the assumption that every decision is already prescribed by the law, and hence the administrator of law only has to derive it from the law by logical processes, is not tenable. It was above all the theory of free law[282] which entered the fray at the beginning of our century, and sharply attacked any juggling with the *argumentum analogiae* on the one hand and the *argumentum a contrario* on the other, in order to show the inadequacies of 'juridical logic'. This had the consequence, that, if the majority of juridical decisions were not to appear as mere products of arbitrary opinion, other sources of knowledge had to be sought for the evaluation of the correctness of such decisions. These, so it seemed, could be furnished by speculation on natural law; and thus it came about that, under the cloak of allegedly pure knowledge, admission to jurisprudence was granted to certain ethical-political postulates.

The critique of these doctrines forms the dominant theme of the '*pure theory of law*'[283] most closely linked with the name of Hans Kelsen, which, in my judgment, constitutes the most important contribution to contemporary investigations directed toward clarifying problems in legal theory. We now want briefly to present, the center piece of this doctrine, in which the concept of the *legal order* is analyzed, in order to show that even here, in this immense achievement of rational reconstruction, the essential methodological core has not been completely extracted from the speculative shell, so that a rational reconstruction of a *second order* proves to be required. The results of these reflections will then form the basis of our analysis of the concept of *positive law*.

In the previous section we have shown, in the analysis of marginal utility theory, that not infrequently the justification of this method, excellent in itself, was sought

---

[280] See note 105 and also note 106 Part One, section "The Concept of Value" on the 'Concept of Value'.

[281] See note 280.

[282] Compare for instance E. Ehrlich, *Freie Rechtsfindung und freie Rechtswissenschaft* (Free jurisdiction and free juridical science) 1903, *Die juristische Logik 1918.*

[283] An overview over the range of the theory, both in a material and a personal respect, is given in the *Bibliographie der Reinen Rechtslehre* compiled by R. A. Metall. It is published in Kelsen's book *Reine Rechtslehre* (Pure Theory of Law) (op.cit. pp. 155–222). With respect to the following presentation, it need only to be mentioned further that the theory of the relationship of legal sentences within a legal order, the 'stage theory' of law was developed by Kelsen, departing from conceptions of A. Merkel (see, for instance, 'Das doppelte Rechtsantlitz' (The double face of law), Juristische Blätter, 57th year, 1918) and in collaboration with the latter.

on the basis of false grounds; it is similar in the case of the pure theory of law. The chief difference between the two consists in that in the former, the justification of an empirical method of science was undertaken, while here the justification of results from the philosophy of science in a narrower sense (methodological critique) is attempted.

As the fundamental thought of the pure theory of law can be designated to be the insight that the juridical concept of law, i.e., the meaning of that which the jurist as interpreter of law *actually* understands by law, can only be found by *analyzing the procedure of legal science*. Thus one must ask oneself under which circumstances a legal scholar would declare a sentence to be a sentence of a specified legal order, and in which cases he would deny it that character. Now this question can be divided into two parts, namely: (1) What conditions with respect to content must sentences satisfy in order to qualify as legal sentences? (2) What are the criteria for ascertaining that sentences which satisfy these conditions are the sentences of a specific legal order? According to this, two main disciplines can be distinguished within the pure theory of law, namely, the *theory of legal sentences* and the *theory of the structure of law*. The theory of legal sentences does not stand in a direct connection with the problem that has occupied us primarily, that of the *unity of the legal order*, as here the legal character of the sentences to be investigated is presupposed. Therefore with respect to this we want to confine ourselves to the brief observation that Kelsen conceives legal norms as *hypothetical judgments* of the form. 'When certain circumstances occur, then compulsion (punishment or execution) toward a certain person is to take place,' and also we want to emphasize that from this determination of legal sentences, he draws decisive conclusions for his *critique of juristic concept formation*. For, in view of the correlation between legal concepts and legal sentences, there follows, according to his exposition, the incorrectness (inadequacy) of every definition of a legal concept which does not contain an '*ought*' element. But with that, judgment has already been rendered on a sociological (generic) definition of legal concepts.

In such definitions, the question of the *essential nature* of law, of the state, of the legal person, of property, etc., are confounded in an impermissible way with the question of their *origin*, and consequently in the definition of, say, the 'law' and the 'state' the typical causal conditions are introduced under which the formation of the law and the state takes place. Thus in the '*power theory*' and the '*consent theory*' it is examined how the power relations have to be constituted in order to secure the formation and preservation of a legal state, or else, what kind the basic attitude of the members of the community in law is required for this. The results of these investigations are then reflected in the genetic definition of law. The decisive argument on the part of the pure theory of law against these definitions lies in this: that in them, there are no conclusive criteria for the decision of a jurist as to whether a sentence is correct. For the law as a jurist understands it – so it is argued – is *not* a *state* of affairs of a certain kind, but an *epitome of sentences* of a specific character, the validity of which, as *normative* validity, can never be derived from facts. The 'is' and the 'ought' lie, as Kelsen says, in 'different planes', and 'is' can never be derived from 'ought' nor 'ought' from 'is'.

But due to these results of methodological critique, the question of the origin of legal validity, which even though 'positive' cannot remain without reference to empirical facts, becomes doubly important; and therefore that part of the pure theory of law which is to present the solution to these questions has increasingly been elaborated as the centerpiece of this theory.

Kelsen's solution is based on his concept of the basic norm, and we want to quote from his recent theoretical writings the especially striking formulation of the argument that led him to the formation of this concept[284]:

> This basic norm establishes the validity of positive law and expresses the hypothetical-relative character of a system of norms clothed only with the validity of positive law. It is not just the hypothesis of a special theory of law, merely the formulation of the assumption necessary for any positivistic grasp of legal materials. It merely raises to the level of consciousness what all jurists are, even unconsciously, doing when, in the comprehension of their subject, they *reject natural law* (i.e., limit themselves to positive law) and yet consider the data of their cognition not as mere facts of *power*, but as *laws*, not as mere *facts*, but as *norms*. They ordinarily understand the legal relationships with which they are concerned not as the natural relation of cause and effect, but as the normative relations of obligations and rights. But why is a human act, occurring in time and space and perceptible by the senses, interpreted as a *legal* act (a legal transaction or a judicial decision) within the meaning of any positive German or French law? Why should such an act be considered a *norm* and not simply a mere event in reality? Why should the *subjective* meaning of this act also be given an *objective* meaning? Why, in other words, does one not simply say that a certain human individual demands that another act in a specified way, but actually contends that one is *entitled* to prescribe and the other *obligated* to act in accordance with the prescription? Why do we assume that what the act in question subjectively conveys must be done objectively, *by law*. The answer of the *positivist* jurist is: because this individual act is in accordance with a 'higher', a more general act, a *law*, because the law prescribes that one is to act as the parties have agreed in their legal transaction, or as the judge has ordered in his decision. One may inquire further, why this law represents a *norm*, why it is *objectively valid*. Prima facie, the law is a mere factual matter, namely, the event of several people having expressed their will that other people should henceforth act in a certain way. But why should the will expressed by these people under these particular circumstances signify a 'law', while, if it were done by others under other circumstances, it would by no means have the same significance? Here the answer will be: The event which we interpret as the making of a law is in accordance with a still higher norm, the *constitution*, because these persons have been entrusted by the constitution with the power of making laws. This 'constitution' is, in turn, nothing else but a prima facie factual event whose normative meaning can only be found by recourse to a prior constitution according to whose rules it has been created. This recourse must ultimately end in the *original* constitution which can no longer be derived from a still earlier one. The positivistic jurist, who cannot go beyond the fundamental facts, *presupposes* that this original historical fact has the meaning of 'constitution', that the resolution of an assembly of men or the order of a usurper has the normative significance of a fundamental law. Only by making this assumption can he demonstrate the normative meaning of all other acts which he comprehends as legal acts simply because he ultimately traces them all back to the original constitution. The hypothetical basic norm which established the original legislator expresses this assumption; it consciously formulates it, nothing more. This means that legal positivism does not go beyond this original

[284] 'Die philosophischen Grundlagen der Naturrechtslehre und des Rechtspositivismus', (The philosophical foundations of the theory of natural law and of positivism of law), *Philosophische Vorträge*, Kant-Gesellschaft, Charlottenburg 1928.

> constitution to produce a material and absolute justification of the legal order. It stops at that point. The basic norm is an indispensable assumption because without it, the normative character of the basic historical event could not be established. This ultimate act, to which the positivistic jurist takes recourse and beyond which he does not proceed, is interpreted as an establishment of norm as it is expressed in the basic norm, which in turn is *not* justified by a higher norm and therefore itself transmits only hypothetical validity.
>
> The essential characteristic of positivism, as contrasted with natural law theory, may be found precisely in the difficult renunciation of an absolute, material justification, in this self-denying and self-imposed restriction to a merely hypothetical, formal foundation in the basic norm[285].

According to Kelsen, the result of this is especially that legal science, by establishing the basic norm, 'generates' its object – where the word 'generates' is understood in the sense attributed to it by the *Marburg* school of neo-Kantians.

Now the methodological controversy about the pure theory of law has almost always gone astray because its material critics turned chiefly against its *philosophical pre-suppositions*, and thus against the *neo-Kantianism* at its base; and failed to examine its decisive methodological content. Had this been done, then it would have had to be conceded at the outset that a large part of the results of the pure theory of law concerning the critique of method is irrefutable, and from this the conclusion would have followed – insofar as the untenable character of the philosophical doctrines it opposed is demonstrable – that Kelsen's undoubtedly appropriate *critique of method* can be *detached* from the *philosophical pre-suppositions* on which it is based. We now wish to show that actually the significant *methodological results* at which Kelsen arrives in his critique of method can also be attained *without the assumption of a dualism of the 'is' and the 'ought'* and *without* the *doctrine* based on this assumption, of the *basic norm*.

The pure theory of law intends to be the theory of *dogmatic legal science* – the theory of the *interpretation of law* – and accordingly its quest for the meaning of the concept of law (for the essential nature of law) aims to comprehend, by rational reconstruction, what is understood by 'law' in the procedure of this interpretive science, on what *criteria* the judgment depends whether a certain *sentence* belongs to the *material to be interpreted* or not.[286] Now here we observe, in the first place, that dogmatic legal science, as the theory of the interpretation of law, never asks its questions concerning the law as such, but always about the '*law of a certain legal order*'. However, the statement that a sentence belongs to a certain legal order means, to the interpreter of law *qua* interpreter of law, nothing else but that it forms *material for interpretation*. All secondary conceptions of the power that stands behind the sentence, of the social effectiveness of the moral obligations generated by the sentence, remain external to the subject matter; they play no role in providing insights for evaluating the legal character of a sentence.

Instead, for this evaluation it is decisive whether the sentences in question are the contents of legislative acts which stand in a relationship, to be specified with

---

[285] *op. cit.* p. 12 ff.

[286] *Compare* above Part I "Proposal for a Universal Methodological Schema".

certain 'supreme' legislative acts. It is this relationship which is indicated by the words '*on the grounds*', when it is declared that an administrative act is valid on the grounds of an ordinance, an ordinance on the grounds of a law, a law on the grounds of a constitution. Now because for legal dogmatics, the assertion of the 'legal validity of a sentence' means nothing else than that the sentence is a component of the material to be interpreted, then the specification as to which legislative acts are to be regarded as ultimate sources of validity, is nothing else than a *selection of topics*; thus considered from the viewpoint of formal logic, a *definition*. When one declares, for instance, that the constitution of the year 1867 must be considered as the ultimate source for Austrian law, then thereby it is established, by definition, what is to be understood by 'Austrian law'. It is therefore no more admissible to speak of a hypothetical assumption concerning the validity of the 1867 constitution, than it would be correct to say it was a hypothesis that the modern era began in the year 1492. For here, just as in the former case, it is not an *assertion that is made*, but a *stipulation that is set*. The one delimits the material of the dogmatics of Austrian law, the other the material of a history of modern times. The only assumption that can be designated as a hypothesis is that 'Austrian law', defined in this manner, covers the epitome of those sentences which are of specific social relevance – to be described more precisely by the sociology of law – for a human community, the Austrians. Here our general reflections about the real definition[287] can be applied by analogy.

This insight also provides an understanding of the analogy which Kelsen – in accordance with conceptual tendencies within the Marburg school – draws between the legal order and the hypothetical systems of natural science. According to his formulation, the basic norm (sometimes also called the 'original norm' or also 'original hypothesis') constituted the unity of a certain legal order, analogous to a hypothetico-deductive system, as the most general – i.e., not further deducible, but only hypothetically applicable – principles of physics constitute the unity of the system of physics. Therefore, as every law of nature derives its 'validity' from those ultimate principles, while (just because) a similar question cannot be posed in a meaningful way with respect to those principles, thus every legal norm of a certain legal order derives its validity from the basic norm, while (just because) a similar question with respect to the fundamental norm would be meaningless.

If, however, one speaks of the 'validity' of the basic hypothesis of physics, then this word has to have a different meaning, and with that one would mean here 'heuristic suitability', which is provided when those basic assumptions make it possible to grasp the events of physics within a unity of law. Analogously, the juridical basic norm would be 'valid' when it made it possible to grasp legal events, i.e., those events which, within a certain personal domain are 'generally' regarded as legal events, in a *normative* unity. Accordingly, if one assumes the viewpoint of science *in statu nascendi*, these norms will be selected in such a way that, by means of them, facts of the sociology of law could be understood from a unified viewpoint. But the

[287] Compare above Part One, section "Logical-mathematical Thought".

question of this choice does not fall within the framework of science regarded as a completed system, since this science is itself constituted by that choice.[288]

What requires revision in this conception, is the thesis implicit in it, that it forms the key to the understanding of the specifically 'normative validity' of 'legal order'. This is a consequence of Kelsen's assumption of a dualism in the spheres of 'is' and 'ought', according to which an 'ought' can never follow from an 'is', so that an autonomous origin must be sought for the normative. But as our analysis of the concept of norm has shown,[289] this assumption is not tenable; it owes its origin solely to the ambiguity of the term 'norm', where the aspect of issuing a command is confounded with the aspect of 'correctness' (itself in turn conceived in a confused manner) – which points to unspoken underlying goals. After the elimination of the ambiguity, a specific kind of positing as criteria for the validity of law results – i.e., a positing where the two statements, 'The sentence L – which in its content is shown to be a legal sentence – is valid' ('is a component of a certain legal order', is 'positive law') and 'the sentence L was posited in that specific manner' are *per definitionem* equivalent. Therefore there is no state of affairs at all that would require a justification to be carried out by a hypothetical approach. However, that in spite of this the principle of the logical heterogeneity of the 'is' and the 'ought' has led to the most important results in the critique of method, can be explained, as we have already observed, as due to its having become the guiding conceptual theme for the refutation of false doctrines, i.e., doctrines inadequate for juridical procedure, about the essential nature and origins of the validity of law. For it was believed that legal validity could be identified with factual obedience, or in any case that it could be derived from it, while in general the jurist decides the question of the legal character of a sentence by recourse to the manner of its positing. Thus the legal character of a sentence is certainly justified by facts, but not those facts which the 'sociological definitions of law' regarded as decisive. Consistently, following up this divergence on meaning into the subtlest ramifications of legal concept formation has led the pure theory of law to results of great significance for the critique of method; but to base this distinction on the notion that the criteria of legal validity lie in a sphere transcending reality cannot be supported.

Let us summarize our results up to this point: That a sentence possesses legal validity means, from the point of view immanent in the procedure of the fixed foundations [*Rechtsdogmatik*] of law, nothing else than that it is a component of the material to be interpreted. The criteria of legal validity are thus the conditions established by definition under which a sentence is conceived as being a component of the material.

But just as the question of the origin of legal validity does not lead to the establishment of a specifically normative method, so the structure of the individual legal sentences does not indicate such a necessity either. As this follows directly from our

[288] Compare for instance Kelsen 'Die philosophischen Grundlagen', p. 20.

[289] See above Part Two, section "Value Problem in the Social Sciences".

analysis of the concept of norm, we do not need to occupy ourselves further with this point.[290]

But there is still a third argument in the pure theory of law for a legal science possessing a specifically normative method – namely the assertion that the *complex of relations of validity* of legal sentences (*relations of delegation*), the '*grounded validity*', would be a specifically normative complex; it is with this that we now have to deal. Setting up a logical schema of '*grounded validity*' will make its untenable character clear.

The major premise states: 'If conduct is in accordance with a (particular) command C_1 of a person (or also a majority of persons) P_1, then it is also in accordance with a (particular) command C_2 of a person (group of persons) P_2.'

The minor premise states: 'The conduct of A is in accordance with the command C_1 of P_1'.

From this follows the conclusion: The conduct of A is in accordance with the command C_2 of P_2.

We want to call the obedience of C_2 '*mediative* obedience of C_1', and we will permit an unlimited number of steps of mediation.

Whether in a given case direct obedience to a certain command at the same time is mediative obedience of certain other commands depends on the circumstances under which it was posited, for example, on whether it was issued by certain persons and published in a certain manner. Thus we can construct the following simple relation of delegation: A father commands a child: "You ought to do what your mother orders you to do." The mother commands: "You ought to go to your aunt and do what she orders you to do." The aunt finally commands the child "You ought to go to the store next door and fetch me one kilo of coffee." The child, in doing so, *directly* obeys the command of the aunt and at the same time *mediatively* the commands of the mother and the father. In the same way a person who directly obeys a certain ordinance of a government thereby mediatively obeys the law on the grounds of which this ordinance was issued, and the constitution on the ground of which that legislation was implemented. Now obviously the complex of relations described is not a complex of the contents as such; that – in our example – the child is to fetch coffee is by no means logically derivable from the content of the sentences 'The child ought to do what its mother orders it to do' and 'The child ought to do what its aunt orders it to do.' Rather we have here a *definition of conduct according to command by an empirical coordination* of facts of a certain kind. Definitions of this kind are by no means confined to the sphere of commands. This we want to demonstrate by an example of the same structure drawn from a sphere unrelated to commands:

A says to B: "C has news of your brother in America." C says to B: "The news of your brother in America to which A refers come from D." D says to B: "The news

[290] For this, see Kaufmann, 'Juristischer und soziologischer Rechtsbegriff' (Juristical and sociological concept of law) in *Gesellschaft, Staat und Recht*, op.cit., pp. 14–41, p. 32.

of your brother in America to which C refers is the following": (there follows then the content of the communication).[291]

One can see that here – in complete analogy to the schema given above about delegation relations – the content of the news of B's brother is not determined by the communications of A and of C; only the persons who are capable of giving this news are indicated, and by this indication the way is pointed out to B by which he can arrive at knowledge of the content. Thus it follows that the relationship of mediative obedience of commands characteristic for law can *not* be regarded as a specifically *normative complex of relations*.

With this critical resolution of the idea of a normative method, various questions treated within the framework of the pure theory of law are eliminated, for example that of the '*tension between the "is" and the "ought"*', or the question how it might be possible that in spite of the separation of spheres between the 'is' and the 'ought' *positive law* might be *at the same time an 'is' and an 'ought'* – according to the aspect under which it is considered at the time.[292] It is also possible then to remove the obscurities that are linked to the conception of *positivity* as an alleged bridge linking the two spheres of the 'is' and the 'ought', and to isolate the various levels of meaning within this family of concepts. This shall be clarified by the following:

We have already pointed out that the jurist as interpreter of law never has to answer the isolated question whether a sentence is simply 'law' or 'positive law' or whether it is an element within a certain – i.e., *historically determined* – legal order; the criterion for answering this lies in ascertaining whether it is the content of an act of positing which stands in a more closely characterized linkage with other acts of positing, a linkage whose general structure we have just described. Accordingly, the assertion that a sentence is positive law is incomplete, as long as the historical context is not indicated within which the act [the legislative act] of positing proper to it is contained. For as the concept of a legal sentence – like the concept of a statement in general (a judgment) – is conceived to be invariant as against the occasional moments of its positing, so legal norms as such cannot enter into historical relationships, but only the *issuing of commands* of a certain content. As a consequence, the questions of the legal dogmatists, as to whether a certain legal sentence would be positive law, have a different meaning depending on the reference system in each case, but what is common to them is the formal structure of the relational complexes, defined as *criteria of positivity* which we have called '*relations of delegation*' By this, a *legal theorist* justifies operating with a general *structural type of positivity* transcending the particular order, which is to characterize the belonging to a delegation relation, as just described, the more precise characterization of which remains *open*.

The concept of legal positivity especially takes on this meaning when it is desired to distinguish the 'law which has validity by virtue of human positing' from 'natural law which has validity by virtue of its content'. Understood in this manner, the dichotomy 'positive right' – 'natural right' has an appropriate meaning, and, indeed,

---

[291] *Ibid.*, p. 33.

[292] Compare Kelsen, 'Die philosophischen Grundlagen', p. 10.

it is indispensable for rational reconstruction of juridical procedure. For it cannot be doubted that a great number of legal judgments and administrative decisions are oriented – more or less consciously – according to *principles of interpretation* which contain certain *purposive viewpoints*. For this reason these principles are subject to interpretation in a manner analogous to those sentences which are contents of the pertinent acts of positing.

The error of natural right doctrines, which has been opposed quite properly by schools of positive law, thus does not lie in the rejection of the thesis that all law is valid due to its positing, but only in the claim of *necessary* validity which is raised for certain sentences on the grounds of the assertion that they represent rational truths about the absolutely correct (just) conduct of human beings. The refutation, on principle of this exaggerated claim follows directly from our general analysis of the value concept.[293] However, a thorough critique of method will not be permitted to stop here, but will have to trace out the *underlying ends* in each instance. It will emerge only too frequently that the apparently unambiguous character of concepts which stand at the center of natural law speculation – as for example 'justice' or 'varying degrees of culpability' – does not in fact exist. This will emerge especially strikingly where underlying purposes *collide*; where, for instance, the goal of the uniformity of jurisdiction collides with that of taking into consideration the peculiarities of the case, or where the goal of general prevention collides with that of particular prevention.[294]

Let us clarify this in terms of the core concept of natural law speculation, the concept of *justice*. This concept finds application in extra-legal domains too; thus one speaks for instance of a 'just wage'. In order to recognize the ambiguous nature (need for completion) of this concept, it is sufficient to ask oneself the question, whether it is just that workers who are supporters of families should receive higher – or, a certain percentage higher – wages than those who only need to provide for themselves. In the effort to decide this question one will have to weigh the 'for' and 'against' of this question, i.e., one will have to reflect on the unspoken underlying purposive contexts, and examine what effect the measures would have with respect to these. The decision will then depend on the one hand on the order of preference of the ends to be considered, and on the other hand on the suitability of the measure in question for achieving one or the other of these ends.

In surmounting the prejudices of natural law, whose common root is the absolutization of the concept of justice, the following steps can be distinguished: In the beginning there is the insight that the concept of justice – as a subordinate concept of practical correctness – is a relational concept. To this is linked the observation of the psychology of knowledge, that this circumstance most frequently remains unrecognized because the goals to which the concept of justice relates in each instance are *unquestionable goals* in the sense made precise above.[295] From this observation in turn comes the task of making the underlying ends explicitly clear,

[293] Compare above Part One, section "The Concept of Value".

[294] Compare for instance Kaufmann, 'Strafrechtsschuld', p. 114.

[295] See Part Two, section "The 'Historical' in the Social Sciences".

and in so doing not infrequently their merely conditional (limited) validity, as maxims for action, will become evident, due to their confrontation with conflicting ends. Accordingly the variations in the setting of social goals, which go hand in hand with the changes of civilization and culture, correspond to changes in the idea of justice, i.e., of the idea of the correct social order in general, and the correct legal order in particular. The study of the development of views concerning 'just punishment' from prehistoric times down to '*normative culpability theory*'[296] on the one hand, and to the culpability-free penal law doctrines of the Italian positivists[297] on the other, furnishes an especially good insight into the character of these changes and the way they were conditioned historically and sociologically. In any case, it is possible – disregarding perhaps a few isolated exceptions – to ascertain *one* purposive idea implicitly underlying the concept of justice as historically invariant in the face of all these variations, namely the *idea of the maintenance of communal peace*,[298] and of securing the cohesion of the community. In conformity with this, almost always those measures are declared to be unjust, which particular groups in the community feel to be 'unbearable'. From this perspective, the particularly close coupling of the idea of justice with that of legal security and thus with that of legality (uniformity of jurisdiction) also becomes understandable, so that 'justice' appears precisely to be the opposite of 'arbitrariness'.

However, we cannot pursue these considerations of the sociology of law further, but must now turn to the general consequences of our reflections on natural law for the theory of legal methodology in general and problems of the concept of positivity in particular. What we are dealing with here is the *natural law components within a legal order* which emerge[299] in interpretations of general norms that are carried out in an analysis having, as its aim, the establishment of individual norms (for example, sentences imposed on persons on trial).

As we have already mentioned, it was especially the doctrine of free law – which attained great influence after World War I – which took up the battle against the prejudice of the universality of legal prescription. It uncovered the misuse of the argument from analogies and the *argumentum a contrario* which was made in support of this prejudice, and for the purpose of an allegedly unobjectionable filling in of the '*gaps of the law*'. Consequently one arrives at the treatment of the question, as to whether a judge making *praeter legem* decisions is free of all limitations, or else what limitations on him there may be.

In this connection, we must examine, in the first place, to what extent prescriptions for 'filling in the gaps' exist in the laws themselves. Such additional stipulations (rules of interpretation) are incorporated in the great majority of codifications

---

[296] Compare Kaufmann 'Strafrechtsschuld', p. 113 ff.

[297] Their intellectual leader was A. Ferri. See his *Progetto preliminare di codice penale italiano per i delitti*, Milano 1921. The doctrine was based on the theories of the psychiatrist Lombroso.

[298] Compare for this, Verdross, 'Die Rechtstheorie Hans Kelsens', *Juristische Blätter*, 59th year 1930, pp. 421,423.

[299] See for this F. Schreier, *Die Interpretation der Gesetze und Rechtsgeschäfte* (The interpretation of laws and legal business), Vienna 1927.

of law. In most cases it is prescribed that the judge must decide according to the 'intent of the law' ('according to the intent of the legislator's will'), by which is meant that he is to grasp the tendencies of legal policy which find expression in the law, and then render his judgment in harmony with these tendencies. After our fundamental considerations with respect to this topic, no detailed justification is required for the observation that this presupposition is untenable, and belongs to the type of prejudice we have called '*dogmas of pre-stabilization*'.

These prescriptions contain the presupposition that the will of the legislator was derivable from the legal text under all circumstances, and could form an unambiguous guideline for every judicial decision. Still one has to ask, by which thought-processes the 'will of the legislator' is to be constituted.

At this point, given the lack of precise prescriptions for interpretation within the law, the methodological controversy, evoked by diverging impulses of legal policy, arose. In the section before last, we have already pointed to the controversies between the protagonists of subjective interpretation and those of objective interpretation. We realized that the kind and extent of these divergences were hardly ever grasped with full clarity by the parties to the quarrel. This is already the case for interpretation in the narrower sense – in which it is merely to be decided *what the legislator intended to say*, but not, as well, *what his intention was in saying it*; all the more so for the interpretation in the broader sense, which has to take recourse to the purposes of legislation.

But even today – as at all times past – the attempt is made to present a decision in one sense or the other as the sole correct one, by maintaining that it follows by rational principles from the 'essential nature of the object'. The practical consequence is that the purposive viewpoints which speak for the decision in the one sense are considered to be the sole decisive ones, so that there is a failure to take into consideration conflicting purposes which would suggest another decision. Even in cases where at first glance the legislator appears to have provided for all contingencies, considerations of the purposes of the jurist applying the law play a role as well, which then are frequently established as absolutely valid in the manner characteristic of natural law.

This is true all the more where, by rules of interpretation, specified by legal statute, the task is imposed on the judge, not to comprehend the goals set by others, but to reflect on his own goals. A famous example of a rule of interpretation obliging the judge to such self-reflection is Article 1, formed in emulation of the *categorical imperative*, of the *Swiss Code of Civil Law* by Egon Huber.[300] One should note especially that the judge is not permitted, or instructed, by this prescription to decide wholly arbitrarily, so that in a decision *praeter legem* he could always decide in favor of that party which was more sympathetic to him, or corresponded more closely with his own interests; rather, he has to render judgment in such a manner

[300] It states: the law finds application to all legal questions, for which it contains a decision according to its wording or its interpretation. If no prescription can be drawn from the law, then a judge will have to decide by customary law, and where none exists, according to the rule which he himself would establish as legislator. In this he must follow proven doctrine and tradition.

that it is consistent with his general (pertinent) valuations. All that remains is to point out that this task as well (understood as a cognitive task), by no means always permits of an unambiguous solution, so that the likely procedure would be to remove the doubts that arise by decreeing a *natural law dogma*.

Above all this will be the case where value concepts (for instance, 'moral' or 'immoral') are contained in the legal text, without indication of the criteria or, the persons whose valuations are to be considered decisive. Each of the following can serve as basis: the valuations of the legislator, or those of the jurist applying the law, or those of a selected circle of 'just and righteous thinking men', or, special experts for precisely these questions, or finally the prevailing valuations within a community of law. Each of the alternatives, with respect to the method of ascertaining the valuations, is to be precisely specified. The natural law pre-stabilization dogma hides all these difficulties, and in this lies its seductiveness, and its danger.

Due to the fact that various philosophers of law presuppose different contents of value concepts, methodological controversies, with a background of legal policy, arise, which apparently render all prospect of resolution hopeless. The guidelines for legally correct decision which the philosopher of law, who believes in natural law, furnishes, to those who apply the law, in his opinion, originate in the clear sources of the evident insight of natural law. Thus, when the guidelines of another philosopher of law are incompatible with his own directions, they must, according to his conviction, be false.

In these conceptual errors the *mistaken conception* of the process of *rational reconstruction* plays an important role, as is understandable after our general presentation. In analyzing the procedure of the administrator of law certain teleological principles can frequently be found, which, without being supported by the statutes, function as a basis for interpretation in the same way as if they were legal prescriptions. By comparing the majority of the decisions of any legal administration of law, one can then judge whether these have been *consistently* based on these guiding viewpoints (external to the statutes). Should one observe inconsistencies one can undertake this or that rectification in order to grasp what the legal administrator of law '*really*' intended. All this is rational reconstruction, and thus a process of clarification, and as such a cognitive procedure. However, when the philosopher of law inadvertently substitutes his own teleological (axiological) viewpoint,[301] his findings change into a postulate of natural law, which appears in the guise of knowledge all the same.

After our previous discussions it is not necessary to waste many words about the methodological task which arises from this epistemological situation. What must be done is indeed to carry out the rational reconstruction just designated, i.e., to take one's departure from the (average) typical conduct of the legal administrator and then to rectify this in the manner indicated. Insofar as the rational reconstruction leads to unequivocal results – and one must not let one's expectations soar too high with respect to this – one will thereby gain insight into the actual functioning of the unspoken implicit purposes and their implicit ranking within the procedure of the

[301] Compare above Part One, section "The Concept of Value".

application of law in general, and of making decisions *praeter legem* in particular. One will also obtain points of reference for how this function would be modified with increased axiological consistency, achievable through more profound self-reflection. It can readily be seen that such knowledge can also have considerable significance for *legal policy*, for the legal administrator striving for clarity about his procedure as well as for the legislators who want to estimate the consequences which, on the one hand, are to be expected from the incorporation of this value concept in a text of law or, on the other, by establishing this or that interpretative rule. Such rational reconstruction was the task undertaken by Hans Kelsen in the pure theory of law, and, indeed, accomplished by him and his students in the treatment of many important problems in the theory of law. At other points, however, his analysis could not penetrate to the desired level of profundity because his prejudice of the dualism of the 'is' and the 'ought' stood in the way. This is true above all for the precise specification of the relation between positive law and natural law, and for the delimitation of the dogmatic science of law and theory of law as against the *sociology of law*.

On the first point, all that is essential has already been said. Reflection here leads both to the overcoming of a naïve positivism of law, which assumes that all individual norms are prescribed unequivocally by general norms, as well as to overcoming the naïve natural law doctrine, which believes that legal decisions can by derived deductively from value insights with *a priori* validity.

Secondly, concerning the delimitation between dogmatics of law and theory of law, on the one hand, and the sociology of law, on the other, it follows from what has been said above that this cannot be accomplished by means of the thesis of the dualism of 'is' and 'ought', of a causal method and a normative method. But in spite of this, the struggle against the syncretism of method, carried on within the framework of the pure theory of law, is a methodological achievement of extraordinary significance, because the topics and procedures of the sociology of law, and of the dogmatic determinations of law, are indeed to be kept sharply separate.

We now can characterize the differences in principle. The topic of legal dogmatics is the interpretation of particular legal sentences. *Legal theory* is the *rational reconstruction* of this interpretative procedure; in particular it clarifies the criteria according to which sentences are determined as belonging to one and the same legal order, as the totality of the material to be interpreted. The *sociology of law*, on the other hand, investigates in what manner, to what extent, and under which conditions the *behavior* of socialized human beings *actually* is influenced in a typical manner by legal procedure – i.e., the decreeing and interpretation of sentences in which under certain circumstances the members of a certain social circle are threatened with the application of physical compulsion – and thus arrives at a sociological concept of the positivity of law.

From the confounding of the two questions we have an erroneous conception that the legal question of the legal validity of a commandment – i.e., of whether it belongs to a certain legal order – is to be decided by an investigation of its social effectiveness. But as we have already observed at the beginning of this section, the 'validity' equated with 'social effectiveness' is not the 'validity' for which the interpreter of law is asking. The relation between these two kinds of 'validity', or as one could say in keeping with Kelsen's terminology, between the *juridical* and *sociological concept of law*, can be described as follows:

When one observes the totality of commands which seek to induce human beings of a certain social circle to undertake or to refrain from certain actions by threats of application of typical physical compulsion (Max Weber),[302] one will recognize that their relations typically prove to be *relations of delegation*, of the kind analyzed above. Furthermore, one comprehends that in the 'generation' of such commands a certain *assessment-procedure* which aims at recognizing whether a certain conduct in question violates certain other commands, plays an essential role. For it is dependent on the result of such assessment, whether the decreeing of certain commands *to* the persons whose behavior was being judged, or the decreeing of certain commands about a behavior (punishment, execution) imposed *upon* these persons will be carried out, or not. For reasons of social techniques, those persons who make these assessments will frequently also be competent to issue the commands just named. But this is not always the case (jury trial) and even where it is the case, one must distinguish carefully between the act of assessment and the act of issuing the command connected with it.

Thus this assessment procedure – *the juridical procedure of legal interpretation* – is an integral part of the social process in question, but for the immanent analysis of legal interpretation, its relationship to the total social process should not be taken into consideration, unless it appears in the procedure itself. This relationship can be made readily understandable by analogy with the relations of the rules of a certain game of cards to the stake for which the players are playing. The significance of a game of cards for those participating in it and, possibly, a wider circle of persons, is frequently decisively determined by the amount of the *stake*, and therefore when a treatise on the '*sociology* of games of chance' is written, special weight will have to be placed on the amount of the customary stakes and their relation to the income of the players. But in a *theory* of the various games of chance, the question of the amount of the stakes is not at issue. On the other hand, in a sociology of games of chance, the theory of these games cannot be left out of consideration, for otherwise it would not be possible to distinguish, for example, the pure games of chance from those in which certain skills, mental concentration and self-control, influence the outcome. And from this, in turn, certain inferences can be drawn about the types of players attracted to the various particular games.

Only in this same sense, can one speak of a *primacy of legal theory vis a vis* the sociology of law. In the framework of the pure theory of law its adherents operated with the argument that legal theory had priority *vis a vis* sociology of law, as one first had to know *what* '*law*' was, before one could do sociology of law. But this reasoning is obviously incompatible with the insight that *law* is a *social*

[302] See for this Max Weber, *Wirtschaft und Gesellschaft*, p. 17 f.

phenomenon, for one surely cannot designate the species as logically prior to the genus. One will arrive at the correct core of this thesis, however, as soon as one realizes that by 'sociology of law', Kelsen does *not* understand *descriptive* sociology in Max Weber's sense, but *causal social research*. Thus the assertion of this primacy is justified insofar as it states the *priority* of the *description* of a domain as *against* its *aetiology*. To be sure, this only holds true under the assumption of completely *clear* thinking, for though causal research in a certain domain must presuppose a certain amount of description, which characterizes the facts of the field, it is not dependent on a complete preceding description.

On the basis of the foregoing considerations, we can in conclusion characterize the relations of the most important of the meanings associated with the term 'positivity':

To begin with we have to distinguish whether we are speaking of the positivity of a particular legal sentence or of an entire legal order. In the first case 'positivity' means 'belonging to a historically determined legal order presupposed as a reference system'. In this sense – correlative to the question of juridical validity – the term 'positivity' requires a supplement. Accordingly, one frequently speaks of 'positive Austrian law', 'positive German law' and so forth. Here the words 'Austrian', 'German' and so forth express that the order with respect to which the membership of the sentence in question is to be examined, is relevant for a certain social-historical reality, for which it has 'social validity'. However, this social validity, also called 'positive validity' (in the sociological sense), which is presupposed for the order *en bloc*, remains *outside of the topic* with respect to the question of the membership of a certain sentence to this order. To be sure, the 'historical' element, whose connection with 'positivity' was always perceived, enters into our question in another manner; for the decision about the positivity of a legal sentence will, as we have recognized, be made dependent whether it has been 'posited' in a certain manner, i.e., whether it forms the content of expressive acts which can be incorporated into a relation of delegation. Thus the *structural types* of positive law are obtained, and one arrives at the contrast of sentences of positive law, which have validity by virtue of their positing, with sentences of natural law, which have validity by virtue of their content. We have recognized that the latter must not be neglected in the theory of law either, but their validity within positive law is subsidiary (restricted to cases of gaps in the law) so that the total legal order is not deprived of its character as a complex of statutes, i.e., as 'positive order'.

With that, we have also characterized a third concept of 'positivity': it refers to the total legal order, understood as the material to be interpreted.

Now only a few words have to be said about the positive validity of a total legal order in the sociological sense. This is understood as synonym for social effectiveness (motivational power) and thus we arrive at the discussion, already mentioned, under what conditions positive law originated and remains in effect, questions which are supposed to be answered by the various power – or recognition – theories.

Only those who are familiar with the doctrinal history of theory of law and the state, especially during the past half century, can have some notion of the extent of the confusion which has been produced by the confounding of the various concepts of 'positivity'. That this confusion has reached such an extent, and even today can by no means be regarded as completely overcome, may be traced to the lack of clarity with respect to the essential nature of juridical procedure. It was Hans Kelsen who undertook to come to grips with this clarification most energetically, and who actually carried it out to a great extent. After adopting the changes described here, his pure theory of law can be regarded as a paradigm example of rational reconstruction in the dogmatic social sciences.

* * *

Index of Names

R.S. Cohen and I.K. Helling (eds.), *Felix Kaufmann's Theory and Method in the Social Sciences*, Boston Studies in the Philosophy and History of Science 303, DOI 10.1007/978-3-319-02845-3,

Printed by Printforce, the Netherlands